Olive Production Manual for Oil

Technical Editors
Dr. Selina C. Wang
Dr. Louise Ferguson

University of California Contributing Editor
Dan Flynn

University of California
Agriculture and Natural Resources

Davis, California
Publication 3559

UNIVERSITY OF CALIFORNIA
Agriculture and Natural Resources

To order or obtain UC ANR publications and other products, visit the UC ANR online catalog at https://anrcatalog.ucanr.edu/. Direct inquiries to

UC Agriculture and Natural Resources
Publishing
2801 Second Street
Davis, CA 95618

E-mail: anrcatalog@ucanr.edu

Publication 3559

ISBN-13: 978-1-62711-169-0

Library of Congress Cataloging-in-Publication Data
Names: Wang, Selina C., editor. | Ferguson, Louise, editor. | Flynn, Dan (Olive researcher), editor.
Title: Olive production manual for oil / technical editors, Selina C. Wang, Louise Ferguson ; University of California contributing editor Dan Flynn
Other titles: Publication (University of California (System). Division of Agriculture and Natural Resources) ; 3559
Description: Davis, California : University of California Agriculture and Natural Resources, [2023] | Series: Publication ; 3559 | Includes bibliographical references and index | Summary: "Olive Production Manual for Oil covers a complete array of topics involved in growing olives for the production of olive oil, ranging from site selection and preparation to irrigation management, pests afflicting olive, the olive harvest, and processing virgin olive oil"-- Provided by publisher
Identifiers: LCCN 2023052476 | ISBN 9781627111690 (paperback)
Subjects: LCSH: Olive--California. | Olive oil industry and trade--California.
Classification: LCC SB367 .O45 2023 | DDC 338.4/76413463--dc23/eng/20240109
LC record available at https://lccn.loc.gov/2023052476

Photo credits: Cover—nito100, iStockPhoto.com; p. 1—Selina Wang; pp. 11, 43, 57, 95, 117, 189, 203—Louise Ferguson; p. 23—Bilanol, iStockPhoto.com; p. 33—Corto Olive Co.; p. 75—Luke Milliron; p. 135—Marshall W. Johnson; p. 161—Amanda Hodson; p. 169—Elizabeth Fichtner; p. 225—Leandro Ravetti; p. 239—Hector Amezcua.

Design by Sandra Osterman.

PRECAUTIONS FOR USING PESTICIDES

Pesticides are poisonous and must be used with caution. READ THE LABEL CAREFULLY BEFORE OPENING A PESTICIDE CONTAINER. Follow all label precautions and directions, including requirements for protective equipment. Use a pesticide only on crops specified on the label. Apply pesticides at the rates specified on the label or at lower rates if suggested in this publication. In California, all agricultural uses of pesticides must be reported. Contact your county agricultural commissioner for details. Laws, regulations, and information concerning pesticides change frequently, so be sure the publication you are using is up to date.

LEGAL RESPONSIBILITY. The user is legally responsible for any damage due to misuse of pesticides. Responsibility extends to effects caused by drift, runoff, or residues.

TRANSPORTATION. Do not ship or carry pesticides together with foods or feeds in a way that allows contamination of the edible items. Never transport pesticides in a closed passenger vehicle or in a closed cab.

STORAGE. Keep pesticides in original containers until used. Store them in a locked cabinet, building, or fenced area where they are not accessible to children, unauthorized persons, pets, or livestock. DO NOT store pesticides with foods, feeds, fertilizers, or other materials that may become contaminated by the pesticides.

CONTAINER DISPOSAL. Dispose of empty containers carefully. Never reuse them. Make sure empty containers are not accessible to children or animals. Never dispose of containers where they may contaminate water supplies or natural waterways. Consult your county agricultural commissioner for correct procedures for handling and disposal of large quantities of empty containers.

PROTECTION OF NONPEST ANIMALS AND PLANTS. Many pesticides are toxic to useful or desirable animals, including honey bees, natural enemies, fish, domestic animals, and birds. Crops and other plants may also be damaged by misapplied pesticides. Take precautions to protect nonpest species from direct exposure to pesticides and from contamination due to drift, runoff, or residues. Certain rodenticides may pose a special hazard to animals that eat poisoned rodents.

POSTING TREATED FIELDS. For some materials, reentry intervals are established to protect field workers. Keep workers out of the field for the required time after application and, when required by regulations, post the treated areas with signs indicating the safe reentry date.

HARVEST INTERVALS. Some materials or rates cannot be used in certain crops within a specific time before harvest. Follow pesticide label instructions and allow the required time between application and harvest.

PERMIT REQUIREMENTS. Many pesticides require a permit from the county agricultural commissioner before possession or use. When such materials are recommended in this publication, they are marked with an asterisk (*).

PROCESSED CROPS. Some processors will not accept a crop treated with certain chemicals. If your crop is going to a processor, be sure to check with the processor before applying a pesticide.

CROP INJURY. Certain chemicals may cause injury to crops (phytotoxicity) under certain conditions. Always consult the label for limitations. Before applying any pesticide, take into account the stage of plant development, the soil type and condition, the temperature, moisture, and wind direction. Injury may also result from the use of incompatible materials.

PERSONAL SAFETY. Follow label directions carefully. Avoid splashing, spilling, leaks, spray drift, and contamination of clothing. NEVER eat, smoke, drink, or chew while using pesticides. Provide for emergency medical care IN ADVANCE as required by regulation.

WARNING ON THE USE OF CHEMICALS

Pesticides are poisonous. Always read and carefully follow all precautions and safety recommendations given on the container label. Store all chemicals in their original labeled containers in a locked cabinet or shed, away from foods or feeds, and out of the reach of children, unauthorized persons, pets, and livestock.

Recommendations are based on the best information currently available, and treatments based on them should not leave residues exceeding the tolerance established for any particular chemical. Confine chemicals to the area being treated. THE GROWER IS LEGALLY RESPONSIBLE for residues on the grower's crops as well as for problems caused by drift from the grower's property to other properties or crops.

Consult your county agricultural commissioner for correct methods of disposing of leftover spray materials and empty containers. Never burn pesticide containers.

PHYTOTOXICITY. Certain chemicals may cause plant injury if used at the wrong stage of plant development or when temperatures are too high. Injury may also result from excessive amounts or the wrong formulation or from mixing incompatible materials. Inert ingredients, such as wetters, spreaders, emulsifiers, diluents, and solvents, can cause plant injury. Since formulations are often changed by manufacturers, it is possible that plant injury may occur, even though no injury was noted in previous seasons.

CONTAINER DISPOSAL. Dispose of empty containers carefully. Never reuse them. Make sure empty containers are not accessible to children or animals. Never dispose of containers where they may contaminate water supplies or natural waterways. Consult your county agricultural commissioner for correct procedures for handling and disposal of large quantities of empty containers.

PROTECTION OF NONPEST ANIMALS AND PLANTS. Many pesticides are toxic to useful or desirable animals, including honey bees, natural enemies, fish, domestic animals, and birds. Crops and other plants may also be damaged by misapplied pesticides. Take precautions to protect nonpest species from direct exposure to pesticides and from contamination due to drift, runoff, or residues. Certain rodenticides may pose a special hazard to animals that eat poisoned rodents.

Contents

Acknowledgments

This first edition of the *Olive Production Manual for Oil* was produced by many hands.

The Olive Oil Commission of California, through grower assessments, provided funding to the University of California Division of Agriculture and Natural Resources (UC ANR) and the UC Davis Olive Center. The funding was essential to producing the manual. The Commission also supplied valuable input on content and presentation.

The UC Davis Olive Center founding executive director, Dan Flynn, with our assistance, initiated this manual, recruited authors, and directed completion of the first draft. We gratefully acknowledge Firmin Berta for financially supporting much of Dan Flynn's time, and we also thank Dan for his many volunteer hours of supervision and review of this manual.

Authors of the chapters are UC ANR academics and members of the olive oil industry. Experts at Boundary Bend Olives, California Olive Ranch, Corto Olive Co., and Agricultural Advisors—as well as Pablo Canamasas, an independent milling consultant based in Argentina—coauthored several chapters. The combination of academic research and decades of industry experience produced a stronger, more relevant manual.

Peer reviewers included academic and industry experts. Their insights and questions greatly improved the manual. Jim Downing, former UC ANR director of publishing, ensured rapid publication.

This manual builds upon the work of our predecessors. Dr. Hudson Hartmann produced the inaugural reference book for olive growers, *Olive Production in California*, in 1953. It was revised as the *Olive Production Manual* by Steven T. Sibbett and Louise Ferguson in 1994 and 2005, and followed by Paul Vossen's *Organic Olive Production Manual* in 2007. We hope that this *Olive Production Manual for Oil* will be equally long-lived.

Dr. Selina C. Wang
Associate Professor of Extension,
Department of Food Science and Technology,
University of California, Davis

Dr. Louise Ferguson
Professor of Extension,
Department of Plant Sciences,
University of California, Davis

California Olive Oil Industry and Standards

Selina C. Wang, Associate Professor of Extension in the Department of Food Science and Technology at UC Davis

Dan Flynn, Executive Director Emeritus, UC Davis Olive Center

Highlights

- Acreage of oil olives has reached its highest in the past decade, surpassing acreage of table olives.

- Driving the industry's acreage growth are larger, intensive, mechanically harvested orchards, but there is also a significant number of smaller growers producing a wide range of varieties and oil flavor profiles.

- The International Olive Council's (IOC's) chemistry and sensory standards set the foundation for the California olive oil industry to assess olive oil quality; the California industry took several actions to align state law and federal regulations with IOC standards.

- The California Department of Food and Agriculture (CDFA) adopted California olive oil standards based on the recommendations of the Olive Oil Commission of California (OOCC) in 2014.

- California olive oil standards apply solely to producers of 5,000 gallons (18,927 L) or more, but smaller producers can opt to follow state standards.

- State quality standards include stricter limits and more descriptive grades than international standards. California requires two tests that assess oil freshness and potential adulteration that are not included in international standards.

- California standards require producers to pay an annual assessment, test every lot, keep extensive traceability records, and include on their labels a best-before date that is supported by technical evidence.

- Best practices for achieving high-quality extra virgin grade are growing healthy fruit, harvesting efficiently, minimizing material other than olives in harvest bins, controlling time and temperature of malaxation, and storing the oil properly.

live oil has been produced in California for more than 200 years, but it has only been in recent decades that the oil olive crop has achieved significant growth. California industry organizations have adopted standards that apply solely to olive oils produced in the state. Understanding how olive oil grades and standards assess quality will help producers recognize the importance of implementing best practices to maximize oil quality.

Industry beginnings

As researched by Judith M. Taylor (Taylor 2000), olives arrived in California when Spanish missionaries first planted cuttings at San Diego around 1775. The Franciscans valued olive oil for soap, cooking, lighting, and the Catholic sacraments. The Mexican government secularized the missions in 1833, 12 years after winning independence from Spain. The Franciscans left, and the olive groves were abandoned.

A significant push to establish a state olive oil industry in the nineteenth century was driven by Elwood Cooper in Santa Barbara County and Frank Kimball in San Diego County. Both men took cuttings from the missions to plant large orchards, reasoning that domestic olive oil would appeal to a growing California population of Italian immigrants. Their business plan collapsed when importers offered plentiful and cheap olive oil in the 1890s, although a number of small producers continued to produce California olive oil (fig. 1.1).

The emergence of California-style "ripe" olives in the early twentieth century led many olive oil growers to shift their harvest to table olives because the canners paid a better price than the millers. When the Spanish Civil War and then the Second World War halted olive oil imports from Italy and Spain, many California table olive growers shifted their harvest to oil production. Most olive growers moved their harvest back to table olives after 1950, when imported olive oil returned to the United States. For the next several decades, California olive oil was a specialty product made by a small number of processors.

Figure 1.1

Olive oil company in San Diego, early 20th century.

The modern industry

Around 1990, the California olive oil industry began emerging into its modern form with a surge of artisan production. Interest was particularly strong in Northern California wine country, where enthusiasts were inspired to produce fresh oils like they had tasted in the Mediterranean region. Some planted high-density (HD), sometimes called medium-density, orchards of about 200 to 300 trees per acre (500 to 750 trees per ha) primarily with Italian varieties, while others worked with older orchards of table olive varieties planted at about 35 to 90 trees per acre (85 to 225 trees per ha).

To share information on methods to produce high-quality oil, these producers formed the Northern California Olive Oil Council, which evolved into the California Olive Oil Council (COOC) by 1992. More than 500 oil olive growers were established in California by 2004, with 42 percent in Napa, Sonoma, Marin, and Mendocino Counties. State acreage was estimated at 6,168 acres (2,496 ha), with an average orchard size of 14 acres (5.7 ha) (Vossen and Devarenne 2005).

The arrival of the super-high-density (SHD) system in California in 1999 brought a paradigm shift to the state industry, which until then was constituted of mostly small, hand-harvested orchards. The SHD system had been pioneered in Spain 4 years earlier, and the first California SHD orchard and processing facility were backed by Spanish investors. The system featured densely planted hedgerows (600 to 900 trees per ac, 1,500 to 2,225 trees per ha) that allowed for efficient harvest with over-the-row equipment (fig. 1.2).

More than 11,500 acres (4,650 ha) of SHD oil olives were planted in the Central Valley between 2005 and 2008, with an average orchard size of about 200 acres (80 ha). In 2013, California oil olive acreage hit 30,000 acres (12,000 ha), nearly all of it in the Central Valley, surpassing California table olive acreage for the first time. In 2014, Australia's largest olive oil producer established acreage and a processing facility in California, expanding the planting of mechanically harvested HD orchards. Companies affiliated with major imported olive oil brands also established processing facilities and olive orchards in the state.

California producers joined with grower associations outside the state to establish the American Olive Oil Producers Association (AOOPA) in 2012 to address trade issues. The following year, with the industry's sponsorship, the California legislature approved the establishment of the Olive Oil Commission of California (OOCC) (California Legislature 2013), a state government agency to fund research and develop standards through grower assessments (sidebar 1.1).

By 2020, California had an estimated 37,000 acres (15,000 ha) of oil olives, producing about 4.25 million gallons (16.1 million L) of oil, equivalent to about 4.4 percent of U.S. olive oil consumption. California olive oil was widely available nationwide, and California artisan producers consistently received top awards in international competitions, with high-volume California producers also garnering accolades. The American olive oil industry outside of California totaled 11,000 acres (4,450 ha) in 2020; most of those plantings were in Georgia, Texas, and Arizona, with much smaller plantings in Oregon, Florida, Alabama, and Hawaii.

International standards

The California olive oil industry's knowledge of olive oil standards and grades was shaped in the 1990s and 2000s by International Olive Council (IOC) standards. IOC standards govern most olive oil imported into the United States, which is more than 95 percent of olive oil consumed in the country.

The IOC was established by the United Nations in 1959 primarily to adopt trade standards and encourage international cooperation on research. IOC standards are also the foundation

Figure 1.2

Approximate comparative spacing in traditional, HD, and SHD systems.

for standards established by other countries and the European Union.

IOC standards include the following designations and grades: (IOC 2021):

- **Virgin olive oils fit for consumption as they are:** These are oils produced from olives by mechanical extraction without high heat or chemicals. Grades in this category are, in descending order of quality, extra virgin, virgin, and ordinary, although the only grade commonly available to U.S. consumers is extra virgin.
- **Virgin olive oils that must undergo processing prior to consumption:** These are the lowest-quality virgin oils, known as lampante grade, which must undergo a refining process to remove poor flavor and aroma and elevated free fatty acidity.
- **Refined olive oil:** These oils are physically refined with vacuum distillation at high heat or chemically treated with sodium hydroxide. They are generally labeled as "light olive oil" in the United States.
- **Olive oil composed of refined olive oil and virgin olive oils:** These oils generally are very high in refined olive oil and low in virgin olive oils and are marketed in the United States as "pure olive oil."

- **Olive-pomace oil:** These oils are extracted with hexane from olive pomace. Grades in this category are crude olive-pomace oil, refined olive-pomace oil, and olive-pomace oil composed of refined olive-pomace oil and virgin olive oils. Olive-pomace oils generally are used in foodservice and food processing and may be available in some retail markets focusing on imported specialty foods.

All olive oils and olive-pomace oils are subject to chemistry standards, which evaluate oil quality and purity:

Quality: The chemical standards determine the grade of the oil, detecting negative chemical changes that occur as it ages or because of the quality of fruit, processing, or storage.

Purity: The chemical standards determine whether the oil may be adulterated with other, less-expensive oils. Purity tests generally cost more than quality tests and typically are deployed by regulators and buyers to prevent fraud.

Additionally, grades within the "virgin olive oils fit for consumption as they are" category are subject to sensory standards. Sensory evaluation is conducted by a panel of at least eight tasters who are trained by a panel leader to identify and assess positive sensory attributes, which include fruitiness, bitterness, and pungency, and negative sensory attributes caused by substandard fruit, processing, storage, or packaging conditions. IOC sensory standards require extra virgin olive oil to have no detectable negative flavors. The IOC assesses the proficiency of the panel through an annual recognition process.

While the United States is not a member country of the IOC, the California olive oil industry aligned itself with IOC standards to some extent as the industry developed:

- In 1997, the University of California began training a sensory panel for the California Olive Oil Council (COOC) using IOC protocols (Gage 2009).
- In 1998, COOC adopted a variation of IOC standards for extra virgin grade, with the COOC establishing a stricter 0.5 percent standard for free fatty acidity compared to the IOC standard of 0.8 percent.
- In 2008, COOC sponsored California legislation that adopted IOC standards for all olive oils sold in the state, the first update of California standards since 1947 (California Legislature 2008).
- In 2010, after several years of collaboration with COOC, the United States Department of Agriculture (USDA) adopted voluntary nationwide standards largely based on IOC standards (USDA 2010). These standards apply only when a producer opts to qualify for a USDA seal of quality. USDA standards diverge from IOC standards in limited instances to reflect the chemical characteristics found in California olive oil.
- In 2011, COOC sponsored California legislation that made the USDA standards mandatory for olive oil sold in California (California Legislature 2011).

California standards

In 2014, the California Department of Food and Agriculture (CDFA) adopted olive oil grades and standards ("California standards") that were recommended by the Olive Oil Commission of California (OOCC) based on California olive oil data produced over several years by the University of California. The CDFA issues the standards for each market season, which may be revised based on OOCC recommendations (CDFA 2021).

The CDFA standards are required for every California olive oil "handler," defined by the CDFA as a person who engages in the operation of marketing 5,000 gallons (18,927 L) or more "of olive oil that he or she has produced, or purchased or acquired from an olive producer, or that he or she is marketing on behalf of an olive producer, whether as an owner, agent, employee, broker, or otherwise." Producers of less than 5,000 gallons annually are exempt from these standards. California standards differ from international standards in several important respects.

Grades

In contrast to IOC standards, California standards do not include ordinary virgin or lampante grades and instead require that oils of similar quality to those oils be graded as crude olive oil. For refined oils, California standards prohibit the term, and variants of, "light" and require that these oils be graded as refined olive oil. For blends of refined and virgin olive oils, California standards prohibit the term "pure" and its variants and require that these oils be graded as refined olive oil blend comprised of refined olive oil and virgin or extra virgin olive oil.

Purity standards

California standards for purity are generally similar to international standards, with some exceptions to accommodate the natural variability in California olive oil chemical composition that is related to variety and growing conditions. California standards prevent the grading and sale of lots that fail chemical purity testing unless traceability documentation (as described below) is provided to, and accepted by, the OOCC as evidence that the lot is comprised of authentic California olive oil.

Quality standards

Several California standards for quality have stricter limits than international standards (table 1.1). The quality tests use chemistry and sensory evaluation to assess the quality of fruit, processing, storage, and packaging conditions. California standards also include two chemistry tests

Table 1.1

California quality standards include stricter limits than international standards, and additional tests

Tests	CDFA	IOC	CDFA	IOC	CDFA	IOC	
	Extra virgin		Virgin		Crude	Ordinary	Lampante
Free fatty acids (%, g oleic acid/100 g oil)	≤ 0.5	≤ 0.8	≤ 1.0	≤ 2.0	≤ 1.0	< 3.3	> 3.3
Peroxide value (meq O_2/kg oil)	≤ 15	≤ 20	≤ 20	≤ 20	≤ 20	< 20.0	no limit
Absorbency in UV K_{232}	≤ 2.4	≤ 2.5	≤ 2.6	≤ 2.6	≤ 2.6	< 2.6	no limit
Absorbency in UV K_{270}	≤ 0.22	≤ 0.22	≤ 0.25	≤ 0.25	≤ 0.25	< 0.25	no limit
Absorbency in UV ΔK	≤ 0.01	≤ 0.01	≤ 0.01	≤ 0.01	≤ 0.01	< 0.01	no limit
Diacylglycerols (%)	≥ 35	N/A	N/A	N/A	N/A	N/A	N/A
Pyropheophytins (%)	≤ 17	N/A	N/A	N/A	N/A	N/A	N/A
Organoleptic median of fruitiness	> 0	> 0	> 0	> 0	N/A	N/A	N/A
Organoleptic median of defects	0	0	0 < Me ≤ 2.5	0 < Me < 3.5	> 2.5	3.5 < Me < 6.0	> 6.0
Insoluble Impurities (%, g/100 g oil)	≤ 0.1	≤ 0.1	≤ 0.1	≤ 0.1	≤ 0.2	< 0.1	≤ 0.2
Moisture and volatile matter (%, g/100 g oil)	≤ 0.2	≤ 0.2	≤ 0.2	≤ 0.2	≤ 0.3	< 0.2	≤ 0.3

(diacylglycerols and pyropheophytins) that provide information on both quality and purity; IOC standards do not include these tests. Additional information on the quality tests required by California standards is in sidebar 1.2.

Mandatory testing

California standards require annual olive oil quality testing by both the OOCC and each handler, with slightly different requirements for the two parties. The California standards require the OOCC to annually request the CDFA or an independent sampling party to randomly sample five lots from each handler by February 1. The sampling party must send the samples to an accredited laboratory designated by the OOCC, request the quality tests identified in sidebar 1.2, and report testing results and grades to the OOCC by March 1. The OOCC requires that a random number of the lots also be tested for purity. The OOCC quality tests include induction time, which is not required under California standards but is a useful predictor of the oxidative

stability of the oil. The OOCC distributes each handler's testing results to them by March 15. A handler may request that the commission retest samples.

The California standards require each handler to test every lot (not just five random lots as required for OOCC testing) for the quality parameters identified in sidebar 1.2 and report the results and grades for each lot to the OOCC by March 1; they can request a 30-day extension from the OOCC for samples produced after February 1. Handlers are not required to test for purity.

Traceability records

Each handler must maintain traceability records for 3 years. The records must include identification of the olive oil lot number, the assessor parcel number where the olives were grown, the pesticide records for that location, the harvest company name, the transportation company name, the weight of olives delivered for processing, and the weight or volume of the oil produced from those olives.

SIDEBAR 1.2

Key quality tests in California olive oil standards

Free fatty acids

Free fatty acids (FFA), which are flavorless, come from the breakdown of triacylglycerols through a chemical reaction called hydrolysis. Factors that can lead to high FFA in an oil include poor fruit quality, olive fruit fly damage, diseases, delays between harvesting and milling, poor extraction methods, and improper storage of the oil (such as on sediment). Processors may measure FFA when olives are delivered to the processing plant to determine the extent of initial fermentation. FFA of a properly made and stored oil do not increase much over time. High FFA in the oil right after extraction leads to a shorter shelf life as high FFA accelerates the oxidation rate.

Peroxide value

Peroxide value (PV) is a crude measurement of initial oxidation in the oil. Oxidation causes peroxides to transform into aldehydes and other compounds that are responsible for rancid flavors. Oxidation is a natural process, and PV is expected to increase as the oil ages, especially if the oil is exposed to light, high heat, and oxygen. As oxidation advances, PV can decrease as the primary oxidation products transform into secondary oxidation products.

Ultraviolet absorbance

There are three ultraviolet (UV) absorbance tests. K_{232} measures initial oxidation products in the oil (similar to PV). K_{270} (sometimes expressed as K_{268}) measures secondary oxidation products, indicating whether oxidation has advanced past initial oxidation. ΔK measures the difference between the absorbance at 270 nanometers and at 266 to 274 nanometers, which is useful in detecting the presence of refined or pomace oil.

Diacylglycerols

A diacylglycerol (DAG) is formed when a triacylglycerol molecule with three fatty acids undergoes hydrolysis, losing a fatty acid from the glycerol. The resulting DAG contains two fatty acids on a glycerol backbone in a 1,2 position (fig. 1.3). As oil ages or is heated, these DAG molecules equilibrate, in a predictable and linear manner, so that the fatty acids move to a 1,3 position, which is more stable thermodynamically and therefore favored. The DAGs test assesses the extent of aging and heating by analyzing the ratio of 1,2- and total DAGs;

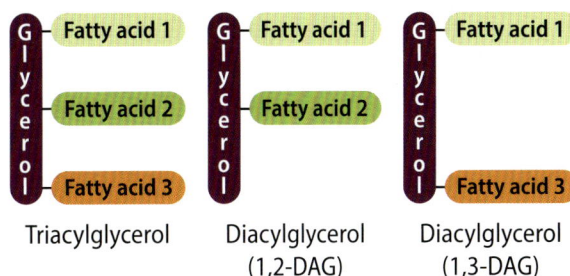

Fig 1.3.

A diacylglycerol (DAG) is formed when a triacylglycerol molecule with three fatty acids undergoes hydrolysis and loses a fatty acid. The remaining fatty acids attach to glycerol in the 1,2 position (1,2-DAG), then convert to the 1,3 position (1,3-DAG) as the oil ages or is heated.

the ratio decreases with age and heat. In addition, DAGs, like FFA, are related to the hydrolysis reaction and can be used as quality markers for olives and post-harvest practices.

Pyropheophytins

Pyropheophytins (PPP) are degradation products of chlorophyll a as a result of aging or heating. Chlorophyll a converts to pheophytin a and then to pyropheophytin a. The ratio of pyropheophytin a to total PPP increases linearly with time, which makes it useful for detecting oils that are aged. The ratio is also useful for detecting the presence of refined oils as high temperatures during refining hasten the conversion of pheophytin a to pyropheophytin a.

Sensory evaluation

Sensory testing is required by a panel certified by the IOC or the American Oil Chemists Society. The panel must include at least eight trained tasters to assess the median intensity of fruitiness and defects. Defects include those caused by oxidation (rancid); fermentation (fusty/muddy sediment, musty, winey); fruit condition (grubby, frozen); processing (cooked, burnt), storage (fusty/muddy sediment), and packaging (metallic).

Other tests

California standards also require testing for insoluble impurities as well as for moisture and volatile matter.

Best-before date

Each handler is required to include a best-before date on the label, and the best-before date must be supported by technical evidence. Shelf life is directly related to olive oil's oxidation rate. All olive oil oxidizes, even oil that starts out as very high quality. An extra virgin olive oil that initially tastes highly complex and fruity eventually will be transformed by oxidation into stale and ultimately rancid oil that no longer meets extra virgin standards. The OOCC permits the handler to select among several methods identified in a literature review by University of California, Davis (Wang 2017), to provide the required technical evidence. One of these methods estimates shelf life based primarily on testing data already collected by handlers as required by California standards (sidebar 1.3).

Best practices for quality

A key objective of California standards is to gain greater consumer and trade confidence in the consistent, high quality of California olive oils (CDFA 2021). Almost all California olive oil is verified by testing as extra virgin grade. The following best practices support the production of high-quality extra virgin oil, as reflected in the chemical and sensory tests required by the California standards.

Grow healthy fruit

The best quality olive oil comes from healthy trees, with fruit that has not been damaged by pests such as olive fruit fly, diseases such as olive knot, or environmental conditions such as drought and frost.

Harvest efficiently

Minimize fruit damage during harvest by proper hedge alignment and harvester calibration. Track the moisture and fat content of the fruit to determine the best time for harvesting. To produce high-quality extra virgin olive oil, do not harvest fruit from the ground. Minimize the amount of time needed to get olives from the field to the processing plant (see chapter 15, The Olive Harvest).

Minimize foreign material

Keep leaves, rocks, twigs, and other material other than olives (MOO) below 5 percent in harvest containers. Although equipment at the processing plant usually removes much of the MOO, MOO that gets processed with the olives can have a negative impact on the quality of the oil and may damage processing equipment (for more information on processing best practices, see chapter 16, Processing Virgin Olive Oil).

SIDEBAR 1.3

Five steps to estimate shelf life

1. Test oil.
    Test for FFA, PPP, DAGs, and induction time (at 110°C).

2. Calculate value A.
    $$(DAGs - 35) ÷ FFA\ factor = Value\ A$$
    FFA factor = 1.7 if FFA < 0.4%, 2.1 if FFA > 0.4% and < 0.6%, and 2.5 if FFA > 0.6%.

3. Calculate value B.
    $$(17 - PPP) ÷ 0.6 = Value\ B$$

4. Identify value C.
    $$Induction\ time\ (hours) = Value\ C$$

5. Identify lowest number.
    Compare values A, B, and C; the lowest number is the estimated shelf life, in months.

Example

1. Test oil.
    Sample results: FFA = 0.2, DAGs = 90, PPP = 2, induction time = 30 hours

2. Calculate value A.
    $$(90 - 35) ÷ 1.7 = 32$$
    DAGs is 90 for this sample, and the FFA factor is 1.7 because FFA for sample is < 0.4%.

3. Calculate value B.
    $$(17 - 2) ÷ 0.6 = 25$$
    PPP is 2 for this sample.

4. Identify value C.
    $$Induction\ time\ (hours) = 30$$

5. Identify lowest number.
    For this sample, value B has the lowest number, at 25.

Therefore, 25 months is the estimated shelf life for this sample.

Based on Guillaume and Ravetti 2016.

Process as soon as possible

Minimize time between harvest and processing. Fruit damage, a long transport time, high ambient temperature, and poor container ventilation are factors that contribute to olive fermentation, which degrades quality. These factors become critical when processing lags behind harvest by more than 4 hours. The longer it takes to process the olives, the more critical it is to have little fruit damage, well-aerated containers, and a low ambient temperature.

Control malaxation

Keep malaxation times as short as possible to allow for a steady kneading of the paste and a slow release of the oil. Generally, shorter malaxation times and lower temperatures during malaxation produce more aromatic and complex oils, but they reduce yield if taken to extremes.

Store properly

Use stainless steel tanks, nitrogen blanketing, and temperature control in the storage room to minimize oxidation processes in the oil. The oil is best kept at temperatures between 59° and 64°F (15° and 18°C). Filter the oil or ensure that the storage tanks are drained regularly (racked) to remove water and sediments to minimize hydrolytic reactions that reduce quality and shelf life. Once the oil has fully settled (approximately 30 to 45 days after processing, depending on room temperature), move the oil to a new clean tank.

References

California Legislature. 2008. Senate Bill 634, Chapter 694, Food labeling: Olive oil.

———. 2011. Senate Bill 818, Chapter 567, Food labeling: Olive oil.

———. 2013. Senate Bill 250, Chapter 344, Olive Oil Commission of California.

[CDFA] California Department of Food and Agriculture. 2021. 2021-2022 grade and labeling standards for olive oil, refined-olive oil and olive-pomace oil.

Gage, F. 2009. The new American olive oil. New York: Stewart, Tabori & Chang.

Guillaume, C., and L. Ravetti. 2016. Shelf-life prediction of extra virgin olive oils using an empirical model based on standard quality tests. Journal of Chemistry, article 6393962. https://doi.org/10.1155/2016/6393962

[IOC] International Olive Council. 2021. Trade standard for olive oils and olive pomace oils, COI/T.15/NC No 3/Rev. 17. Madrid.

Taylor, J. M. 2000. The olive in California: History of an immigrant tree. Berkeley: Ten Speed Press.

[USDA] United States Department of Agriculture. 2010. United States standards for grades of olive oil and olive-pomace oil. Washington, DC: USDA.

Vossen, P., and A. Devarenne. 2005. California olive oil industry survey statistics 2004. University of California Cooperative Extension, Sonoma County.

Wang, S. C. 2017. Shelf life of olive oil and useful methods for its prediction. Report submitted to the Olive Oil Commission of California, UC Davis Olive Center.

Suggested reading

Barranco, D., R. Fernandez-Escobar, and L. Rallo, eds. 2010. Olive growing. Translated by Susan E. Hovell, William A. Hovell. Pendle Hill, NSW, Australia: RIRDC. Originally published as El cultivo del olivo. Madrid: Junta de Andalucia, Consejeria de Agricultura y Pesca, and Ediciones Mundi-Prensa, 2004.

2

Olive Botany, Growth, Morphology, and Physiology

Mohamed T. Nouri,
UC Cooperative Extension Orchard Systems Advisor, San Joaquin County

Louise Ferguson,
Professor of Extension in the Department of Plant Sciences at UC Davis

Highlights

- Olive is a member of the Oleaceae family. Commercial olives belong to the species *Olea europaea* L., the only species that produces edible fruit.

- Olive evolved in the Mediterranean region, where periods of drought, temperature extremes, high solar irradiation, and poor-quality saline soils resulted in adaptations that improved its survival.

- Olive has a moderate cold tolerance, a 11°F (-11.7°C) lower temperature limit, and a low dormant chill requirement for bloom.

- Olive has unique leaf, stem, and root structural and physiological adaptations that allow production in a wide range of temperatures on Class II or poorer-quality soils.

- The roots are shallow, in the top 2 feet (0.6 m) of soil, spreading, and highly efficient at extracting water from dry soils, and they respond rapidly to intermittent irrigation. Olive fruit can rehydrate following water stress.

- The leaves are adapted to protect themselves from high solar irradiation and maintain photosynthesis in hot dry conditions. Olive trees store their carbohydrates primarily in the leaves, avoiding large fluctuations in carbohydrate resources.

- The flower buds of olive develop in the leaf axils of the current year's new shoots on the outer sun-exposed parts of the tree canopy, resulting in the crop being borne exclusively on the outer approximately 1.6 feet (0.5 m) of canopy; the shaded interior canopy is nonproductive. Flowers complete differentiation after dormancy.

- Alternate bearing occurs within individual shoots. The induction of the flower buds occurs during the rapid early growth of the fruit. At this time, any stress, particularly irrigation stress, will decrease the following year's crop.

Olive is a member of the Oleaceae family, subfamily Oleoideae. Commercial olives belong to the species *Olea europaea* L., one of about thirty-five species of *Olea* that evolved from tropical and subtropical species. Only *Olea europaea* L. produces edible fruit. Olive evolved in the Mediterranean agro-ecosystem, between latitude 30° and 45°, which was initially tropical, with later periods of drought, glaciation, temperature extremes, high solar irradiation, and poor-quality saline soils. These stresses resulted in structural and physiological adaptations that improved survival of this sclerophyllous species—characterized by hard leaves, short internodes (the distance between leaves along the stem), and leaf orientation parallel or oblique to direct sunlight. The olive's moderate cold tolerance, 11°F (-11.7°C) lower temperature limit, and low dormant chill requirement for bloom evolved under these conditions.

Olive tree growth, morphology, seasonal phenology

The olive is a long-lived, slow-growing, evergreen tree consisting of roots, a stem or trunk, and leaves (fig. 2.1). It can grow untrained as a multi- or single-trunk tree 13 to 30 feet (4 to 8 m) in height (fig. 2.2). For commercial production, it is trained to a single trunk, and its height is controlled to facilitate harvesting, or it is trained into a hedgerow. The olive tree is polymorphic, having juvenile and adult growth stages. The difference between the two stages is in reproductive, fruit-bearing ability; only adult trees bear fruit.

Olive shoots grow from early spring through summer. Vegetative growth in young and mature trees is apically dominant, almost exclusively from the single apical vegetative bud. If not trained or pruned before bearing, the young branches grow upright, eventually bending over and exposing the upper shoot surface to the sun. New vegetative shoots form on this upper sun-exposed surface (fig. 2.3). After two to four seasons of vegetative growth, the tree reaches bearing

Figure 2.1

Two-year-old Arbequina olive trees have the characteristic upright, apically dominant shoot growth until shoot length and cropping cause the shoots to bend. *Photo:* Mohamed T. Nouri.

Figure 2.2

Louise Ferguson and Sergio Castro-Garcia with an 80-year-old Manzanillo olive tree that shows the size potential of a single-trunk olive tree not pruned in decades. *Photo:* Louise Ferguson.

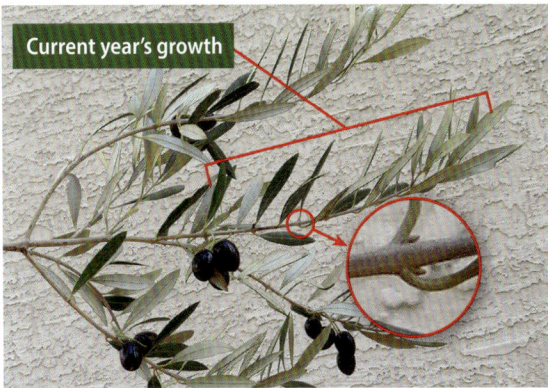

Figure 2.3

New vegetative shoots, and fruit buds, form in the sun-exposed outer canopy. *Photo:* Louise Ferguson.

age, producing flower buds in the opposing lateral leaf axils. When the trees bloom and the olives grow, the weight of the crop bends the long flexible shoots downward, forming the cascading fruit-bearing habit seen in figures 2.3 and 2.4.

The flower buds develop in the current year's new shoots on the outer sun-exposed parts of the tree canopy, resulting in the crop being borne exclusively on the outer approximately 1.6 feet (0.5 m) of canopy; the shaded interior canopy is nonproductive. The flower buds form in the axils of the opposing leaves below the terminal vegetative bud. All flower buds undergo a three-step process before bloom: induction, initiation, and differentiation (see the section "Olive tree flowers," below). Differentiation occurs at the end of winter, which makes olive different from most fruit trees, which fully differentiate their flower buds before winter. The flower buds bloom in late spring, on the now 1-year-old wood.

Figure 2.4

Arbequina trees show the typical cascading growth habit of an olive tree carrying a full crop. *Photo:* Louise Ferguson.

The growth habit of olive and its flower bud development cycle have important implications for orchard management. In the spring, the tree is supporting two crops: the current year's crop on 1-year-old shoot growth and the following year's crop developing as undifferentiated buds on the current year's growth. The induction of the buds occurs during the rapid early growth of the fruit, a period of heavy resource demand within the shoot. At this time, any stress, particularly irrigation stress, will decrease the following year's crop by reducing shoot growth. Figure 2.5 illustrates the timing of an olive tree's vegetative growth, flower development, and fruit development.

Olive tree roots

Roots have five primary functions: anchorage, absorption of water, absorption of nutrients, synthesis of plant compounds, and storage of plant compounds. Olive trees are generally grown on their own roots. Thus far, no rootstock has been developed that can confer resistance to diseases, particularly Verticillium wilt, or pests, nor an advantageous adaptation to climatic, soil, and water factors, or improved productivity or quality, or the ability to control scion vigor.

Having evolved under conditions of scarce and intermittent rainfall, olive roots have a high water uptake capacity in drying soils and immediate water uptake when irrigated. In rain-fed climates, roots have been found growing horizontally 20 to 23 feet (6 to 7 m) from the tree. If olive trees are produced from seed, they form a taproot that dominates root growth for several years. If they are produced from rooted cuttings, adventitious roots, which are roots formed in an unusual anatomical position, serve as multiple smaller taproots.

The proliferation, depth, spread, and branching of olive root systems depend greatly upon soil type, depth, aeration, and water content. As with most other orchard crops, olive generally has approximately 70 percent of its roots in the top 2 feet (0.6 m) of soil, as this is where oxygen and nutrient content are higher. Also, the olive's horizontally spreading root growth habit, common to species that evolved in arid climates, facilitates rapid water uptake. Even when planted in deeper soils, olive trees tend to remain shallow rooted (fig. 2.6).

Olive development cycles and management

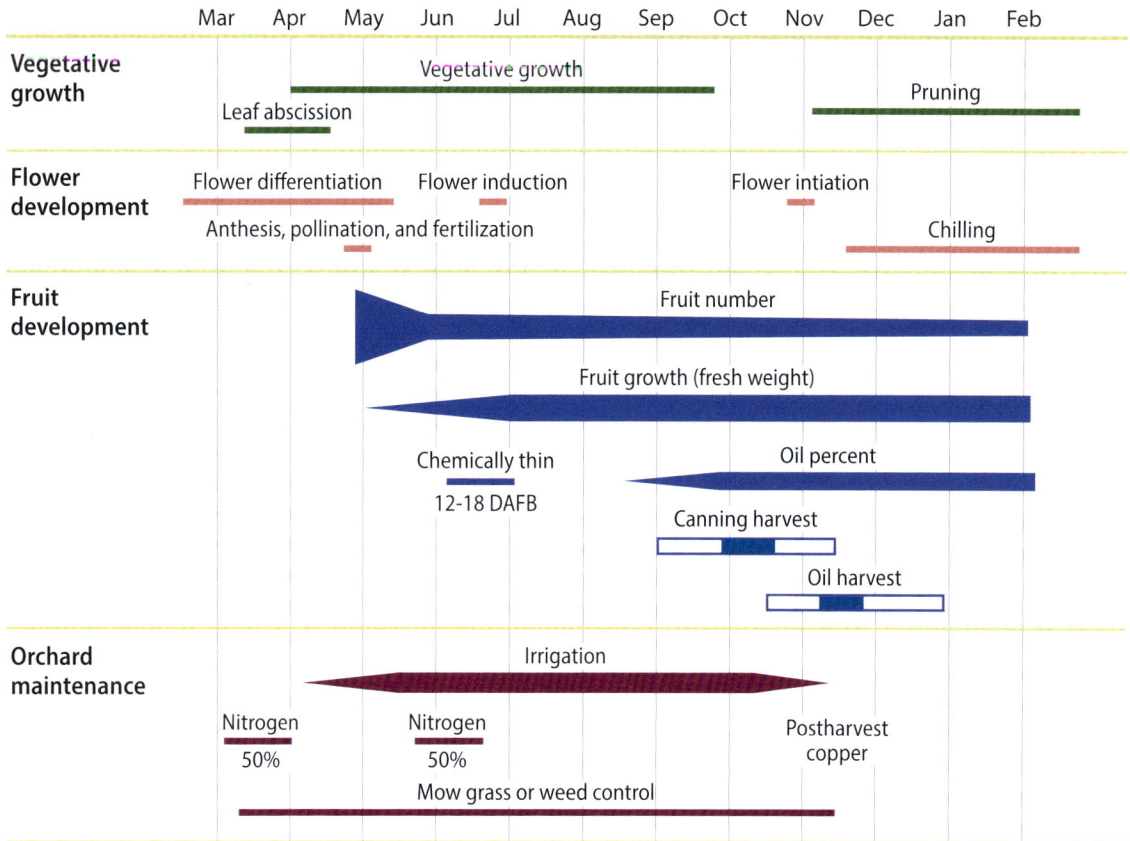

	Mar	Apr	May	Jun	Jul	Aug	Sep	Oct	Nov	Dec	Jan	Feb

Vegetative growth
- Vegetative growth
- Leaf abscission
- Pruning

Flower development
- Flower differentiation
- Flower induction
- Flower intiation
- Anthesis, pollination, and fertilization
- Chilling

Fruit development
- Fruit number
- Fruit growth (fresh weight)
- Chemically thin 12-18 DAFB
- Oil percent
- Canning harvest
- Oil harvest

Orchard maintenance
- Irrigation
- Nitrogen 50%
- Nitrogen 50%
- Postharvest copper
- Mow grass or weed control

Figure 2.5

The timing of olive tree shoot growth and flower and fruit development. *Source:* Paul Vossen.

Water and nutrient absorption occurs in the youngest, white part of the roots, the area immediately behind the root apex. The young root's efficiency is enhanced by root hairs, tubular extensions from the young root surface (epidermis) that effectively increase the actively absorbing root surface area. The youngest most active roots are also most susceptible to fungal and nematode attack. A young root is constantly renewing as its older portion suberizes and becomes less effective at water and nutrient uptake. Inside the epidermal cell layer that generates the root hairs is another layer of cells, the hypodermis, composed of larger, thicker cells with a hydrophobic layer that prevents water loss under stress conditions.

Readily available moisture to the entire root system is necessary for optimal root activity. Roots do not grow through dry soil; their exploration is limited by lack of water and the increasing mechanical resistance to penetration characteristic of dry soils. Uptake of water and nutrients is less efficient and slower in dry soils, and that can cause nutrient deficiency. For example, if

Figure 2.6

The typical horizontally spreading root growth habit of olive is evident in this 3-year-old tree. Even at maturity, an olive tree's most active roots are within the top 2 feet (0.6 m) of soil. *Photo:* Louise Ferguson.

potassium, which is immobile in the soil, is not in the wetted zone, it is not absorbed; potassium deficiency can sometimes be corrected by improving irrigation practices. A dry soil profile, by suppressing root activity, also reduces vegetative and reproductive growth in spring. The nutrient storage capacity of roots, and therefore supplies to the shoots, is reduced with restricted root growth.

However, a greater danger to olive roots than dry soil is the hypoxia created by soil saturation (Grattan et al. 2006; Shepherd et al. 2008). In a saturated soil, terminals of new roots can be killed in 1 to 4 days. Even short periods of waterlogging inhibit shoot growth, causing leaves to pale and yellow. In extreme cases, entire root systems can be killed.

The danger of root damage from excessive moisture is greater with fine-textured or compacted soils, because these soils have slower oxygen reentry after saturation. Deficient aeration is partly responsible for the shallower root systems and smaller trees seen in fine-textured soils, even without excessive water. The effects of waterlogging are exacerbated as temperatures increase. Better root survival in water-saturated soils during winter is a function of lower soil temperatures, and therefore lower root respiration rates (requiring less oxygen).

Olive tree trunk

The function of the tree trunk is to support the tree's branches, leaves, and fruit and connect them to the roots. It is composed, from the inside out, of concentric rings of xylem, cambium, phloem, and bark. The xylem is a ring of vertical tubes that conduct water, nutrients, and other minerals up from the roots to the rest of the tree. It is the sapwood, and a new ring forms on the outside of the previous year's ring every year. Over time, the multiple old rings of xylem cells in the center of the tree become inactive and die, forming the heartwood, the inner core of dead wood that supports the tree.

The cambium is a one-cell living layer that produces the xylem on its inner surface and the phloem on its outer surface. The cambium is what makes the trunk, branches, and roots grow larger in diameter. The phloem, also called the inner bark, transports nutrients and sugar from the leaves down through the branches and trunk to the roots. The annually produced new ring of phloem is interior to the previous year's phloem ring. The outer bark, which is mostly dead tissue, is produced by cork cambium. It protects the tree from injury, disease, insects, and weather. In summary, a mature tree trunk is composed of both living and dead components.

Olive tree leaves

The function of leaves is to produce the carbohydrates that the tree uses to construct itself and produce fruit. The carbohydrates are produced in the green tissues of the leaves by photosynthesis, which uses the sun's radiant energy, atmospheric carbon dioxide, and water to synthesize the carbohydrates. Oxygen is generated as a by-product.

Photosynthesis:

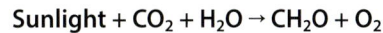

$$\text{Sunlight} + CO_2 + H_2O \rightarrow CH_2O + O_2$$

Carbohydrates are the structural building blocks of plant components. They also can be stored in the leaves and translocated, as sugar alcohol (mannitol), via the phloem to the root tips, shoots, and fruit. There, cellular respiration, using oxygen, breaks down the carbohydrates to energy that the plant can use. Carbon dioxide is generated as a by-product.

Respiration:

$$CH_2O + O_2 \rightarrow \text{roots, tree, and crop} + CO_2 + \text{energy}$$

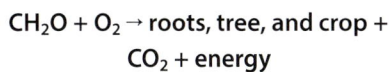

The energy produced during respiration is used to maintain the tree functions (maintenance respiration) and to produce new vegetative and reproductive growth (growth respiration). Both respiration rates are low during the dormant season and high during the growing season.

Within the growing season, the use of the carbohydrates shifts. Early in the growing season, the young, half-expanded leaves are a highly efficient source of photosynthesis but are also a carbohydrate sink, needing carbohydrates to continue constructing themselves. Later, when the leaves mature, the carbohydrates they produce support fruit growth and ripening, which are the major midseason carbohydrate sinks. In a heavy crop year, the developing fruit creates a high demand for carbohydrates, which can reduce the carbohydrates available for vegetative growth.

Competition develops in the period of simultaneous fruit and shoot growth (see fig. 2.5).

Olive leaves are evergreen. They are silvery green, thick and leathery, oblong, and 1.5 to 4 inches (4 to 10 cm) long and 0.5 to 1.25 inches (1 to 3 cm) wide. They are borne in opposite pairs on the shoot, with successive pairs at right angles to one another. New leaves are produced annually on the current year's shoot growth. Leaves grow to final size within 2 weeks and live for 2 or 3 years. Like the leaves of other evergreens, olive leaves abscise in spring. Yellowing leaves on 2-year-old and older growth in spring signal imminent abscission (or physiological or nutritional problems discussed in later chapters).

The process of photosynthesis occurs in the chloroplasts, organelles within the leaves that contain the photosynthetic pigment chlorophyll. Photosynthesis is temperature sensitive. The leaf maintains temperature by allowing ambient air into the leaf's inner layers, where the chloroplasts are located, through stomata, openings on the lower leaf surface. Water from the roots also provides cooling and maintains the aqueous environment the chloroplasts require for photosynthesis. Photosynthesis is generally inhibited above 95°F (35°C). However, olive leaves can maintain 70 to 80 percent of their photosynthetic capacity at 104°F (40°C) (Angelopoulos et al. 1996).

Olive leaves are adapted to hot, high-transpiration environments. The upper, dark green surface has a thick cuticle that protects the leaf from heat, and a layer of silver-white peltate trichomes—overlapping shieldlike structures that maintain a layer of still air on the leaf surface, preventing excess water loss by transpiration. The stomata, the porelike openings that facilitate the entry of carbon dioxide and exit of oxygen during photosynthesis, are on the lower leaf surface, which also decreases water loss. These leaf adaptations increase olive's drought resistance.

As noted above, to facilitate photosynthesis in hot dry environments, the chloroplasts are located in the inner palisade layer of the leaf as they function best in an aqueous environment. However, the low light levels in the inner palisade layer and the thick leaf cuticle result in a relatively low photosynthetic capacity. In a typical olive leaf with good sun exposure on a summer day, the stomatal opening increases as light intensity increases, reaching its maximum by midmorning, and then narrows or closes to prevent water loss in excessive afternoon temperatures.

Olive tree flowers

In olive, the process of flowering starts with the induction of the flower buds approximately 6 weeks after the current season's full bloom. Depending on variety, the flowers self- or cross-pollinate.

Flower induction, initiation, differentiation

Olive flowering involves three developmental events: induction of flower buds, initiation of flowers, and differentiation into flower parts. During induction, in July, the axillary vegetative buds make the developmental changes that result in some becoming flower buds. Floral initiation occurs in November, followed by the final differentiation into flower parts 8 months after induction, in March, after winter dormancy.

Flower bloom

At bloom, each 1-year-old shoot leaf axil contains an inflorescence, called a panicle, with fifteen to thirty flowers, which makes for hundreds of flowers per shoot and maybe 500,000 flowers per tree (fig. 2.7). In years of heavy flowering, less than 1 percent of flowers successfully set fruit. Olive flowers are small, yellow-white, and inconspicuous. Olive trees have two types of flowers: perfect and imperfect. Perfect flowers have functioning stamens, the male pollen-producing part, and a functioning pistil, the female fruit-producing part that consists of an ovary, containing the egg, connected by the style to the superior pistil (fig. 2.8). Imperfect flowers have functional stamens that produce pollen but degenerated pistils incapable of producing fruit (fig. 2.9). The total number of flowers, particularly the proportion of perfect to imperfect flowers, varies as a function of the previous season's crop load. A heavy previous year's crop will limit both shoot growth and the percentage of perfect flowers.

Flower pollination

Olive can self- and cross-pollinate. Cross-pollination is the transfer of pollen from one flower to the stigma (the pollen-receiving part of the pistil) of another flower. Wind facilitates both self- and cross-pollination and is the primary agent of the latter. While insects can facilitate self- and cross-pollination, they are not necessary.

Olive flower geometry facilitates self-pollination. Pollen from the anthers drops onto the stigma between and below them. The pollen germinates, producing a pollen tube containing the sperm that grows through the style to the ovary. The pollen tube penetrates the ovary, and the sperm fertilizes the egg in the ovule, precipitating fruit set. Multiple pollen grains may germinate and grow through the style to the ovary, but only one can successfully fertilize the egg.

Pollination and fertilization occur during what is called the effective pollination period. At this time, four factors are key to fruit set, and they are dependent on temperature and relative humidity: anther dehiscence, which stimulates the pollen to drop onto the stigma; stigma receptivity, which affects its ability to germinate pollen; pollen germination and pollen tube growth rate; and ovule viability.

The optimum temperature for these processes is 86°F (30°C) with 50 percent relative humidity. Temperatures below this level slow pollen germination and pollen tube growth, decreasing fruit set. The tube either fails to reach the ovary or grows too slowly to reach the ovary before the latter degenerates. This may occur when temperatures during bloom are above 95°F (35°C), especially when they are accompanied by dry, windy conditions that reduce fruit set by desiccating the stigma or ovary. Rain harms pollination and fruit set by hindering pollen transport and by altering the osmotic relationship between a pollen grain and stigmatic surface secretions that lead to pollen germination.

Compatibility and self-incompatibility

California oil olive varieties may be self-compatible, able to regularly set commercial crops with their own pollen. Or they may be partially to largely self-incompatible, only self-pollinating and setting regular commercial crops when climatic conditions for bloom and fruit set are

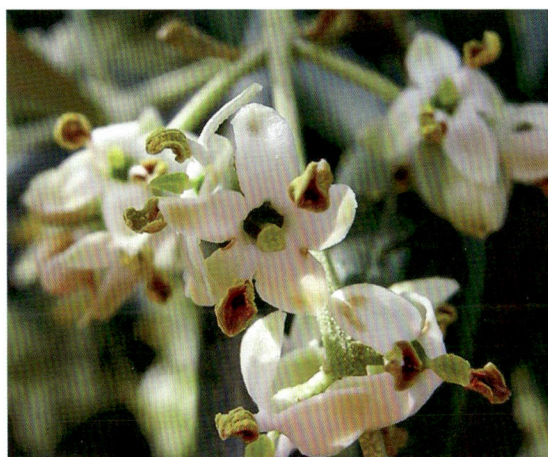

Figure 2.8

Perfect flowers contain functioning stamens, with pollen-producing anthers, and a central functioning pistil. *Photo:* Louise Ferguson.

Figure 2.7

Olive shoots produce hundreds of flowers, but less than 1 percent of the flowers may produce a mature fruit. *Photo:* Louise Ferguson.

Figure 2.9

Imperfect flowers contain functional stamens that produce pollen but have a degenerated pistil. *Photo:* Louise Ferguson.

optimal. Self-incompatibility can decrease olive oil yield. Self-compatible plants evolved self-incompatibility to enhance outbreeding, genetic variability, and consequently evolutionary diversification (Igic et al. 2008; Montemurro et al. 2019). Research in both Spain and California has demonstrated better yields in orchards planted with multiple cross-pollinating varieties.

Fruit set

The objective of pollination is fruit set—the process of a fertilized ovary becoming a fruit. After the pollination period, any flower, pollinated or not, that is not destined to become a fruit, abscises naturally. Imperfect flowers, those with no ovary to fertilize, are the first flowers to drop. Unfertilized perfect flowers and fertilized young fruit may also drop, the latter often due to competition for resources among young fruit. Varieties vary, but most fruit abscission occurs soon after full bloom, and final fruit set is achieved within a month after full bloom.

Olive tree fruit

The olive fruit is a drupe, botanically similar in form but unrelated to almond, apricot, cherry, nectarine, peach, and plum. It is made of two main parts: the flesh, called a pericarp, and the hard seed, called a pit. The pericarp consists of the skin (exocarp), a cuticular lipid layer with stomata; the flesh (mesocarp), tissue that is eaten and contains the oil; and the lignified shell (endocarp) enclosing the seed (embryo) (fig. 2.10).

The shape and size of the fruit and pit, oil content, and flavor vary greatly among varieties. Fruit and pit size and shape are the most reliable morphological features for making a visual identification of a variety. However, genetic analysis is now increasingly used for variety identification.

Olive fruit growth occurs in a typical sigmoid fashion (fig. 2.11). The pit enlarges to full size and is completely hardened approximately 90 days after flowering. During that time, inside the pit, the endosperm (food source for the seed) begins to solidify and seed development takes place, leading to embryo maturity 60 to 90 days after flowering. The flesh and skin continue their gradual growth. Fruit color changes from green to yellow-green (straw), and the purple anthocyanin blush develops. Individual fruit growth is dependent upon temperature and tree crop load. Heavier crops have smaller fruit and mature later. Oil accumulation is complete 130 to 210 days after flowering, depending on variety.

Impacts of abiotic environmental factors

The major objective in producing any tree fruit crop, including oil olive, is to establish and manage the orchard in a climate and environment that optimize the tree's ability to consistently flower and mature fruit. A climate suitable for commercial olive production has hot, dry summers and winter temperatures that fulfill the dormant chill requirement. The early season irrigation requirement may be provided by winter rain; however, for good fruit and oil production, supplemental irrigation is required through the growing season (see chapter 7, Irrigation Management). Several abiotic stress factors affect olive tree productivity.

Temperature stress

Temperature is the most important environmental factor in determining where olives can be grown and successfully produce; freeze limits the geographical range of production, and winter chill and summer heat greatly affect fruit production.

The most limiting temperature factor is cold damage (Vossen 2007). Winter low temperatures will determine where to establish an orchard as mature olive trees will not survive below 11°F (-11.7°C). The previous season's growth is damaged by winter temperatures below 22°F (-5.5°C). Fruit is damaged by fall temperatures below 29°F (-1.7°C). Young trees are damaged by winter temperatures below 25°F (-3.9°C).

The second most limiting temperature factor is insufficient dormant chill for the tree to break dormancy and produce bloom. Olive requires chilling for successful bloom (Martin et al. 2005). Dormant chill requirements differ among varieties. Most varieties require at least 10 weeks below 54° to 56°F (12.2° to 13.3°C), 70 to 80 days fluctuating between 36° and 59°F (2.2° and 15°C), or 200 to 300 hours total below 45°F (7.2°C). Temperatures constantly below 45°F (7.2°C) or above 59°F (15°C) may inhibit flowering, and midwinter temperatures above 68°F (20°C) for 2 to 3 weeks may delay bloom. In contrast to flower buds, vegetative buds of olive seem to have little

Exocarp (skin)

Endocarp (stone)   Pericarp

Mesocarp (flesh)

Seed
(embryo)

Figure 2.10

Botanically the olive is a drupe. The pericarp produces the flesh from which the oil is extracted. *Source:* Mohamed T. Nouri.

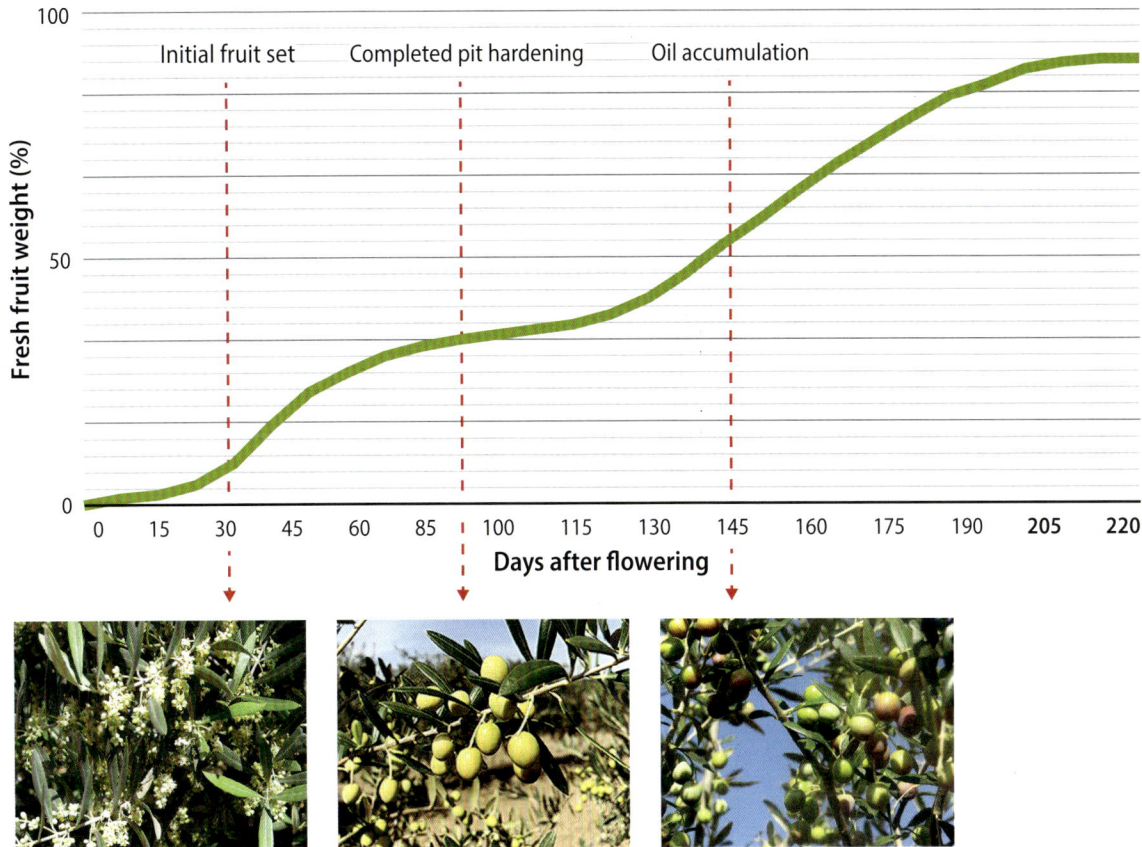

Figure 2.11

The seasonal development cycle of an olive fruit has three stages. *Source:* Mohamed T. Nouri.

if any dormancy requirement, growing whenever the temperature is 70°F (21.1°C) or higher.

The most temperature-sensitive period for olives is flowering. Low temperatures, below 60°F (15.5°C), can delay or prolong flowering and fruit set. High temperatures, above 95°F (35°C), can limit pollination and pollen tube growth, and if extremely high temperatures occur after flowering, they can eliminate the crop.

The third most limiting factor is insufficient heat to successfully mature the fruit. There is limited information on the hours of heat

accumulation required to mature the different oil olive varieties. Rather, orchard establishment considerations are based on average annual temperatures and average low and high temperatures. For good olive production, the average low temperature from April to October should be between 50° and 65°F (10° and 18°C) and the average high temperature between 75° and 95°F (24° and 35°C) (fig. 2.12).

Photosynthesis is optimal in olive leaves at temperatures between 59° and 86°F (15° and 30°C). Until recently, high temperatures during the growing season, except for extreme highs during bloom that can destroy the crop, were not thought to limit productivity or quality. However, Nissim et al. (2020) demonstrated temperatures between 104° and 113°F (40° and 45°C) 60 to 120 days after full bloom, the period after pit hardening through fruit ripening, have a variety-dependent deleterious effect on fruit weight, oil concentration, and oil quality in Koroneiki, Coratina, and Picholine varieties.

Temperatures high and low, depending upon the season, have been demonstrated to have significant effects on olive tree mortality, tree and fruit damage, and productivity. Given the warming temperatures due to human-caused climate change, defining the effects of temperature on the physiology of olive tree growth and particularly on the reproductive cycle should be the focus of future olive tree physiological research. If the optimal temperature ranges and chill and heat accumulations, and how they interact with one another, could be defined for the different varieties, determining where to plant and how to manage olive orchards could be improved. And knowing the mechanisms of how the plants cope with temperature extremes could inform future plant improvement programs.

Light stress

Only leaves on the outer edge of the canopy are exposed to full sun, and these only for a portion of the day as the sun crosses the sky. Therefore, for most of the day, photosynthesis in most of the leaves is not light saturated. The photosynthetic capacity of olive leaves drops to approximately 70 percent if the leaves develop in the shade. Light management is important for two reasons: First, shading decreases flower induction. Second, evergreen olive trees store more of their carbohydrates in their leaves than do deciduous trees; leafy shoots, even if they are not bearing, contribute to the tree's carbohydrate storage. Increasing light penetration into the interior canopy to maximize photosynthesis, and increase carbohydrate storage, is an important part of maximizing yields.

The best way to improve light penetration within the canopy is pruning, which is covered in chapter 9, Canopy Management. As most modern olive orchards are high- or super-high-density hedgerows designed for mechanical pruning, the three most important factors in canopy management are maintaining a high hedgerow (the maximum height is determined by the harvest machinery), avoiding shading of adjoining hedgerows, and maintaining canopy porosity to facilitate light penetration (Conner et al. 2014).

Water stress

In the Mediterranean climates where olive evolved, precipitation occurred primarily in the winter and was usually sufficient to produce bloom and fruit set. The final fruit yield and quality was highly dependent on the water status through the growing season. Both drought and soil saturation cause water stress. Drought stress precipitates stomatal closure and decreases photosynthesis. Because fruit production is a 2-year process, a season of drought stress affects not only the flower differentiation, bloom, fruit set, growth, and ripening of the crop in the current year but also next year's crop by decreasing shoot growth and bud induction. In drought-stressed trees, the current year's fruit exert a stronger demand on available water supplies than does shoot growth. If olive fruit are stressed to the point of shrivel, they can recover if trees are irrigated. To learn more, see chapter 7, Irrigation Management.

Water stress also occurs when excess soil moisture produces hypoxia in the root zone during the growing season, depriving the actively respiring roots of oxygen. Hypoxia decreases the roots' water uptake efficiency, signaling the stomata to close, which slows photosynthesis, decreasing carbohydrate production. Excess soil moisture is less lethal during winter, when the roots are less active. Saturated soils, particularly during cool wet springs, also make hypoxic roots susceptible to root fungal pathogens, particularly Verticillium wilt. Saturated soils also increase humidity within the olive canopy, exacerbating the risk of foliar fungal pathogens and insect pests during the growing season.

Salinity stress

There are two forms of salinity stress: osmotic stress and specific ion toxicity. Osmotic stress occurs first, subtly and slowly. High concentrations of sodium and/or chloride ions in the soil water decrease the soil water potential, decreasing the tree's ability to extract water from the soil. The symptoms are slower growth, smaller leaves,

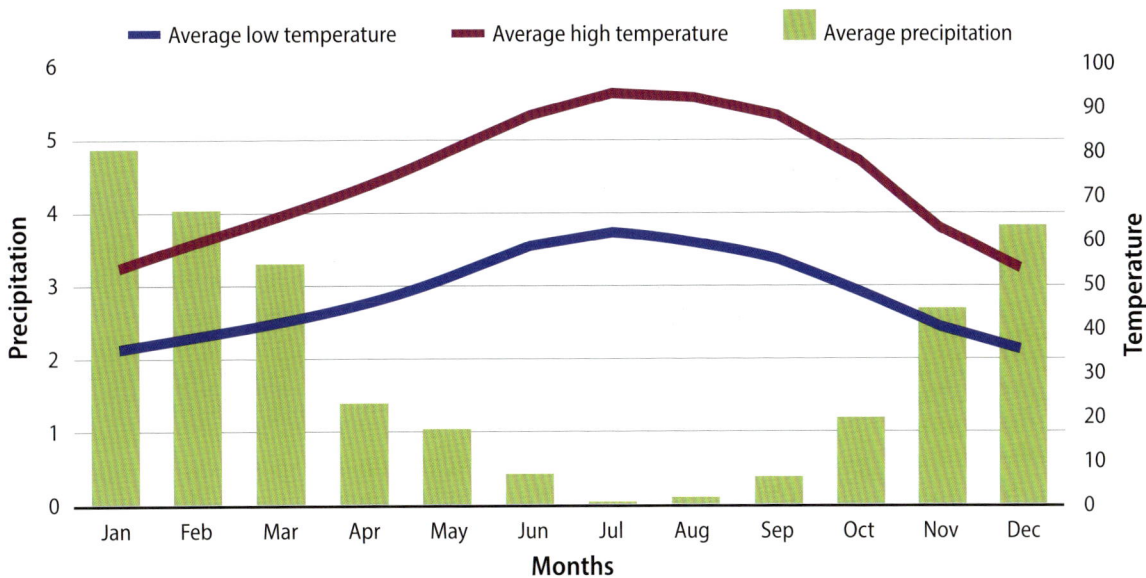

Figure 2.12

Temperature is the most important factor in assessing whether a climate is suitable for good olive oil production. The summer and winter temperatures, and precipitation amounts, shown here are ideal. *Source:* Mohamed T. Nouri.

smaller trees, and less fruit. The second form of salinity stress is specific ion toxicity. When water extracted from the soil contains sodium or chloride ions, those ions accumulate in the leaves after the water evaporates through the stomata, destroying leaf tissue and decreasing photosynthetic capacity. The symptoms are more pronounced late in the season.

Olive is considered moderately tolerant of salinity. Leaf concentrations of 0.5 percent chloride or 0.2 percent sodium on a dry-weight basis are the designated critical toxicity limits. If leaf toxicity symptoms do not manifest, the subtler symptoms of slower growth and lower vitality are often missed. Reduced shoot and root growth and lower oil production in fruit due to salinity have been reported without the appearance of leaf symptoms.

Some varieties are more tolerant of salinity than others. Unfortunately, the relative salinity responses among California varieties are not known. Marin et al. (1995) reported Picual and Lechin de Sevilla displayed greater salinity tolerance than other varieties. The difference in tolerance among varieties is the degree to which their roots either sequester salt ions or exclude salt ions; exclusion protects the leaves from damaging ion accumulations. For more information on salinity in olive, see chapter 8, Soil and Nutrient Management.

Alternate bearing

Like apple, pear, mango, orange, pistachio, and pecan, olive produces alternating large and small crops in successive years. Alternate bearing is not harmful to the tree, but it is disruptive for marketing and returns. Alternate bearing is a within-shoot phenomenon. It is initiated with the first heavy crop, which inhibits the current year's apical shoot growth, making the shoot shorter, with fewer buds and a higher percentage of imperfect flowers, thus reducing the following year's crop. The specific mechanism of alternate bearing has not been elucidated, but evidence suggests it is precipitated by low shoot carbohydrate levels. At the time of the most active shoot growth and bud induction, carbohydrates are also needed for the rapid growth of the current year's fruit (see fig. 2.5). Alternate bearing can manifest at a tree, orchard, or regional level.

Any cultural practices that diminish olive tree vigor, such as a decrease in nutrition or water, or exert a heavy demand on the tree's resources, such as leaving a large crop on the tree late in the season, exacerbate alternate bearing. Environmental stress and lack of pest control can also increase alternate bearing. Freezes or unseasonably high temperatures during bloom can synchronize alternate bearing within a region by destroying the current season's crop. Mechanical pruning, with its indiscriminate heading cuts, has been observed to moderately mitigate alternate bearing by decreasing the ratio of bearing and nonbearing shoots (Albarracín et al. 2017). No fertilization practice has been demonstrated to decrease alternate bearing.

References

Albarracín, V., A. J. Hall, P. J. Searles, and C. M. Rousseaux. 2017. Responses of vegetative growth and fruit yield to winter and summer mechanical pruning in olive trees. Scientia Horticulturae 225:135–194. https://doi.org/10.1016/j.scienta.2017.07.005

Angelopoulos, K., B. Dichio, and C. Xiloyannis. 1996. Inhibition of photosynthesis in olive trees (*Olea europaea* L.) during water stress and rewatering. Journal of Experimental Botany 47(301):1093–1100. https://doi.org/10.1093/jxb/47.8.1093

Conner, D. J., M. Gomez-del-Campo, M. C. Rosseaux, and P. S. Searles. 2014. Structure, management and productivity of hedgerow olive orchards: A review. Scientia Horticulturae 169:71–93. https://doi.org/10.1016/j.scienta.2014.02.010

Grattan, S. R., M. J. Berenguer, J. H. Connell, V. S. Polito, and P. M. Vossen. 2006. Olive oil production as influenced by different quantities of water. Agricultural Water Management 85(1–2):133–140. https://doi.org/10.1016/j.agwat.2006.04.001

Igic, B., R. Lande, and J. R. Kohn. 2008. Loss of self-incompatibility and its evolutionary consequences. International Journal of Plant Science 169(1):93–104. https://doi.org/10.1086/523362

Marin, L., M. Benlloch, and R. Fernández-Escobar. 1995. Screening of olive cultivars for salt tolerance. Scientia Horticulturae 64(1–2):113–116. https://doi.org/10.1016/0304-4238(95)00832-6

Martin, G. C., L. Ferguson, and G. S. Sibbett. 2005. Flowering, pollination, fruit, alternate bearing, and abscission. In G. S. Sibbett and L. Ferguson, eds., Olive production manual, 2nd edition. Oakland: UC Agriculture and Natural Resources Publication 3353.

Montemurro, C., G. Dambruoso, G. Bottalico, and W. Sabetta. 2019. Self-incompatibility assessment of some Italian olive genotypes (*Olea europaea* L.) and cross-derived seedling selection by SSR markers on seed endosperms. Frontiers in Plant Science 10:451. https://doi.org/10.3389/fpls.2019.00451

Nissim, Y., M. Shloberg, I. Biton, Y. Many, A. Doron-Faigenboim, H. Zemach, R. Hovav, Z. Kerem, B. Avidan, and G. Ben-Ari. 2020. High temperature environment reduces olive oil yield and quality. PLoS One 15(4):e0231956. https://doi.org/10.1371/journal.pone.0231956

Shepherd, G., F. Stagnari, M. Pisante, and J. Benites. 2008. Olive orchards: Visual soil assessment. Rome: Food and Agriculture Organization of the United Nations.

Vossen, P. M., ed. 2007. Organic olive production manual. Oakland: UC Agriculture and Natural Resources Publication 3505.

3 — Site Selection and Preparation

Dan Flynn, Executive Director Emeritus, UC Davis Olive Center

John Post, President, Agricultural Advisors

Ciriaco Chavez, Vice President of Agriculture, Cobram Estate Olives

Paul Busalacchi, Director, Agricultural Operations, Corto Olive Co.

Highlights

- Site selection may be the most important factor in determining the ultimate profitability of an olive orchard.

- Many growers have wasted large amounts of money by selecting sites that were not suitable for consistently producing high yields of quality oil.

- Evaluate the soil texture, depth, and stratification to determine how best to prepare the soil for planting.

- Hardpan and abrupt changes in soil texture can impede the movement of moisture (and salts) through the profile, resulting in saturated layers that can damage or kill olive roots.

- Send soil samples from at least two depths for chemistry testing, and do not intermix the samples.

- While olive is drought tolerant, the best production results from delivering 24 to 30 inches (61 to 76 cm) of irrigation annually.

- Send water samples for chemistry testing, as water quality impacts orchard productivity and longevity.

- Isolated orchards incur significant additional costs that may be justifiable for large orchards but not small orchards.

- Design the block with north-south orientation of the tree rows when possible to promote the highest productivity, and ensure that there is sufficient room for a mechanical harvester to turn at the ends of rows.

- Soil preparation may include land grading, deep tillage, surface tillage, berms, soil amendments, fumigation, Verticillium wilt control, and weed control.

- The irrigation system and tree support structure usually are installed prior to planting.

- Olives are planted from March through October in California, with growers identifying April, May, September, and October as the best months.

Orchard profitability starts with selecting a suitable orchard site and preparing that site well for planting. A mature olive orchard is more likely to be profitable if it consistently produces strong fruit yields with high oil content, and achieving those yields is easier with optimal climate, soil, and water. To maximize profitability, and ensure crop income exceeds costs as soon as possible, site preparation must facilitate a rapid development of the canopy to increase yields during the orchard's early years.

Site selection

Site selection may be the most important factor in determining the ultimate profitability of an olive orchard, but many growers do not devote adequate time to assessing a site's environmental conditions. California growers have lost large numbers of trees and potential yield by planting in the wrong places.

Climate

Olive is a subtropical species, and the best olive oil production and quality occur in areas with mild winters and long, warm, dry summers. Some cold winter weather is needed to supply the vernalization—winter chilling—that all olive varieties need for flower development. About 200 hours below 45°F (7°C) during winter induces proper bloom, though major California growers believe that in super-high-density (SHD) systems olives produce better with 300 chill hours. Winter daily temperatures fluctuating between 35° and 65°F (2° and 18°C) are ideal.

Excessively cold temperatures can injure or kill trees, although the impact depends on the variety's cold-hardiness as well as the suddenness and duration of the cold (table 3.1). Temperatures between 32° and 23°F (0° and -5°C) can cause shoot tip dieback and slight shoot cracking. Temperatures below 25°F (-4°C) can damage young trees. Temperatures between 23° and 18°F (-5° and -8°C) can crack or kill the large limbs of older trees. Temperatures below 18°F (-8°C) can kill cold-sensitive trees and if they drop below 15°F (-9°C) can injure or kill large limbs and whole mature trees of cold-hardy varieties.

Freezing-induced cracks can also be easily invaded by olive knot bacteria, which can girdle, defoliate, and kill twigs and branches.

Flower bud development, pollination, and fruit set are influenced by climatic conditions. Prolonged cold below 55°F (13°C) and wet weather during bloom development reduce fruit set by producing fewer perfect flowers and more imperfect (staminate) flowers than under ideal conditions (see chapter 2, Olive Botany, Growth, Morphology, and Physiology). The same conditions hinder pollination and thus reduce fruit set. Conversely, hot, dry, windy conditions during bloom shorten the tree's receptive period for pollination by drying out the flowers.

Cold climates also make it difficult for the fruit to accumulate oil, as do temperatures above 93° to 95°F (34° to 35°C). If the temperature falls below 30°F (-1°C) prior to harvest, it can lead to frost-damaged olives, which produce off-flavored oil prone to rapid oxidation. Growers should avoid a site with unfavorably cold conditions or, if there is no alternative site, plant cold-hardy varieties.

Significant summer rainfall favors the development of peacock spot fungus and olive knot bacteria, which diminish productivity and increase management expenses. Wet soil from rainy conditions during harvest can also impede operation of heavy mechanical harvesters and delay harvest.

Consult historic weather data for the orchard site. Pay particular attention to low areas where cold air accumulates. Ideally, develop multiyear, site-specific data using a weather station, electronic data logger, or remote monitoring service. In the absence of site-specific data, there are free online sources of regional weather data from the federal

Table 3.1

Seasonal climate challenges for olive orchards

Winter	• Minimal tree damage: 32° to 23°F (0° to -5°C) • Young tree damage: < 25°F (-4°C) • Tree cracking and large limb death: 23° to 18°F (-5° to -8°C) • Tree death of cold-sensitive varieties: < 18°F (-8°C) • Large limb and mature tree death of cold-hardy varieties: < 15°F (-9°C)
Spring	• Reduced fruit set: prolonged temperatures < 55°F (13°C), very high humidity, or hot, dry, windy conditions during bloom
Summer	• Reduced oil accumulation: > 93° to 95°F (34° to 35°C) • Development of peacock spot fungus and olive knot bacteria: significant rain
Fall	• Frost damage and oil flavor defects: < 30°F (-1°C) • Harvest delays: rain

National Oceanic and Atmospheric Administration (NOAA) and the California Irrigation Management Information System (CIMIS).

Soil

Although olive is adapted to a wide variety of soils, production is best where the trees can develop roots without adverse impacts from soil physical or chemical conditions. There is a strong relationship between the productivity of an orchard and the soil's physical condition and chemistry. See chapter 8, Soil and Nutrient Management, for a detailed discussion of soil diagnostic methods.

Soil physical condition

The physical condition of a soil describes its

- **texture:** the relative amounts of sand, silt, and clay making up the soil
- **depth:** measured to the point where there is a barrier to root growth
- **stratification:** layers of different textures

Olive trees are shallow rooted and prefer a moderately fine soil texture known as loam. Loam soils contain mostly sand and silt with a smaller proportion of clay. Depending on the specific sand, silt, and clay content, they are classified into several categories: sandy loam, loam, silt loam, clay loam, and silty clay loam (fig. 3.1). Loam soils provide aeration for root growth, are quite permeable, and have adequate water holding capacity. Sandy soils do not have good nutrient or water holding capacity, and clay soils often do not have adequate aeration for root growth. Sandy and clayey soils are difficult to manage for maximum production.

Soils with an unstratified profile of 4 feet (1.2 m) are suitable for olive, although trees in an SHD system can be planted in soils with a depth as little as 2 feet (0.6 m) if there are not difficult soil textures, unfavorable soil chemistry, or perched water (water pooling above a layer of impermeable rock) below this level.

Evaluate the soil texture, depth, and stratification to determine how best to prepare the soil for planting. Start the soil evaluation with

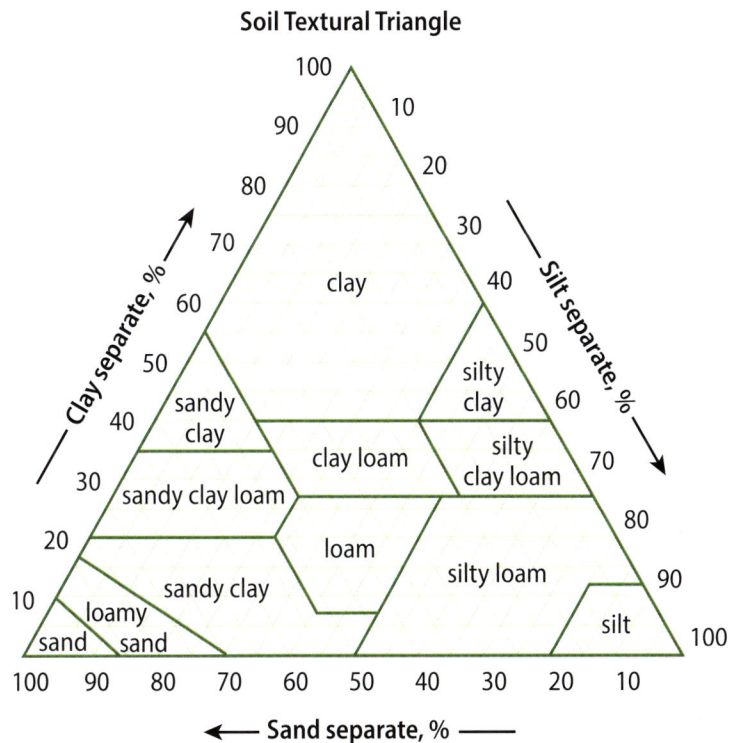

Soil Textural Triangle

Figure 3.1

Olive grows best in loam soils, which contain sand, silt, and clay.
Source: USDA Natural Resources Conservation Service.

a review of existing soil survey data. The UC Davis California Soil Resource Lab offers free soil information online, including detailed soil survey data compiled from multiple government sources at the United States Department of Agriculture National Cooperative Soil Survey.

Excavating soil inspection holes is a useful way to evaluate soil conditions for their suitability for olive growing and to identify conditions requiring modification before planting. If more than one soil type is present at the site, dig one or more inspection holes in an area of each type. Growers typically excavate two to four sites per 50 acres (20 ha), more if multiple soil types are present. Holes should be dug to a depth of 6 feet (1.8 m). Many growers use a backhoe, which can dig the holes rapidly and easily.

For each inspection hole, observe soil texture and structure to assess soil water holding capacity (clay soil holds about double the water sandy soil holds.) Check for stratification (fig. 3.2). Hardpan and abrupt changes in soil texture can impede the movement of moisture (and salts) through the profile, resulting in saturated layers that can damage or kill olive roots.

Observe the health of any roots present, and the depth to which they have penetrated. Dig holes at the drip line of any existing trees to observe the roots and gain insights on root growth conditions. Look for indicators of iron oxidation (rusty orange mottling), which illustrate past saturated soil conditions, and a possible salt accumulation zone (fig. 3.3). These conditions would impede olive tree growth.

Check for the presence of perched water; it may be rainwater or irrigation failing to drain because of hardpan or changes in soil texture, or it may be subsurface water from a nearby river, rice field, irrigation ditch, or uphill source (fig. 3.4). Even if perched water is not present, past occurrences could be indicated by iron oxidation. Flooding adjacent to the site (for example, for rice production) can cause seasonal perched water. It may be necessary to inspect the soil more than once to determine if perched water is present.

Perched water that rises to within 4 feet of the soil surface at any time of year is detrimental to root health and tree survival. The conditions causing it must be addressed before planting. Tile drains may be needed, which involves a contractor installing drainage pipes about 6 feet (1.8 m) below the soil surface in a grid pattern designed specifically for the water quantity and soil texture at the site; it is critical that the diversion pipe has an outlet, such as an empty ditch, that has sufficient capacity for the drainage water.

Figure 3.2

A soil inspection hole reveals three soil layers: silt (top), sand (middle), and clay (bottom). *Photo:* John Post.

Silt loam soil with signs of iron oxidation

Salt accumulation

Stratified sandy soil

Figure 3.3

A soil inspection hole reveals indicators of iron oxidation and salt accumulation. *Photo:* John Post.

Soil chemistry

Olive tolerates soils of varying chemical quality. The best soils for olive production are moderately acid to moderately alkaline (pH between 6.5 and 8.5) (table 3.2). A pH below 5.5 can induce aluminum and manganese toxicity in trees; soils with a pH above 8.5 often have poor structure and may be high in sodium. High sodium can be toxic as well as cause water penetration and drainage problems that kill olive roots.

Figure 3.4

A soil inspection hole reveals perched water in the rooting zone. *Photo:* John Post.

Avoid soils with salinity (electrical conductivity, EC) higher than 4 dS/m, an exchangeable sodium percentage (ESP) above 4, or a sodium absorption ratio (SAR) above 15. Avoid soils with excessive boron ($\geq$ 2 ppm), excessive chloride ($\geq$ 10 meq/L), or a ratio of calcium to magnesium plus sodium of less than 1:1, as these conditions may reduce productivity. Test soil also for nitrogen, phosphorus, potassium, and zinc to assess future fertilization needs.

Take soil samples from a soil inspection hole; if there is more than one type of soil at the site, take soil samples from a hole of each soil type. Sample the soil in the hole at these depths: between the soil surface and 6 inches (15 cm), and between 6 inches (15 cm) and 24 inches (61 cm) below the soil surface. If soil investigations have revealed a potential problem, such as salts, iron oxidation, or hardpan, consider taking an additional sample between 24 inches (61 cm) and 48 inches (122 cm) below the soil surface.

Keep the samples from each depth and each hole separate: do not mix soil from different layers in a hole, and do not make composite samples from multiple holes. Lab results will provide information on each layer, which may identify undesirable properties that may be exacerbated by deep ripping or slip plowing. Do not rely on analysis of soil samples from a home soil testing kit.

If there is a question about the history of the site, if existing vegetation shows odd symptoms or poor growth, or if other soils in the area have known toxic levels of excess nutrients, then more soil tests may be justified. For example, testing for

Table 3.2

Soil chemistry test parameters for olive sites

Test parameter	Desirable level
Acidity/alkalinity (pH)	6.5–8.5
Electrical conductivity (EC)	< 4 dS/m
Exchangeable sodium percentage (ESP)	< 4
Sodium absorption ratio (SAR)	< 15
Boron (B)	< 2 ppm
Chloride (Cl-)	< 10 meq/L
Calcium to magnesium + sodium ratio	> 1:1 (better at 3:1 or higher)
Nitrogen (N), phosphorus (P), potassium (K), and zinc (Zn) to assess future fertilizer needs	

Verticillium wilt is advisable if the site was previously planted with cotton, tomatoes, cucurbits, eggplant, peppers, potatoes, or any other potential host. Nondefoliating strains of Verticillium wilt present at less than 2 microsclerotia per gram of soil may pose less risk. Many growers do not plant olives on a site previously planted with cotton.

Water

While olive is drought tolerant, maximizing oil productivity in Central Valley orchards requires between 24 inches (61 cm) and 30 inches (76 cm) of irrigation water per acre annually (rain during the growing season counts toward the irrigation requirement). Surface and groundwater availability (and, if present, well and pump capacity) need to be evaluated to determine whether there is adequate volume and quality of irrigation water available throughout the growing season to make the orchard economically competitive.

The irrigation requirement is based on the climatic demand of the growing site after subtracting the amount of water stored in the soil from winter rainfall and the amount of any likely effective rainfall during the growing season (rainfall within the root zone that would be used by the tree). Climatic demand information can be obtained from the California Irrigation Management Information System (CIMIS), an online source for evapotranspiration data derived from weather stations located throughout the state. Once the irrigation requirement is calculated, knowing the soil type and rooting depth and the application efficiency of the irrigation system (see chapter 7, Irrigation Management) can then help determine irrigation timing and application rates.

Irrigation water quality impacts orchard productivity and longevity. Compared to other trees, olive trees tolerate water that is relatively high in boron, but water that is high in nitrogen promotes excessive vegetative growth, which hinders fruit production. Excess sodium in irrigation water can accumulate in the soil, causing sodic soil conditions and water penetration problems.

Send a water sample to a lab and request standard agricultural water testing. The most important parameters to check are pH, electrical conductivity (ECw), sodium, bicarbonate, sodium absorption ratio (SAR), chloride, boron, and nitrate nitrogen. Table 3.3 shows the desirable levels for key water quality criteria. Help with evaluation of water analysis test results is available from University of California Cooperative Extension farm advisors.

Elevations and land features

Evaluate the site for slope and topographical features that may hinder orchard operations. Safe operation of the harvester and other equipment limits planting to areas with a slope no greater than 15 to 30 percent, depending on harvesting method (determine the harvester's maximum slope capacity with the dealer or manufacturer). Contour features that could direct cold air or water into a low-lying orchard pose a risk to the health of the trees and fruit quality.

Natural features such as vernal pools and native species habitat may limit planting. In California, agricultural operations are subject to laws and regulations designed to protect designated environmentally sensitive areas. Consider contracting with an environmental engineer to do a site evaluation to determine whether any specially protected areas are present and to advise on required protection measures.

Site planting history

Assess the site planting history to determine whether olive-damaging nematodes (see chapter 11, Nematodes of Olive), Verticillium wilt, or Armillaria root rot (see chapter 12, Diseases of Olive) may be present and warrant further investigation or action. Consider avoiding affected areas entirely.

Proximity to supplies and services

Isolated orchards may incur significant additional costs to access transportation, power, and water infrastructure; essential supplies and services; labor; and processing facilities. These costs may be financially justifiable for large orchards but less so for small orchards.

Block design

Design the planting blocks so that each block has uniform soil physical and chemical properties, which will facilitate management practices. The irrigation system should have the capacity to deliver water uniformly throughout a block.

Table 3.3

Water chemistry test parameters for olive sites

Test parameter	Desirable level
Acidity/alkalinity (pH)	6.5–8.5
Electrical conductivity (ECw)	< 3 dS/m
Exchangeable sodium percentage (ESP)	< 4
Bicarbonate (HCO$_3$.)	< 3.5 meq/L
Sodium absorption ratio (SAR)	< 6
Chloride (Cl$^-$)	< 3 meq/L
Boron (B)	≤ 1–2 mg/L
Nitrate nitrogen (NO$_3$-N)	< 5 ppm

Sunlight and row orientation

Olive trees bear fruit on 1-year-old shoots, mainly on the canopy's outer periphery, where they are exposed to adequate sunlight. If shoots are shaded, production decreases. In full shade, shoots do not form flowers or produce fruit. A mature orchard is considered at optimal bearing potential when trees are spaced and managed at the greatest density that still lets them intercept enough sunlight for annual productive shoot growth.

Orient rows north-south when possible. The sun exposure on the east side of the row in the morning and the west side in the afternoon produces a symmetrical canopy with good production potential on both sides. An east-west orientation, with more sun exposure on the south side of the rows than the north side, produces an asymmetrical canopy with lower production on the north side. Research on a variety of tree crops has shown that north-south hedgerow orchards typically produce about 20 percent more yield than east-west hedgerow orchards.

Issues with harvester access, row length, drainage, or crosswinds may determine the row orientation. High-quality production is still possible in east-west plantings, but more research is needed to better understand how to maximize their productivity.

Row length

Row lengths in modern planting systems can exceed 2,600 feet (800 m), which increases efficiency by minimizing the time needed for turning equipment around, but a very long row makes it harder to service a harvester that has a mechanical breakdown, and the harvest yield may exceed the capacity of the receiving gondola, potentially delaying the harvest until another gondola arrives. Block design also includes allowing sufficient space at the row ends for the harvester to make turns, up to 40 feet (12.2 m), depending on the equipment.

Preparing the soil

Once a site has been selected and the blocks designed, the soil must be prepared for planting. Implementing the following operations will facilitate efficient growth of the newly planted trees and long-term management of the orchard.

Leveling

The site does not need to be completely flat but ensure that all areas of the orchard have good drainage with no pockets where water can accumulate. Where needed, promote drainage of excess winter rainfall by roughly leveling the site to achieve a minimum 10 percent slope. Contractors are available to evaluate site topography using GIS measurements and develop a drainage plan.

Deep tillage

Once the land has been leveled, the next step is deep tillage, which is most often accomplished with a ripper or slip plow. Deep tillage breaks up compacted, cemented, and textural layers that impede water movement in the soil profile; olive trees are killed when poor drainage creates saturated soil conditions in the root zone. Deep tillage is not usually required for soils with a 4-foot (1.2-m) homogenous profile unless they have been compacted by cultivation. Summer is the best time for deep tillage, as dry soils are easier to fracture and manipulate.

Ripping, which involves pulling a 3- to 5-foot (0.9 to 1.5 m) shank through the soil profile with a tractor, is used to break up compacted soils. Rip the entire site, both in and between the proposed tree rows. The number of passes depends on the soil texture and density. Ripping is best suited to homogenous soils.

Slip plows, which pull a steel shoe about 2 feet long through the soil, are used for deep tillage of soils with varying textures in the profile. They mix the layers and homogenize the soil profile above the plow blade, improving rooting depth and enhancing even water penetration throughout the rooting zone (fig. 3.5). Do not use a slip plow if the lower soil profile has toxic levels of salt, because it will bring the salt up into the rooting zone.

Surface tillage and berms

After deep tillage, the field typically is disked or chiseled and then "floated" to level the surface. At this stage, some growers also rototill the planting row to ensure that the soil there is broken up very finely.

Some growers create berms in the planting row. These raised strips of soil, typically about 10 inches (25 cm) high and 18 inches (46 cm) wide, ensure good drainage for the tree crown and provide more soil depth for root development (particularly if the soil evaluation shows impediments to growth such as salts and hardpan lower in the profile). In side-sloped plantings, mounds can be used instead of berms to allow drainage.

Soil amendments

Some amendments are best applied along the planned tree row and incorporated just prior to planting. Lime is an example; it is not sufficiently soluble to be applied with irrigation so it is applied down the tree row and disked into the soil or berm prior to planting. Also, organic amendments, like compost, which help improve soil water and nutrient holding capacity, can be mixed into the soil at this time. If trees are being planted

on berms, a good time to apply and incorporate into the root zone any needed preplant amendments is during berm formation.

Most fertilizers are applied after planting in the irrigation system—phosphorus in one of the initial irrigations to help with root growth, followed by nitrogen to help with growth of the entire plant. Gypsum is often applied, before or after planting, if there is an imbalance with the calcium and magnesium ratio or excess sodium in the soil chemical analysis. Potassium can be applied before planting but more often is applied through the irrigation system after trees are established, and for the life of the orchard (see chapter 8, Soil and Nutrient Management).

Fumigation

If fumigation is needed, it is applied after leveling and deep tillage. Fumigation can suppress plant-parasitic nematodes (see chapter 11, Nematodes of Olive) and provide some suppression of weeds, and also Armillaria root rot, although it doesn't eradicate the pathogen. (Thorough removal of

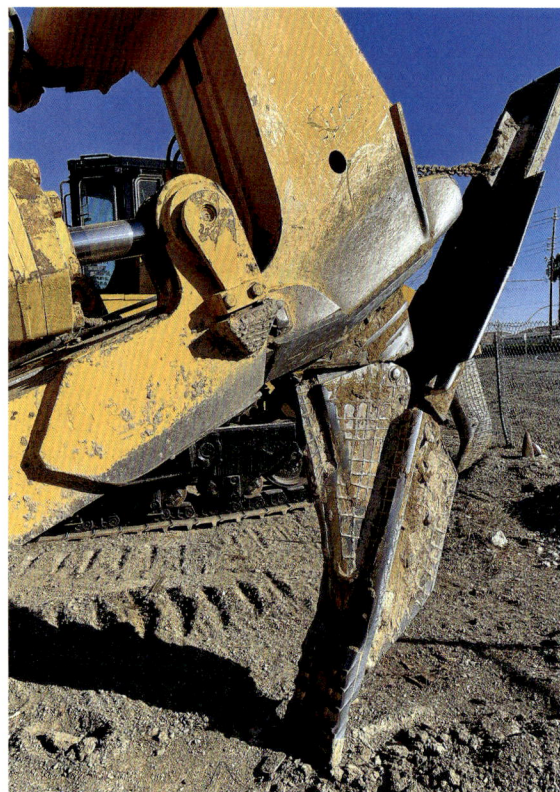

Figure 3.5

The purpose of a slip plow is to mix layers in the soil profile. *Photo:* John Post.

any roots of infected plants that remain after deep tillage, before planting, can also help slow progress of Armillaria root rot.)

In general, California olive growers have not found it necessary to fumigate before planting. Fumigation is an expensive practice and must be done correctly to be effective. It is essential that the soil be prepared and at the proper temperature and moisture conditions for treatment. Consult commercial fumigant applicators or product manufacturers for guidance on proper prefumigation soil preparation.

Verticillium control

The microsclerotia (fungal survival structures) of Verticillium wilt fungi have been documented to survive for over 30 years in the soil after an infected perennial crop. As mentioned above, olive growers may want to avoid such a site. Verticillium inoculum levels can be reduced by removing (during ripping or tillage) as many roots as possible of previous trees or vines, by summer flooding, by growing several seasons of grass cover crops (especially rye or sudangrass), or by a combination of these treatments, but it is unknown whether these techniques result in significant disease reduction.

Weed control

Weed management starts before the orchard is planted. Reducing weed pressure in the season or two prior to planting may be time and money well spent because it allows the young trees to grow with less competition and lessens the need for herbicides and thus the risk of herbicide damage to the young trees. At a minimum, do not allow weeds to go to seed in the fallowed orchard site prior to planting.

Control bermudagrass, dallisgrass, and johnsongrass with repeated disking and drying during summer (if the site is not irrigated); seedlings of these weeds can be controlled after the orchard is planted. Field bindweed can also be reduced by disking and drying during summer, but it is best controlled by first irrigating to produce a vigorous plant and then treating with glyphosate or 2,4-D (2,4-D is not registered for use on olive in California but can be used prior to planting), followed in 10 days by disking and drying the soil. Field bindweed is not completely controlled by

any method, but seedlings can be managed with cultivation or contact herbicides.

After planting, preemergence herbicides registered for use in young orchards can control seed-propagated weeds. Weeds that emerge later can be managed with cultivation or postemergence herbicides.

In organic production systems, chemical weed control options are limited and expensive. As such, weed management is based largely on cultural, physical, and mechanical measures. Soil solarization is one nonpesticidal method of pest and weed control before planting. Burying irrigation drip lines in the soil reduces weed growth but must be planned for in the orchard design; weed flaming and in-row cultivation are other control methods, but irrigation lines need to be suspended above the soil surface, and that also needs to be part of the orchard design. If cover crops are part of a weed management strategy, they may require more irrigation than olives and that should be planned for; it is cheaper and easier to ensure adequate irrigation capacity during the orchard design than retrofitting an undersized system later. Also plan ahead for whether mulches, another weed reduction strategy, will be installed before or after planting. See chapter 13, Weed Management in Olive, for detailed information on weed types and control methods.

Preparing for planting

Well in advance of tree arrival, growers should have scheduled the best time for planting, installed the irrigation system, and probably also installed the support structure for the trees. Planning and completing these operations will maximize early growth, minimize tree losses, and ensure proper training.

Timing of planting

Olives are planted from March through October in California, with growers identifying April, May, September, and October as the best months. Schedule planting when there is a low probability of frost and the ground is not saturated with moisture. To avoid Phytophthora root rot problems in clay-rich soils that drain slowly, plant later in the spring, when ground temperature is sufficient to maximize root development and growth. Planting after late spring reduces the first-year growth potential. Planting in the fall is an option in areas that have a low likelihood of winter temperatures falling below 30°F (-1°C),

Irrigation system

Install the irrigation system before planting, so the orchard can be irrigated prior to and immediately after planting (see chapter 7, Irrigation Management). Growers with SHD orchards typically install a single-line drip system, although some prefer a double-line system to help trees get established, deliver sufficient water when summer heat is at its peak, and provide more flexibility in adjusting the size of the wetted area as trees grow. Growers with HD orchards usually install a double-line drip system. For specifications on installing the irrigation system, see chapter 4, Establishing a Super-High-Density Olive Orchard; chapter 5, Establishing a High-Density Olive Orchard; and chapter 6, Microirrigation Systems and Fertigation.

Support structure

A tree support structure is usually installed prior to planting in hand-planted orchards where over-the-row harvesters will be used. In mechanically planted orchards, the support structure is installed after planting. The support structure holds the trees upright, prevents wind-misshapen trees, and ensures a straight tree row to accommodate over-the-row harvesters. Options for the support system include a small trellis, heavy stakes, and a large trellis. For specifications on installing a support structure, see chapter 4, Establishing a Super-High-Density Olive Orchard, and chapter 5, Establishing a High-Density Olive Orchard.

Suggested reading

Elmore, C. L., J. J. Stapleton, and C. E. Bell. 1997. Soil solarization. Oakland: UC Agriculture and Natural Resources Publication 21377.

Rius, X., and J. M. Lacarte. 2010. The olive growing revolution: The super high density system. Barcelona: Private-sponsored publication.

Sibbett, G. S., and L. Ferguson, eds. 2005. Olive production manual. 2nd edition. Oakland: UC Agriculture and Natural Resources Publication 3353.

Vossen, P. M., ed. 2007. Organic olive production manual. Oakland: UC Agriculture and Natural Resources Publication 3505.

4

Establishing a Super-High-Density Olive Orchard

Dan Flynn, Executive Director Emeritus, UC Davis Olive Center

John Post, President, Agricultural Advisors

Paul Busalacchi, Director, Agricultural Operations, Corto Olive Co.

Lizandro Magana, Director of Farming North, California Olive Ranch

Highlights

- Compared to the high-density (HD) system, the super-high-density (SHD) system has the advantages of higher yields in the early years of the orchard, availability of contract harvesters, and greater capability of harvesters to manage steep slopes.

- The disadvantages of an SHD system include higher establishment costs and the fact that fewer varieties are suitable for the SHD system, which limits a grower's options to match a variety to agronomic conditions and provides less diversity in the oil sensory and phenolic profiles.

- Many California SHD growers have found that stronger yields come from rows spaced at 12 feet (3.7 m).

- California growers have reported better yields with SHD varieties when they are planted with pollenizers.

- Growers can stipulate desired tree size in the contract with the nursery, and that the trees be free of pests and diseases.

- Arbequina is the most widely planted oil olive variety in California. Growers con- sider it the easiest to grow and earliest to ripen of the three major SHD varieties, and it produces the mildest oil.

- Arbosana has the least vigor and the highest, most consistent fruit yield, but it is sensitive to cold and olive knot, and unharvested fruit is prone to linger as mummies.

- Koroneiki has the most vigor, making it suitable for marginal soils, and produces the most robust oil.

- Most California SHD growers install a small single-wire trellis, which is less expensive than a larger two-wire trellis.

- Vigilant care is needed after planting to minimize tree losses, especially losses due to inadequate soil moisture.

- Avoid excessive tree pruning because it delays the onset of bearing.

- Control weeds so that they do not com- pete with the new trees for water and nutrients.

Properly establishing a super-high-density (SHD) orchard, which can range from about 600 to 900 trees per acre (1,500 to 2,225 trees per ha), has impacts on the profitability of the orchard. The grower must plant, nurture, and train the trees to minimize replants, quickly establish a hedge, and produce profit as soon as possible. For long-term profitability, it is also essential during the establishment period that the trees are trained carefully into straight hedgerows that will accommodate the mechanical harvester, minimizing tree damage during harvest.

The SHD production system has dominated California oil olive plantings since 2005. Pioneered in Spain in 1995, the SHD system significantly changed olive growing by allowing hedgerows of densely spaced trees to be harvested with over-the-row mechanical harvesters similar to the modern harvesting systems developed for other tree crops and vineyards (fig. 4.1). Approximately 70 percent of California oil olive acreage is planted in the SHD system, compared to about 4 percent globally.

Compared to the high-density system (200 to 300 trees per ac, 500 to 750 trees per ha; see chapter 5, Establishing a High-Density Olive Orchard), the SHD system has the advantage of higher early yields in the initial years of the orchard. Other advantages of using the SHD system are the availability of contract harvesters and harvester capability to manage steeper slopes. Disadvantages include higher establishment costs and few suitable low-vigor varieties, which limits the grower's options to match a variety with agronomic conditions and provides less diversity in the oil sensory and phenolic profiles. The SHD system tends to be favored by large growers rather than small growers given that the cost of the straddle harvester usually necessitates larger plantings, and, also, small growers typically seek to differentiate their flavor profile from the widely available SHD-produced olive oil.

Tree spacing

The primary objectives in designing the tree spacing are to quickly create a hedge and to facilitate sufficient light interception in the canopy to maximize olive number, size, and oil accumulation. The vast majority of California growers with SHD systems have used a row spacing of 12 feet (3.7 m) or 13 feet (4 m). In the early years of the SHD system in California, 13 feet (4 m) was the predominant row spacing, but many growers since have found that stronger yields come from orchards with rows spaced at 12 feet (3.7 m). Spanish research confirms that higher SHD yields are found in orchards with narrower row spacing. Industry authorities in Spain report that Arbosana, which is less vigorous than the other primary SHD varieties, can be grown with row widths as low as 10 feet (3 m), assuming the availability of suitable harvest equipment.

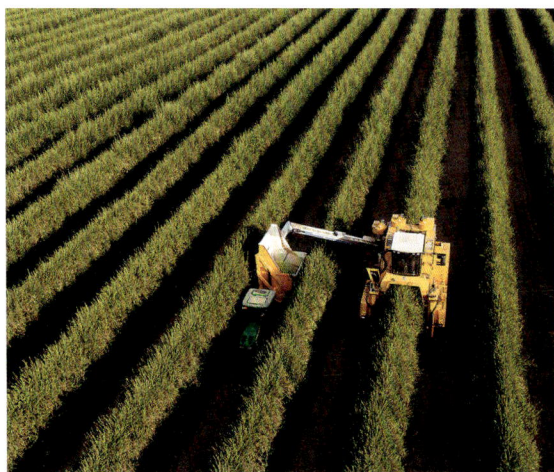

Figure 4.1

Olives grown in an SHD system are harvested with over-the-row equipment. *Photo:* Corto Olive Co.

SHD trees in California are generally spaced 5 feet (1.5 m) apart in the row. In areas of higher rainfall or deeper soil, growers may increase tree spacing in the row to 6 feet (1.8 m) to accommodate expected higher tree vigor, particularly for Koroneiki, which is more vigorous than the other primary SHD varieties. In shallower or less-fertile soil, growers may reduce tree spacing to 4 feet (1.2 m), particularly for Arbosana.

In general, higher tree densities mean the orchard comes into production earlier, but higher tree densities also mean added costs for tree purchase and early orchard maintenance. Higher tree densities can also increase pest and disease pressure and limit canopy light interception. Lower densities reduce the tree costs, but hedge establishment is slower, which delays production and the return on investment.

Support structure

A support structure usually is installed before the trees are planted if the trees are being manually planted, and after planting if the trees are being mechanically planted. The three options that have been used in California are a small trellis, heavy stakes, and a large trellis.

Small trellis

A small trellis has been the most common support structure used for California SHD olives in recent years, as it costs less than the large trellis that once dominated California SHD orchards. The small trellis requires these materials:

- end posts of treated wood with a diameter of 5 inches (12.7 cm) or metal with a spade and a diameter of 2 ⅞ inches (7.3 cm) and 6 to 6.25 feet (1.8 to 1.9 m) long, installed at about a 65- to 75-degree angle at the ends of each row, anchored 3 feet (0.9 m) into the ground
- metal T-posts 4 to 4.5 feet (1.2 to 1.4 m) long, anchored 2 feet (0.6 m) into the ground along the row at intervals of 50 to 100 feet (15.2 to 30.5 m), each T-post equidistant between two tree locations
- bamboo canes 3.5 to 5 feet (1.1 to 1.5 m) long and ⅜ to ⅝ inch (10 to 16 mm) in diameter, pushed 6 to 16 inches (15 to 41 cm) into the ground along the tree row, one at each tree position, so that at least 3 feet (0.9 m) of the cane is above ground

- one 12- to 13-gauge galvanized wire wrapped around the end posts, strung with adequate tension the length of the row at a height of 18 to 30 inches (46 to 76 cm), and clipped to the T-posts and bamboo canes

Some growers do not want the end posts or T-posts more than 1 inch (2.5 cm) above the wire; others allow up to 3 inches (7.6 cm). Some growers do not want the bamboo canes more than 2 inches (5 cm) above the wire; others allow up to 12 inches (30 cm), depending on where the tree will be topped. The central leader is tied to the cane at 8- to 14-inch (20- to 36-cm) intervals as the leader grows (fig. 4.2).

Heavy stakes

Heavy stakes have become the most common choice in European SHD olive orchards in recent years. To build the support, push a single 5-foot-long (1.5 m) wooden stake with a diameter of 1 inch (2.5 cm) 20 to 24 inches (50 to 61 cm) into the ground at each tree location and tie the tree to the stake at 8- to 14-inch (20- to 36-cm) intervals as the leader grows. Heavy stakes are less expensive than a trellis but some California growers believe that a trellis is stronger, particularly in windy locations, and that the stakes can be damaged by the harvester when young trees are being harvested.

Large trellis

A large trellis was the favored support globally for about 15 years after the SHD system was

Figure 4.2

A small trellis with trees tied to bamboo stakes is the most common tree support in California. *Photo: Agromillora California.*

introduced (fig. 4.3). A number of growers still use a large trellis, but most California growers have moved to a small trellis because of its significantly lower costs. The large trellis requires these materials:

- end posts of treated wood with a diameter of 5 inches (12.7 cm) or metal with a spade and a diameter of 2 ⅞ inches (7.3 cm) and 8 feet (2.4 m) long, installed at about a 65- to 75-degree angle at the ends of each row, anchored 3 feet (0.9 m) into the ground
- metal T-posts 8 feet (2.4 m) long, anchored 2 feet (0.6 m) into the ground along the row at intervals of 50 to 55 feet (15.2 to 16.8 m), each T-post equidistant between two tree locations
- bamboo canes 7 feet (2.1 m) long and ⅝ to ⅞ inch (16 to 22 mm) in diameter, pushed 4 to 6 inches (10 to 15 cm) into the ground along the tree row, one at each tree position (the central leader is tied to the cane at 8- to 14-inch (20- to 36-cm) intervals as the leader grows)
- two 12- to 13-gauge galvanized wires wrapped around the end posts, strung with adequate tension the length of the row, and clipped to the T-posts and bamboo canes—the lower wire at 3 to 3.5 feet (0.9 to 1.1 m) and the upper wire at 5 to 6 feet (1.5 to 1.8 m)

Nursery tree orders

Nurseries propagate SHD trees from cuttings under mist irrigation in greenhouses (fig. 4.4). The buyer can specify a minimum tree size and that the trees be free of pests and diseases. An appropriate minimum tree size is 12 to 14 inches (30 to 36 cm). Various factors impact nursery production, so nurseries do not guarantee a delivery date at the minimum size. Growers should order trees well in advance of planting, as far as 2 years prior, as deliveries are prioritized on a first come, first served basis. Expect to put down a deposit of 25 to 50 percent.

SHD trees generally arrive from the nursery with a single central leader. Growers report that they have received trees from 6 to 28 inches (15 to 71 cm) tall. Larger trees cost more but may produce a crop in the second year from planting. The nursery may also have a limited quantity of taller trees in larger pots. As long as taller trees are not root-bound, they have a better survival rate than small ones. Some states may require that a California nursery provide a phytosanitary certificate for olive trees shipped to orchards in their state.

The nursery grows the trees in approximately 3-inch (7.6-cm) pots, which may be biodegradable, and ships the potted trees in cartons. Open the cartons as soon as they arrive, as the trees can defoliate after several days without light. When

Figure 4.3

A large trellis features tall canes and two wires. *Photo:* Corto Olive Co.

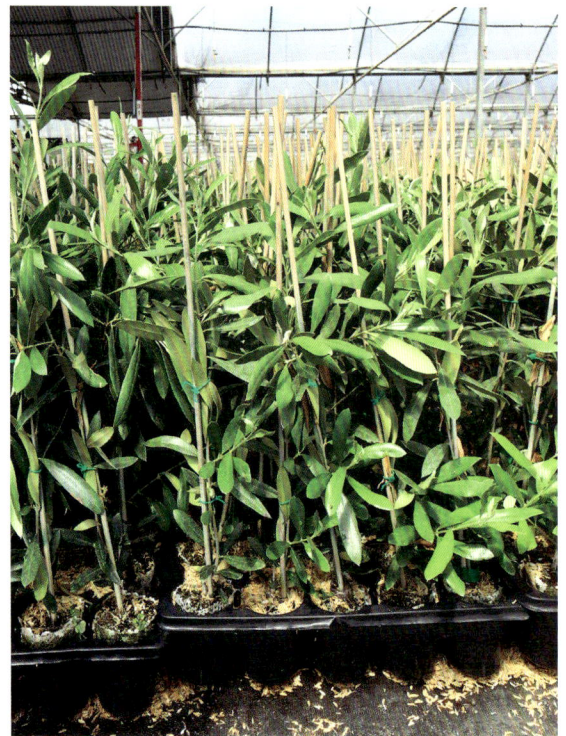

Figure 4.4

Trees are raised in greenhouses to a size that can be specified by the grower. *Photo:* Dan Flynn.

the trees are staged at a central location in the field, some growers apply a precautionary spray for black scale. Check the moisture of the root ball to determine if gentle misting is needed to increase moisture prior to planting.

Varieties

The SHD system relies on clonal selections of olive varieties that have reduced vigor, which allows the tree canopy to be maintained in a hedge. Growers should not attempt to establish an SHD orchard with varieties other than those specifically developed for the SHD system because other varieties have too much vigor to be consistently productive and fit the harvester.

The three major SHD olive varieties are Arbequina, Arbosana, and Koroneiki. They have dominated SHD plantings in California since the system was introduced in the state in 1999. California growers have experimented for more than a decade with other varieties that can be accommodated by the SHD system, and while they see promise in a couple of those varieties, no major grower has invested significantly in them.

The three major SHD varieties are precocious, coming into bearing 2 or 3 years after planting, and have high fruit productivity. They ripen at different times in the season, which can help a grower stagger harvest times.

All three varieties are purportedly self-fertile, but California growers have reported better yields when they are planted with a pollenizer. Any of the three varieties can serve as a pollenizer to the others. A different pollenizer, not one of these three main varieties, might be chosen, but it should

- be of sufficiently low vigor to grow in an SHD system
- produce high oil accumulation in the fruit and a desirable flavor profile in the oil
- be compatible with the main variety (California data on compatibility of SHD varieties as pollenizers is lacking)
- have a flowering time that overlaps the flowering time of the main variety
- provide consistent pollen annually (that is, not be highly alternate bearing)

California table olive research shows that the most efficient wind pollination occurs within a 100-foot (30-m) radius of the pollenizer, achieved by a pollenizer row every 200 feet (60 m); however, SHD growers generally favor a tighter pollenizer spacing, based on a 60- to 75-foot (18- to 23-m) radius. There are two primary industry practices for planting pollenizers: a pollenizer row every ten to twelve rows or a pollenizer tree about every 125 feet (38 m) within each row. Planting whole pollenizer rows is a good option for a grower who owns harvest equipment because the pollenizer row can be harvested at the optimal ripeness period, rather than with the rest of the block.

Arbequina

Arbequina has been propelled by the popularity of the SHD system into becoming one of the major olive varieties globally (fig. 4.5). Native to Catalonia in northeastern Spain, Arbequina is now grown in more than twenty countries. It accounts for about 80 percent of California SHD olive plantings. The oil generally has low bitterness and pungency, and the mild flavor profile is agreeable to consumers. Arbequina generally is lower in antioxidants and oleic acid than other varieties and has a shorter shelf life.

Growers consider Arbequina to be the easiest to grow of the three major varieties, with

Figure 4.5

Arbequina accounts for about 80 percent of California SHD plantings. *Photo:* Dan Flynn.

medium to strong vigor compared to the other two varieties, which enables it to thrive even in heavier clay soils. Arbequina is the most resistant to cold and olive knot but is the most sensitive to peacock spot.

The fruit ripens before the fruit of the other two major varieties. The force required to remove the olives is less than the force required for the others, making Arbequina the easiest to harvest. Fruit that does not come off the tree during harvest tends to drop off before the next crop, minimizing the presence of mummies on the trees, which can significantly degrade the oil quality of the next crop.

Arbosana

Arbosana once was a minor variety in Catalonia, in northeastern Spain, but has achieved international prominence in the SHD system (fig. 4.6). The fruit produces an oil with a medium flavor profile with moderate shelf stability. Arbosana has the lowest vigor of the three major varieties and will not thrive in marginal soils, but it is a good choice for highly fertile soils. Arbosana comes into its first fruiting earlier than the other varieties and has the most consistent fruit yield;

Figure 4.6

Arbosana has the most consistent fruit yield of the three main SHD varieties. *Photo:* Dan Flynn.

even with its smaller canopy, Arbosana produces as much fruit as Arbequina, with a higher oil yield. The trees can be planted closer together due to the low vigor.

Arbosana is sensitive to cold and highly susceptible to olive knot; it should be planted in warm, dry areas that will not get frost during the late fall harvest period. Arbosana should not be planted in areas with a history of hard winter freezes. It is the most resistant to foliar diseases such as peacock spot, but it may be susceptible to the Neofabraea diseases, based on preliminary findings in California.

Of the three major varieties, Arbosana ripens last and requires the most force to remove the olives during harvesting. Olives that are not harvested tend to remain on the tree. If a significant amount of fruit remain, it is generally removed in late winter or early spring with the same equipment used for harvesting, so that mummies do not compromise the oil quality of the next crop.

Koroneiki

Koroneiki is the primary oil variety of Greece, particularly in the Peloponnese and Crete, accounting for 50 to 60 percent of the oil olive acreage there (fig. 4.7). The fruit produces a robust oil with significant bitterness, pungency, and antioxidants. The oil adds fruity intensity and extends shelf life to oil blends. Koroneiki has the most vigor of the three major SHD varieties and may be difficult to control in deep, fertile soils. It is better suited to marginal soils with low water retention capacity. Some growers find Koroneiki to have more consistent bearing than other SHD varieties, while others find it to be more alternate bearing.

Koroneiki is sensitive to cold and susceptible to olive knot and should not be selected for areas where winter freezes are common. It is resistant to peacock spot. It ripens after Arbequina and before Arbosana. The force required to remove the olives is high, but the unharvested fruit tends to fall off the tree before the next harvest.

California research has found that oil from Koroneiki trees grown in the Central Valley or inland deserts of Riverside and Imperial Counties can exceed California's maximum limit for campesterol (UC Davis Olive Center 2019). Campesterol is among the sterols evaluated in

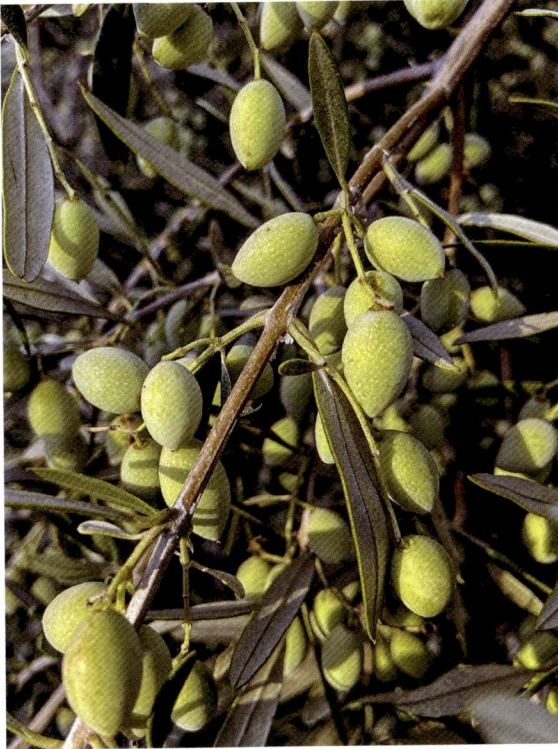

Figure 4.7

Koroneiki has the most vigor of the three main SHD varieties. *Photo:* Dan Flynn.

chemical standards for olive oil purity, and exceeding the limit could potentially require the oil to be marketed as low-value vegetable oil rather than as olive oil.

Planting

Before planting, ensure that the root balls are moist. Irrigate the field for 2 days prior to planting to supply adequate soil moisture. After planting, trees need to be watered daily for the first several weeks, so it is essential to have the irrigation system immediately operational. Do not dig holes in advance of the tree delivery; it is unnecessary if the soil has been prepared properly (see chapter 3, Site Selection and Preparation).

Ensure that each tree is planted within 2 inches (5 cm) of an emitter. Some growers pull the drip line tight down the row and plant each tree at the emitter closest to the bamboo cane. Others leave some slack in the drip line to allow flexibility in positioning the emitters at the trees, or install a drip line with an emitter every 12 inches (30 cm) to ensure that the emitters will properly align with the 5-foot (1.5-m) tree spacing.

To plant the tree, push a shovel about 6 to 8 inches (15 to 20 cm) into prepared soil, tilt the shovel to one side to open a hole, and drop the tree into the hole. Remove the shovel, fill any gaps with loose soil, and tamp the soil to remove air pockets. Check the planting depth: the tree crown should be level with or no more than 0.5 inch (1.3 cm) below the native soil. The potting mix surrounding the root ball is porous and can dry out quickly at the soil surface, so place up to 1 inch (2.5 cm) of soil over the root ball after planting to keep the root ball moist. Tie the tree to the bamboo cane, either manually or using a tool designed for the purpose.

In windy areas, plant the tree on the side of the bamboo cane that will ensure that the prevailing wind does not cause friction damage to the tree. Do not put fertilizer, compost, or planting mix in the planting hole.

When planting larger trees, dig a hole about twice the size of the tree container to accommodate the root system and accompanying soil. If necessary in clay soils, rough up the smooth sides of the hole with a shovel to facilitate lateral root growth and water movement. Then finish planting the tree as indicated for smaller trees, above.

Most California growers prefer manual planting because they have found it more efficient than mechanical planting. Manual planting also allows proper soil tamping after planting to remove air pockets. A worker can plant about 1 acre (0.4 ha) of trees per day, depending on the planting density. Some growers favor mechanical planting because of difficulties in obtaining labor for manual planting. Mechanical planting machines work with GPS, and most plant a single row at a time, although some can plant two or three rows simultaneously. The machines plant the trees, install bamboo canes, and in some cases place the drip lines in one pass.

Trunks of newly planted trees sunburn readily and can also be damaged by animals and herbicides if left exposed. There are several options for trunk protectors. Growers consider the best option to be sturdy, tube-shaped, translucent plastic protectors that allow light to reach the tree and are reusable. Longer protectors offer more protection than shorter ones against herbicide sprays. If the protector does not permit light to reach the tree, it is essential that the tree is taller than the protector, or the tree may not survive. Remove the protector the second year after planting as it can become a habitat for black scale, ants, wasps, and rodents. After planting, begin drip irrigation immediately so the newly planted tree does not dehydrate.

Care after planting

The main objectives after planting are to promote root growth from the porous potting soil into the native soil, establish a strong tree structure to facilitate development of a productive hedge, and develop robust canopy growth so that the tree comes into bearing as quickly as possible. Attaining these objectives demands vigilance. Inattentive orchard management during this critical period risks excessive tree losses and unnecessarily limited yields.

Promote root growth

The most critical time in the olive tree's life is the 6 weeks after planting, when the roots must grow beyond the root ball into the field soil (fig. 4.8). This requires careful irrigation, and neglect will result in a poor tree survival rate.

The amount and frequency of irrigation needed are influenced by soil composition and weather. Heavy soil retains water longer than sandy soil, and more irrigation is needed in hot weather than in cool weather. At planting, the tree's root system is confined to the potting soil, which has a limited ability to retain water and can dry out quickly in high temperatures. Irrigate daily for the first 6 weeks so that the root ball does not dry out. Avoid prolonged soaking because the roots are sensitive to asphyxiation and Phytophthora root rot, particularly in cold soils in early spring.

Fertigate with soluble phosphorus once a week for the first 2 weeks and nitrogen weekly throughout the first growing season to encourage root and shoot development. Excessive fertilizer can cause tree death, so manage the dosage carefully. Use calcium ammonium nitrate, CAN-17, in the first 2 years of orchard establishment.

Once the roots go beyond the root ball, reduce irrigation to two or three applications per week because now the roots are in soil that likely has a higher water retention capacity than the potting soil. Continue fertigating with nitrogen weekly or as needed by the trees. Monitor field soil moisture

Figure 4.8

Successfully establishing the orchard requires vigilant care after planting. *Photo:* Agromillora California.

regularly with a shovel or auger, two or three times a week in summer. When using sensors, place them close to the trees, given the small size of the root systems, and always combine sensor information with physical inspection of soil moisture (see chapter 7, Irrigation Management, and chapter 8, Soil and Nutrient Management).

After weather cools in the fall, reduce irrigation to stimulate hardening of the tree bark and to help guard against freeze damage. Manage irrigation and fertigation based on site-specific needs. In areas with cold winter temperatures, discontinue nitrogen in early September to reduce vegetative development that will not harden in time for the colder temperatures. Applying phosphate and potassium in late summer and early fall might help strengthen the plant's resistance to low temperatures.

Control weeds in the tree row so that the tree does not have to compete for water and nutrients (see chapter 13, Weed Management in Olive). Keep a 3- to 4-foot (0.9-to 1.2-m) strip of soil (or the entire width of a berm) free of weeds throughout the growing season. Watch for signs of damage from pests and diseases (see chapter 12, Diseases of Olive, and chapter 10, Arthropod Pests of Olive).

Establish tree structure

The objective in establishing the tree structure, which occurs in the first year, or two, after planting, is to facilitate future canopy management, encourage strong yields, and achieve high harvest efficiency. Training the tree requires progressive removal of lower shoots to make a clear trunk for the harvester. It requires regularly tying the tree to the bamboo cane or heavy stake with stretch tie tape, which maintains the vigor of the central leader and an upright trunk. Growers generally find that hand-tying is quicker and more reliable than using a tie gun. In very windy areas, it may be necessary to use additional stretch tie tape or other flexible material to secure the tree to the bamboo cane, looping the tie between the cane and the tree to avoid friction. Training also requires topping the central leader to finish the shaping of the tree.

Small trellis and heavy stakes training

Establishing the structure of a young tree growing on a small trellis or heavy stakes begins at planting, when the tree is tied to the bamboo cane or stake. As the central leader grows, continue to tie it to the bamboo cane or stake every 8 to 14 inches (20 to 36 cm). When the central leader reaches 24 inches (61 cm), cut off the lower shoots, up to about 12 inches (30 cm). When the central leader reaches the wire at 30 inches (76 cm), California growers typically choose among three training options:

- **Option A:** Continue tying the central leader to the bamboo cane or stake, and when the central leader reaches 36 inches (91 cm), remove any lower shoots, up to 22 to 24 inches (56 to 61 cm). Top the central leader when it reaches 4 to 4.5 feet (1.2 to 1.4 m), which may not occur until the second year after planting. Double-tie with stretch tie tape near the top of the bamboo cane or stake.
- **Option B:** Continue tying the central leader to the bamboo cane or stake, and when the central leader reaches 36 inches (91 cm), remove any lower shoots, up to 22 to 24 inches (56 to 61 cm). Then allow the central leader to grow, without tying it, so that it bows, which encourages lateral shoots to emerge. Top the lateral shoots when they reach about 5 feet (1.5 m), which will occur the second year after planting. To secure the tree, double-tie a vertical shoot close to the trunk with stretch tie tape to the top of the bamboo cane or stake.
- **Option C (known as the "smart tree" option):** Cut the central leader at 26 to 30 inches (66 to 76 cm), which will encourage vigor in nearby lateral shoots. Choose four or five of the shoots to develop a scaffold structure, and remove about 4 inches (10 cm) from the tips of those shoots when they reach a length of about 12 inches (30 cm). Progressively remove lower shoots, up to 22 to 24 inches (56 to 61 cm). Top the tree when it reaches 4 feet (1.2 m), which may not occur until the second year after planting. Secure the tree to the bamboo cane or stake just below the scaffold shoots with double-wrapped 1-inch (2.5-cm) stretch tie tape.

Whichever option is chosen, should it be necessary to replace the main trunk due to damage or death, select a replacement shoot and trim away any vigorous shoots nearby that could compete with it.

Large trellis training

Growers who choose a large trellis generally continue tying the central leader to the bamboo cane every 8 to 14 inches (20 to 36 cm) until it reaches the top of the cane, at 5 to 6 feet (1.5 to 1.8 m). As the tree grows, remove shoots from the lower third of the tree trunk. By the time the central leader reaches the top wire, the first 22 to 24 inches (56 to 61 cm) of the trunk should be clear of shoots.

Develop robust canopy growth

Developing a robust canopy is the goal of early trimming and pruning. Too much trimming or pruning delays the onset of bearing, which generally occurs in the third year after planting and sometimes in the second year. However, it's important to stay on top of early trimming and pruning because allowing shoots to grow too long before cutting them leads to open, unproductive areas of the hedge.

Growers should use a hedging tool to remove 4 inches (10 cm) of growth from the sides of the tree when the branches have grown 20 inches (51 cm) from the trunk, and again when the branches have grown 30 inches (76 cm) from the trunk. Growers following options A and B should top the tree when it reaches the height that can be accommodated by the harvester. Option C requires more frequent topping—at 6 feet (1.8 m), 8 feet (2.4 m), and then to accommodate the harvester. Suckers should be removed, as well as excessively low-hanging shoots.

To get a growth response from pruning, prune between March and August. Do not prune after August because the tree will have limited ability to respond with new growth, and it may make the tree more susceptible to freeze. Do not prune during the rainy season as it makes the tree more susceptible to freeze injury and olive knot.

Regardless of the approach in developing the canopy, the trunk of the tree should be clear of branches up to 24 to 30 inches (61 to 76 cm) as the canopy is prepared for its first harvest. Maximum canopy dimensions are determined by the harvest equipment and guidelines suggested in chapter 9, Canopy Management.

Suggested reading

Díez, C. M., J. Moral, D. Cabello, P. Morello, L. Rallo, and D. Barranco. 2016. Cultivar and tree density as key factors in the long-term performance of super high-density olive orchards. Frontiers in Plant Science 7:1226. https://doi.org/10.3389/fpls.2016.01226.

International Olive Council. 2015. International olive oil production costs study—2015. Madrid: IOC.

Rius, X., and J. Lacarte, eds. 2012. The olive growing revolution: The super high density system. Barcelona: Agromillora.

Sibbett, G. S., and L. Ferguson, eds. 2005. Olive production manual. 2nd edition. Oakland: UC Agriculture and Natural Resources Publication 3353.

Vossen, P. M., ed. 2007. Organic olive production manual. Oakland: UC Agriculture and Natural Resources Publication 3505.

UC Davis Olive Center. 2019. Evaluation of fatty acid and sterol profiles, California olive oil 2018/19 season. Submitted to the Olive Oil Commission of California.

5

Establishing a High-Density Olive Orchard

Dan Flynn, Executive Director Emeritus, UC Davis Olive Center

Ciriaco Chavez, Vice President of Agriculture, Cobram Estate Olives

Highlights

- Compared to the super-high-density (SHD) system, the high-density (HD) system has the advantages of lower establishment costs and offers more options for growers to select varieties to suit agronomic conditions and provide diverse oil sensory and phenolic profiles.

- The disadvantages of an HD system are lower yields in the early years of the orchard, unavailability of contract harvesters, and less capability of harvesters to manage steep slopes.

- Varieties with high vigor on a site with deep soil merit wider spacing, and low-vigor varieties on marginal soil, closer spacing.

- Growers should order trees from nurseries well in advance, up to 2 years prior to planting.

- Growers can stipulate desired tree size in the contract with the nursery, and that the trees be free of pests and diseases.

- Growers should select a variety that suits the agronomic characteristics of the site, as well as production goals, processor preferences, and market acceptance of the variety's oil profile.

- A support structure is not necessary for HD growers planning to use a trunk shaker at harvest or hand-picking unless the orchard is in a very windy location. Growers using a straddle harvester favor trellis installation.

- The most critical time in the life of an olive tree is the 6 weeks after planting, when it needs irrigation and attention to ensure survival.

- Fertigating at each irrigation allows the tree to absorb more of the nutrients, so less nutrients leach below the root area.

- Avoid excessive tree pruning because it delays the onset of bearing.

- Control weeds so that they do not compete with the new trees for water and nutrients.

Long-term profitability demands a high level of management in the first months of the orchard. The grower must properly plant, nurture, and train the trees to minimize replants, quickly fill out the canopy, and produce profit as quickly as possible.

In California, the high-density (HD) system (also known as the medium-density system) averages about 200 to 300 trees per acre (500 to 750 trees per ha) (fig. 5.1). Initial interest in the HD system came from California table olive growers, as a strategy to transition from hand-picking to mechanical harvesting. Small California oil olive growers began experimenting with HD planting densities in the 1990s, and new plantings have accelerated in recent years. An estimated 15 percent of California oil olive orchards are planted in the HD system, which is commensurate with the global percentage.

Compared to the super-high-density (SHD) system, with 600 to 900 trees per acre (1,500 to 2,225 trees per ha), the HD system has the advantage of lower establishment costs and gives the grower more options to select varieties that suit agronomic conditions and provide diverse oil sensory and phenolic profiles. Disadvantages include lower yield in the early years of the orchard; HD straddle, or canopy contact; harvesters' having

less capability to manage steep slopes; and, in California, a lack of contract harvesters suitable for HD orchards—though the expansion of HD plantings may make contract harvesting more available in the future. SHD straddle harvesters are suitable only for the first few crop years of an HD orchard.

Tree spacing

The primary objectives in designing the tree spacing are to quickly create a hedge with a manageable canopy that can be accommodated by harvest equipment and to facilitate sufficient light interception of the canopy to maximize production of olives and oil accumulation in the fruit. Tree spacing should consider harvester requirements, soil characteristics, and the vigor of the variety. Row widths in California HD systems have ranged from 16 to 22 feet (4.8 to 6.7 m), with trees spaced in the row at 8 to 12 feet (2.4 to 3.7 m). Varieties with high vigor on a site with deep soil would merit wider spacing, and low-vigor varieties on marginal soil, closer spacing. Ensure that the tree spacing will accommodate the anticipated harvest equipment, assuming that the canopy width will be half the alley width (see chapter 9, Canopy Management).

Support structure

A trellis system or heavy stakes are necessary for orchards harvested with a straddle harvester because a straddle harvester requires a straight row alignment to avoid damaging the trees. For HD orchards using a straddle harvester, the following trellis was found to be cost effective by a California company that has analyzed numerous trellis configurations (fig. 5.2):

Figure 5.1

A high-density California orchard has about 200 to 300 trees per acre (500 to 750 trees per ha). *Photo:* Boundary Bend Olives.

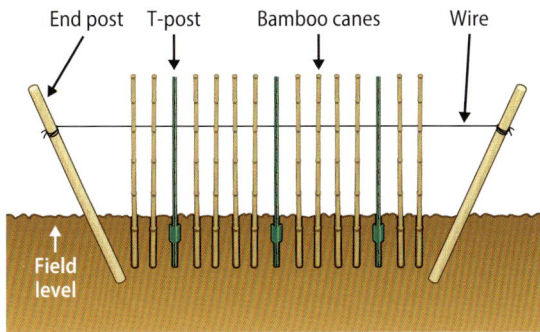

Figure 5.2 labels: End post, T-post, Bamboo canes, Wire, Field level

Figure 5.2

A trellis system is needed if a straddle harvester will be used.

- end posts of either treated wood with a diameter 5 to 6 inches (13 to 15 cm) or metal with a spade and a diameter of 3 inches (7.6 cm) and 8 feet (2.4 m) long, installed at about a 65- to 75-degree angle at the ends of each row, anchored 4 feet (1.2 m) into the ground
- metal spaded T-posts 6 feet (1.8 m) long, anchored 2 feet (0.6 m) into the ground along the row at intervals of 50 to 60 feet (15 to 18 m), each T-post equidistant between two tree locations
- bamboo canes 7 feet (2.1 m) long and ¾ inch (19 mm) in diameter, anchored 2 feet (0.6 m) into the ground along the tree row, one at each tree position
- one 12- to 16-gauge galvanized wire wrapped around the end posts and strung with adequate tension the length of the row at a height of 32 inches (81 cm), and clipped to the T-posts and bamboo canes with a metal tie

Heavy stakes are an alternative to the trellis. Push a single wooden stake 4 to 5 feet (1.2 to 1.5 m) long with a diameter of 1 inch (2.5 cm) 20 to 24 inches (51 to 61 cm) into the ground at each tree location. Heavy stakes are less expensive than the trellis, but they do not provide as much tree support, which would be a concern in windy locations. The stakes also may be damaged by a mechanical harvester when the trees are young and the canopy has not fully developed. The stakes would need to be removed if a trunk-shaking harvester is used, to allow the shaker head to firmly grip the trunk and avoid trunk damage from the stake abrading the bark.

The central leader is tied to the bamboo cane or stake at 8- to 14-inch (20- to 36-cm) intervals as the leader grows.

Pollenizers

California table olive research shows that the most efficient wind pollination occurs within a 100-foot (30 m) radius of the pollenizer, achieved by a pollenizer row every 200 feet (61 m); however, HD growers favor spacing pollenizers within a radius as low as 50 feet (15 m) to increase pollination potential. There are good reasons for providing ample pollenizers—research shows that close proximity to a pollenizer makes for higher yields and that a greater number of pollenizers increases the chances of pollination if there is unfavorable weather during bloom.

The options for placing pollenizers are to plant isolated pollenizer trees within rows of the primary variety or to plant full pollenizer rows or groups of several pollenizer rows. An even number of pollenizer rows facilitates harvest because the harvester can collect the same variety after a turnaround at the end of a row.

Pollenizers should
- produce high oil accumulation in the fruit and a desirable flavor profile in the oil
- be compatible with the primary variety
- have a flowering time that overlaps that of the primary variety
- provide consistent pollen annually (that is, not be highly alternate bearing)

When the compatibility of pollenizer varieties with the primary variety is not known, it is advisable to plant two potential pollenizer varieties to ensure good pollination. California research indicates that Manzanillo is not compatible with Mission; other varieties may have compatible pollenizers identified in their countries of origin, but California research on pollenizers is lacking.

Nursery tree orders

Growers with HD orchards typically order larger, more mature trees than the trees growers order for SHD orchards. Usually the tree is 20 to 24 months old with a height of 2 to 3 feet (0.6 to 0.9 m). Larger trees have a higher survivability rate and the onset of bearing is earlier, typically in the third year after planting. Larger trees cost more, but, with fewer trees planted, the tree cost per acre is lower in an HD orchard than in an SHD orchard.

Some commercial olive nurseries offer a wide range of HD olive varieties, while others focus on a limited number of the most popular varieties. Nurseries propagate most HD trees from cuttings under mist, although there are some varieties that propagate more successfully when grafted onto a rootstock.

Growers can work with the nursery to specify a minimum tree size and also specify that the tree be true to type and free of pests and diseases. Various factors impact nursery production, so some nurseries do not guarantee a delivery date at the minimum size. Growers should order HD trees at least 2 years in advance of planting, keeping in mind the period for nurseries to root the cuttings (that is, start the order) is roughly June to September. Some states may require that a California nursery provide a phytosanitary certificate for olive trees shipped to orchards in their state.

Varieties

Any olive variety can grow in an HD system. Choose varieties based on site conditions, production goals, and oil marketability. Consider agronomic factors such as cold hardiness, pest and disease resistance, vigor, yield, fruit removal force required, and bearing consistency as well as processing factors such as oil content, extraction efficiency, shelf stability, flavor profile, and processor preferences. Table 5.1 summarizes some characteristics of select olive varieties. Nurseries may offer additional varieties to meet growers' needs.

The descriptions of varieties below are based primarily on information from native growing areas and modified by reports from California growers and nurseries. Varieties indicated as "alternate bearing" may be so because of the traditional management practices used in the country of origin, such as not irrigating, infrequent management of the canopy, and harvesting late in the season. These varieties may bear more consistently in irrigated orchards with annual canopy management and earlier harvest. Information on how varieties perform in California will increase as HD plantings expand.

Primary varieties for the HD system in California include Arbequina, Coratina, and Picual, which have attractive agronomic characteristics and provide consistently strong yields of high-quality oil. Other HD varieties used

Table 5.1

Agronomic and processing characteristics of select olive varieties

Variety	Oil content	Fruit removal force	Olive fly	Verticillium wilt	Peacock spot	Olive knot	Cold	Ripening	Vigor	Productivity
Arbequina	M	M	S	M	M	R	R	early-mid	L-M	H, C
Ascolano	L	L	S	R	R	R	R	early	H	M, C
Chemlali	H	H	R	R	S	S	M	early-mid	M-H	H, C
Coratina	H	H	S	S	M	M	S	late	H	H, A
Frantoio	H	M	S	R	M	M	M	late	H	M, A
Hojiblanca	L-M	H	S	S	S	S	R	mid-late	M-H	H, A
Koroneiki	H	H	R	M	R	S	S	late	M	H, C
Leccino	M	M	S	M	R	R	R	early-mid	H	M, A
Manzanillo	L-M	H	S	S	S	S	S	early	M	H, A
Mission	M-H	M	S	S	S	R	R	late	H	M, A
Picual	M-H	L	S	S	S	R	R	mid-late	M	H, C
Taggiasca	H	M	S	S	S	S	S	late	M-H	H, C

Key: L = low, M = medium, H = high, S = sensitive, R = resistant, A = alternate, C = constant

Source: Adapted from International Olive Council 2000.

to a lesser extent include Ascolano, Chemlali, Frantoio, Hojiblanca, Koroneiki, Leccino, Manzanillo, Mission, and Taggiasca.

Arbequina

Arbequina is native to Catalonia in northeastern Spain but is now grown in more than twenty countries. The oil generally has low bitterness and pungency, and the mild flavor profile is agreeable to consumers. Arbequina generally is low in antioxidants, which limits the oil's shelf life.

Arbequina has low to medium vigor compared to other variety options for the HD system and can thrive even in heavy clay soils. It is resistant to cold and olive knot. The fruit ripens early to midseason. The force required to remove the olives is moderate, with fruit that does not come off the tree during harvest tending to drop off before the next crop, minimizing the presence of mummies on the trees, which can significantly degrade the oil quality of the next crop. Spanish research shows that Arbequina is compatible with Picual for pollination (fig. 5.3).

Ascolano

Ascolano was introduced to California in 1885, but the variety fell out of favor for table olives because the fruit bruised easily during harvest. Ascolano's distinctive fruit-forward flavor profile makes it popular as a single-varietal oil as well as for blending. Originally from Italy, it has high vigor, and productivity is medium and constant, but oil production is low to medium. Pollenizers include Mission and Manzanillo. The high flesh-to-pit ratio of the fruit makes it attractive to the olive fruit fly. The variety is resistant to cold, peacock spot, and olive knot, and shows some resistance to Verticillium wilt. The fruit may be harvested early, but the distinctive stone fruit and tropical flavors become more pronounced with maturity (fig. 5.4).

Chemlali

Chemlali represents 60 percent of the olive plantings in Tunisia and is showing some promise in California according to growers who have planted it on a very limited scale. Chemlali has high vigor, is early flowering, and produces ample pollen. It is moderately cold hardy. It is tolerant of

Figure 5.3

Arbequina thrives in heavy clay soils. *Photo:* Dan Flynn.

Figure 5.4

Ascolano has distinctive flavors and is popular for single-varietal oils. *Photo:* MerchantAdventurer, CC BY-SA 4.0.

drought and salinity but susceptible to olive knot. The fruit ripens early to midseason, has a high oil content, and reports of the limited acreage in California suggest productivity to be high and consistent (fig. 5.5).

Figure 5.5

Chemlali has a high oil content and is tolerant of drought and salinity. *Photo:* Boundary Bend Olives.

Figure 5.6

Coratina is very early bearing, and the oil has a long shelf life. *Photo:* Boundary Bend Olives.

Coratina

Coratina is the main variety in Puglia, Italy's most productive olive region. The qualities that have made this variety a favorite among HD growers include very early bearing, adaptability to a wide range of growing conditions, high oil content, and long oil shelf life. The tree has high vigor. The variety is sensitive to frost and moderately tolerant of drought and salinity, but sensitive to Verticillium wilt and susceptible to olive fruit fly. The fruit ripens late and works best in areas with a long growing season and a low risk of frost. The force required to remove the olives is high, and unharvested fruit may survive on the tree as mummies, which can degrade the oil quality of the next crop. Historically, Frantoio and Moraiolo have served as pollenizers although other varieties can be used (fig. 5.6).

Frantoio

Frantoio is a prominent variety from Tuscany. It has a high oil content and produces a superior-quality oil. The tree has high vigor. Fruit ripens late, and productivity is medium and alternate. Frantoio is susceptible to olive fly and has medium susceptibility to peacock spot and olive knot. The tree is resistant to cold and Verticillium wilt. Traditional pollenizers for Frantoio include Pendolino, Leccino, and Maurino (fig. 5.7).

Figure 5.7

Frantoio has high vigor, and a high oil content. *Photo:* Patrimonio, Dreamstime.com.

Hojiblanca

Hojiblanca is a primary variety in southern Spain. The oil is prized for its high quality, although oil content is low to medium. The fruit also is grown for table olive processing in Spain. The tree has high vigor and is resistant to cold and drought but susceptible to Verticillium wilt, peacock spot, olive fruit fly, and olive knot. Production is high and alternate, and the fruit ripens late. The force required to remove the olives is high during the early years of the orchard. Picual, Arbequina, and Manzanillo are compatible pollenizers (fig. 5.8).

Figure 5.8

Hojiblanca has low-medium oil content and produces high-quality oil. *Photo:* Boundary Bend Olives.

Figure 5.9

Koroneiki has a high oil content and produces a robust oil. *Photo:* Dan Flynn.

Koroneiki

Koroneiki is the primary oil variety of Greece, particularly in the Peloponnese and Crete, accounting for 50 to 60 percent of the oil olive acreage there. The fruit has a high oil content. California growers have found that production can be consistent when properly managed. Koroneiki has medium vigor and is resistant to peacock spot but is susceptible to olive knot. It is sensitive to cold and therefore not suitable for areas at high risk of frost. The fruit ripens late, and the force required to remove the fruit is high. The fruit produces a robust oil with significant bitterness, pungency, and antioxidants. The oil adds fruity intensity and extends shelf life to oil blends.

California research has found that oil from Koroneiki trees grown in the Central Valley or inland deserts of Riverside and Imperial Counties can exceed California's maximum limit for campesterol (UC Davis Olive Center 2019). Campesterol is among the sterols evaluated in chemical standards for olive oil purity, and exceeding the limit could potentially require the oil to be marketed as low-value vegetable oil rather than as olive oil (fig. 5.9).

Leccino

Leccino is among the most popular varieties in Tuscany. The tree is vigorous, with a dense, upright canopy. Productivity is medium, and fruit tends to ripen prior to the completion of oil accumulation. Leccino is cold hardy and resistant to wind, fog, olive knot, and peacock spot. Oil content is medium and the oil quality is high, with balanced flavor and low bitterness. In Tuscany, it is usually planted with Frantoio and Pendolino (fig. 5.10).

Manzanillo

Manzanillo is the most popular table olive variety in California and the world, and the fruit also makes a decent oil, particularly when harvested

Figure 5.10

Leccino is vigorous and cold hardy. *Photo:* Dan Flynn.

early. Sevillano, a table olive variety low in oil content, has been the traditional pollenizer for Manzanillo in California. Frantoio, Picual, and Arbequina are alternative pollenizers. Research has shown that Mission and Ascolano are incompatible as pollenizers for Manzanillo. The tree has medium vigor, is sensitive to cold, and is susceptible to olive fruit fly, Verticillium wilt, peacock spot, and olive knot. Production is high and alternate, oil content is low to medium, and the force required to remove the fruit is high. It ripens early and has a high flesh-to-pit ratio. Its high moisture content makes Manzanillo particularly susceptible to emulsions during processing, which makes oil extraction difficult (fig. 5.11).

Mission

Mission has the longest history of any olive variety in California; it grew in the gardens and fields of California missions in the late eighteenth century. The tree has high vigor and upright growth. Mission is resistant to cold and olive knot but susceptible to olive fruit fly, Verticillium wilt, and peacock spot. Production is medium and alternate. The fruit is high in oil and phenolic content and may produce excessively bitter oil when harvested early in the season. Manzanillo is not a compatible pollenizer for Mission (fig. 5.12).

Picual

Picual is the most popular oil olive in the world due to its high and consistent productivity of fruit that has medium to high oil content. The tree is easy to grow and comes into bearing early. It is cold hardy and can tolerate soil with high moisture and salinity. Picual is somewhat resistant to olive knot but sensitive to olive fruit fly, Verticillium wilt, anthracnose, and peacock spot. The fruit ripens mid-late season and has a low removal force. The oil has high stability with fruity attributes that range from a tomato leaf flavor early in the season to tropical flavors in midseason to late-harvest "catty" flavors. Hojiblanca and Arbequina are pollenizers of Picual (fig. 5.13).

Taggiasca

Taggiasca is the principal variety of the Liguria region in northern Italy, where it is known to produce an oil of delicate flavor intensity due to its late harvest. California growers tend to harvest

Figure 5.11

Manzanillo has several compatible pollenizers and ripens early. *Photo:* Dan Flynn.

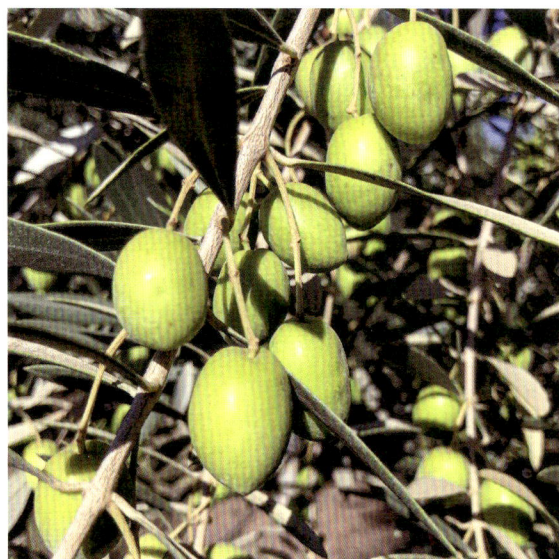

Figure 5.12

Mission is high in oil and phenolic content. *Photo:* Dan Flynn.

the fruit at earlier maturity for a more robust profile. The tree has medium to high vigor and offers consistently high productivity and oil content. It is susceptible to olive fruit fly, olive knot, and cold (fig. 5.14).

Planting

One of the most important considerations before planting is to ensure proper hydration of the field and the root balls of the new trees. Ensure that

Figure 5.13

Picual is easy to grow and has medium to high oil content. *Photo:* Dan Flynn.

Figure 5.14

Taggiasca produces consistent yields and has a high oil content. *Photo:* Dan Flynn.

the root balls have sufficient moisture by fully wetting them before planting. To ensure adequate moisture levels in the soil, growers may irrigate the field prior to planting. This is particularly advisable when planting during warm periods with temperatures exceeding 90°F (32°C). Growers should take care not to oversaturate the soil because planting shovels could create slick walls in the tree holes that can inhibit root expansion in the soil. Root migration may also be impeded when digging holes in advance of tree delivery if

the holes become dry prior to planting (see chapter 3, Site Selection and Preparation).

Usually the irrigation system (with two drip lines) and the end posts for the trellis are in place prior to planting. The bamboo canes, T-posts, and wire can be installed right after planting. If heavy stakes are used instead of a trellis, they are also usually installed after planting. Use drip lines with emitters that are spaced to align with tree spacing so that emitters are on the root ball. Typically, the drip lines are laid on the ground, although some growers may suspend them from the trellis, if a trellis is being used, to ease non-chemical weed management. Nutrient deficiencies should be addressed prior to planting (see chapter 3, Site Selection and Preparation).

Dig a hole about twice the size of the tree container to accommodate the root system. In clay soils, lateral root growth and water movement will be limited if the sides of the hole are smooth, so rough up the sides with a shovel. Place the tree in the hole, with the crown level with the native soil or no more than 0.5 inch (13 mm) below it. Refill the hole with native soil, tamping firmly to prevent air pockets, which could kill the roots. The potting mix surrounding the root ball is porous and can dry out quickly at the soil surface, so place up to 1 inch (2.5 cm) of native soil over the root ball after planting to keep the root ball moist. Tie the tree to the bamboo cane or heavy stake loosely; allowing the tree to move will promote faster and deeper root development.

Trunks of newly planted trees sunburn readily and can also be damaged by animals and herbicides if left exposed. There are several options for trunk protectors. Growers consider the best option to be sturdy, tube-shaped, translucent plastic protectors that allow light to reach the tree and are reusable. If the protector does not permit light to reach the tree, it is essential that the tree is taller than the protector, or the tree may not survive. Longer protectors (18 in [46 cm]) offer more protection than shorter ones against herbicide sprays. Remove the protector the second year after planting as it can become a habitat for black scale, ants, wasps, weevils, and rodents.

Install the bamboo cane, or heavy stake, next to the tree, avoiding damage to the root system. In windy areas, place the tree on the side of the bamboo cane or stake that will ensure that the prevailing wind does not cause friction damage to the tree. For example, in areas with a prevailing north wind, plant the tree on the south side of the bamboo cane or stake. Begin drip irrigation immediately so the newly planted tree does not dehydrate (fig. 5.15).

Care after planting

The main objectives after planting are to promote root growth from the porous potting soil into the native soil, establish a strong tree structure to facilitate the tree's efficient development, and develop robust canopy growth so that the tree comes into bearing as quickly as possible. Attaining these objectives demands vigilance. Inattentive orchard management during this critical period risks excessive tree losses and delayed or limited yields.

Promote root growth

The most critical time in the olive tree's life is the 6 weeks after planting, when the roots must grow beyond the root ball into the field soil. This requires careful irrigation, and neglect will result in a poor tree survival rate.

The amount and frequency of irrigation needed are influenced by soil composition and weather. Heavy soil retains water longer than sandy soil, and more irrigation is needed in hot weather than in cool weather. At planting, the tree's root system is confined to the potting soil, which has a limited ability to retain water and can dry out quickly in high temperatures. Irrigate daily for the first 6 weeks so that the area around the root ball is moist and does not dry out. Avoid prolonged soaking because the roots are sensitive to asphyxiation and Phytophthora root rot, particularly in cold soils in early spring. Fertigating at each irrigation allows the tree to absorb more of the nutrients, so less nutrients leach below the root area; manage the dosage carefully to avoid burning the roots. Fertigate with nitrogen, phosphorus, and potassium throughout the growing season to encourage root and shoot development.

Once the roots grow beyond the root ball, consider reducing irrigation frequency as the roots are in soil that likely has higher water retention capacity than the potting soil (fig. 5.16). Be mindful to maintain adequate levels of available moisture in the soil, without waterlogging the young trees. Continue fertigating with nitrogen,

Figure 5.15

A newly planted tree must be kept well watered. *Photo:* Boundary Bend Olives.

Figure 5.16

Six months after planting, the tree roots have extended into the native soil. *Photo:* Boundary Bend Olives.

phosphorus, and potassium to promote vigorous vegetative growth. Monitor field soil moisture regularly with a shovel or auger, two or three times a week in summer; if desired, combine these observations with data from soil moisture monitoring devices. When installing sensors, place them close to the tree for the most accurate reading of the soil moisture in the tree's rooting zone. Measure the flow from emitters to ensure that flow is uniform throughout the irrigation block.

Reducing nitrogen applications in late summer and early fall can help harden off trees and strengthen their resistance to low winter temperatures. Applying phosphate and potassium in late summer and early fall may also enhance cold hardiness. A foliar copper spray should be applied just prior to fall rains to prevent fungal diseases, such as peacock spot.

Control weeds in the tree row so that the tree does not have to compete for water and nutrients (see chapter 13, Weed Management in Olive). Young olive trees are highly sensitive to weed competition, and careful orchard floor management will support faster root and canopy growth and higher survival rates. Keep a 4- to 5-foot (1.2- to 1.5-m) strip of soil in the tree row free of weeds throughout the growing season. Watch for signs of damage from pests and diseases (see chapter 12, Diseases of Olive, and chapter 10, Arthropod Pests of Olive).

Establish tree structure

The objective in establishing the tree structure, which occurs in the first year after planting, is to facilitate future canopy management, encourage strong yields, and achieve high harvesting efficiency. Training the tree requires the progressive removal of the lower shoots as the tree grows so that the lower trunk is clear for mechanical harvest—a trunk shaker attachment needs to grip the trunk well, or, if a straddle harvester will be used, the plates must close properly under the canopy. Training also requires tying the central leader to the bamboo cane or heavy stake. Maintain a gap of 0.13 to 0.25 inch (3 to 6 mm) between the tree and stake. Ensure the tape is pliable so it does not girdle the tree as it grows. In very windy areas, it may be necessary to use additional flexible material to secure the tree to the bamboo cane or stake, looping the tie between the cane or stake and the tree to avoid friction.

Training the tree begins at planting, when the tree is tied to the bamboo cane or stake. As the tree grows, continue to tie the central leader to the cane or stake every 8 to 14 inches (20 to 36 cm) and remove branches from the bottom half of the tree during every training pass; for example, when the leader reaches 24 inches (61 cm), cut off the lower shoots up to 12 inches (30 cm).

Trim or remove vigorous branches that compete with the central leader. If the competing branch is thinner than the trunk, cut the branch just above a bud or shoot growing in line with the tree row; if the competing branch is thicker than the trunk, remove the branch at the base. Should it be necessary to replace the central leader due to damage or death, select a replacement shoot and trim away any vigorous shoots nearby that could compete with the new central leader. The goals of the initial years of training are to promote upward growth and lift the canopy.

As a general rule, mechanical harvesters require straight and clear trunks up to 32 inches (81 cm); however, the specific clearance height for a desired harvester should be checked and low branches removed accordingly. When the canopy clears the desired height for the harvester, stop removing the low branches during training passes. For example, for a harvester that requires a clear trunk up to 32 inches (81 cm), cease applying the rule to remove branches from the bottom half of the tree once the tree exceeds 64 inches, and simply maintain the clearance of 32 inches (81 cm). As the tree continues to grow, there are two primary options for forming the structure of the tree: free palmette and open center, or vase.

Free palmette

The free palmette is favored by growers using a straddle harvester (fig. 5.17). To create a palmette structure, continue tying the central leader to the bamboo cane until the leader reaches 60 inches (1.5 m) or up to 40 inches (1 m) if using heavy stakes. Once the central leader has reached the top of the bamboo or stake, cease removing competing branches (but continue to eliminate lower shoots for harvester clearance, as mentioned above). Then, on an annual basis, remove large branches growing perpendicular to the row and

Figure 5.17

The free palmette structure of this young tree, 1 year after planting, directs growth within the row, which is favored for straddle harvesters. *Photo:* Boundary Bend Olives.

Figure 5.18

The open center structure is favored for hand-harvesting and shaker harvesters. *Photo:* Louise Ferguson.

also branches growing at less than a 45-degree upward angle in the top quarter of the tree. This will result in a tree with most of the branches growing between the trees within the row, like a fan or palm. As main branches become large, woody, and unproductive, remove one main branch per season to spur new growth and obtain a full renovation of the canopy—the trunk of the tree may be 10 years old, but no branch is older than 4 or 5 years.

Open center, or vase

The open center structure is also called the vase structure (fig. 5.18). It is favored for hand-harvesting and also for shaker harvesters (tree rows should be at least 18 feet [5.5 m] apart to ensure adequate space to maneuver shaker equipment). As the canopy grows above the required harvester clearance, select three to five branches growing laterally but with the tip turning toward upright to be scaffold branches. The scaffold branches should be spaced around the trunk, with sufficient distance between them to later allow a branch shaker harvester access to each branch. Remove the central leader (with its dominant apical tip), making the cut at the junction of the scaffold branches, to promote growth in

the scaffold branches and create the open center, or vase, shape. Over time, prune growth that has accumulated on the scaffold branches, one main branch per season, to spur new growth, maintain an open center, control height, and obtain the full renovation of the canopy.

Develop robust canopy growth

In the first year after planting, the recommended trimming and pruning helps develop the tree structure. In the second and third growing seasons, because olive trees store most of their energy in their leaves, excessive trimming and pruning is avoided to produce a robust, productive canopy. Remove only suckers, water sprouts, low-hanging shoots, branches that could be broken by the harvester, and branches that could girdle from crossing each other. In most other cases, little or no additional pruning is required for the first few years of tree growth, until the trees get about 7 to 8 feet (2.1 to 2.4 m) wide. Do not prune during the rainy season as it makes the young trees more susceptible to freeze injury and the spread of olive knot (see chapter 9, Canopy Management). Rapidly developing the canopy will allow a commercial crop generally in the third year after planting.

Continue irrigating and fertigating as indicated by monitoring data, to maintain adequate moisture and nutrients for root and tree growth. In the second year after planting, begin to expand the wetted strip by moving the drip lines progressively farther apart so that the roots expand accordingly. Send leaf samples to a laboratory for analysis in July of each year, as the nutrient needs of the canopy and the fruit load will need closer management when the trees come into production (see chapter 8, Soil and Nutrient Management). Control diseases, weeds, and pests, particularly black scale, which may hide between the tree and the cane or stake. Once anchoring roots have developed and the tree is firmly established, the cane or stake may be removed.

References

International Olive Council. 2000. World catalogue of olive varieties. Madrid: IOC.

UC Davis Olive Center. 2019. Evaluation of fatty acid and sterol profiles, California olive oil 2018/19 season. Submitted to the Olive Oil Commission of California.

Suggested reading

Connor, D. J., M. Gomez-del-Campo, M. C. Rousseaux, and P. S. Searles. 2014. Structure, management and productivity of hedgerow olive orchards: A review. Scientia Horticulturae 169:71–93. http://dx.doi.org/10.1016/j.scienta.2014.02.010

Díez, C. M., J. Moral, D. Cabello, P. Morello, L. Rallo, and D. Barranco. 2016. Cultivar and tree density as key factors in the long-term performance of super high-density olive orchards. Frontiers in Plant Science 7:1226. https://doi.org/10.3389/fpls.2016.01226

Griggs, W. H., H. T. Hartmann, M. V. Bradley, B. T. Iwakiri, and J. E. Whisler. 1975. Olive pollination in California. University of California, Berkeley, California Agricultural Experiment Station, Bulletin 869.

International Olive Council. 2015. International olive oil production costs study—2015. Madrid: IOC.

Peri, C., ed. 2014. The extra-virgin olive oil handbook. Chichester, UK: Wiley.

Sibbett, G. S., and L. Ferguson, eds. 2005. Olive production manual. 2nd ed. Oakland: UC Agriculture and Natural Resources Publication 3353.

Vossen, P. M., ed. 2007. Organic olive production manual. Oakland: UC Agriculture and Natural Resources Publication 3505.

Khaled M. Bali,
UC Cooperative
Extension (UCCE)
Irrigation Water
Management
Specialist, Kearney
Agricultural Research
and Extension Center

Daniele Zaccaria,
UCCE Associate
Professor and
Agricultural Water
Management
Specialist,
Department of
Land, Air, and Water
Resources, UC Davis

6

Microirrigation Systems and Fertigation

Highlights

- Surface drip is the most common microirrigation system used to irrigate high-density (HD) and super-high-density (SHD) orchards in California. Microirrigation systems include surface drip, subsurface drip, bubblers, microsprinklers, and microsprayers.

- Microirrigation systems apply water more uniformly than surface irrigation systems and other traditional irrigation systems.

- Olive orchards irrigated by microirrigation systems generally use water and nutrients more efficiently and require less weed control.

- Microirrigation systems have higher initial costs and require more frequent maintenance to avoid clogging and rodent damage.

- Microirrigation systems allow easy injection and precise delivery of soluble fertilizers and chemicals, such as soil or water amendments.

- Subsurface drip systems are commonly used on heavy soils but often require a separate irrigation system that applies water on the soil surface (either surface irrigation, such as furrow or border irrigation, or a hand-moved sprinkler system) for tree establishment during the first year.

- Microsprinklers or microsprayers are often used on light-textured soils to improve water application uniformity and reduce water losses due to deep percolation beyond the root zone.

Microirrigation is the slow application of water at low pressure and in discrete locations. It includes surface as well as subsurface systems using drip emitters; surface drip systems can also use microsprinklers, microsprayers, and bubblers. Fertigation is the distribution of fertilizers and water-soluble chemicals (chemigation) through a microirrigation system. While microirrigation and fertigation represent important practices to increase productivity and efficiency in oil olive orchards, their success depends on proper system design, as well as on adequate operation and maintenance. A well-designed and well-functioning system will optimize olive yield and quality.

Reasons for irrigating olive

For centuries, olive trees have been planted on marginal lands in Mediterranean climate conditions and grown widely spaced and typically without supplemental irrigation; they have thrived because of their high resistance to drought, lime, salinity, and heat. While irrigating olive orchards may not be essential under traditional growing practices and with sufficient precipitation, irrigation is a standard practice in California for intensive oil olive orchards to produce consistent yield and good-quality oil.

Intensively planted olive orchards have relatively higher evapotranspiration demands as the soil-water volume available for each tree is more limited. Research findings show that for high planting densities, microirrigation increases olive yields and decreases alternate-bearing behavior (Steduto et al. 2012). Irrigation can lead to faster growth after orchard establishment, earlier onset of fruit production, increased olive yield, and reduced alternate bearing.

Advantages of microirrigation systems

A microirrigation system typically wets 20 to 50 percent of the surface area of the orchard and 30 to 60 percent of the root zone. It allows circulation of machinery in the orchard during irrigation and limits weed growth and weed control costs. The uniformity of a microirrigation system's water application is superior to that of all other irrigation systems, whether on level ground, slopes, or irregular terrain. Microirrigation enables excellent control of the timing and volume of water, and also nutrient applications, because it can provide light and frequent irrigations that maintain adequate soil moisture while minimizing losses due to soil evaporation, waterlogging and surface runoff, and deep percolation.

Microirrigation systems are very well suited to automation, remote control and operation, and implementation of precision irrigation practices, such as plant-based irrigation timing, evapotranspiration-based irrigation scheduling, and volume-based irrigation duration (system starts and shutoffs are based on a predetermined volume of irrigation water delivered per tree), while minimizing irrigation labor. Automation also enables the systematic monitoring of soil moisture sensors and flowmeters, which help evaluate the proper execution of irrigation schedules.

Low-volume, high-frequency irrigations are particularly suited to olive orchards grown either on shallow or low water holding soils, and to high-density (HD) and super-high-density (SHD) orchards with limited root zone. An additional advantage of microirrigation systems is that fertilizers and chemicals injected through the system are delivered near the trees' root zone with high uniformity for more efficient nutrient use and safety. Finally, microirrigation systems minimize the risk of soil structure degradation due to soil erosion, soil surface sealing, and compaction. All of these advantages of microirrigation are important for the majority of olive production regions characterized by high evaporative demand, drought, limited or poor-quality water supplies, and shallow or marginal soils.

Disadvantages of microirrigation systems

Microirrigation systems are pressurized, requiring energy; although operating pressure and power usage are lower than for sprinkler systems, they are significantly higher than those for surface irrigation systems. They also require filtration to eliminate sediment and other suspended materials. These materials decrease system performance by clogging emitters, wearing down components, and causing uneven or decreased water delivery. The clogging of emitters requires periodic chlorine or acid injections accompanied by high-volume flushing to reduce the deposits.

Because design and installation should be done by qualified professionals, microirrigation systems are relatively expensive. Regular system evaluation and maintenance are essential not only to prevent emitter clogging but also to repair leaks

and damage by rodents, other wild animals, and field equipment. A microirrigation system also requires a skilled operator and maintenance labor. Automation increases the initial investment cost, although it decreases labor costs.

Components of microirrigation systems

Microirrigation includes surface drip, subsurface drip, microsprinkler, microsprayer, and bubbler systems. All these systems consist of the following basic components (fig. 6.1):

* pumping system
* vacuum relief valve and other control valves
* primary filters
* injection equipment
* flow control/pressure-regulating valve
* flowmeter
* mainline and submains
* lateral lines
* emitting devices (drip emitters, microsprinklers, microsprayers, bubblers, in-line emitters in drip tape)

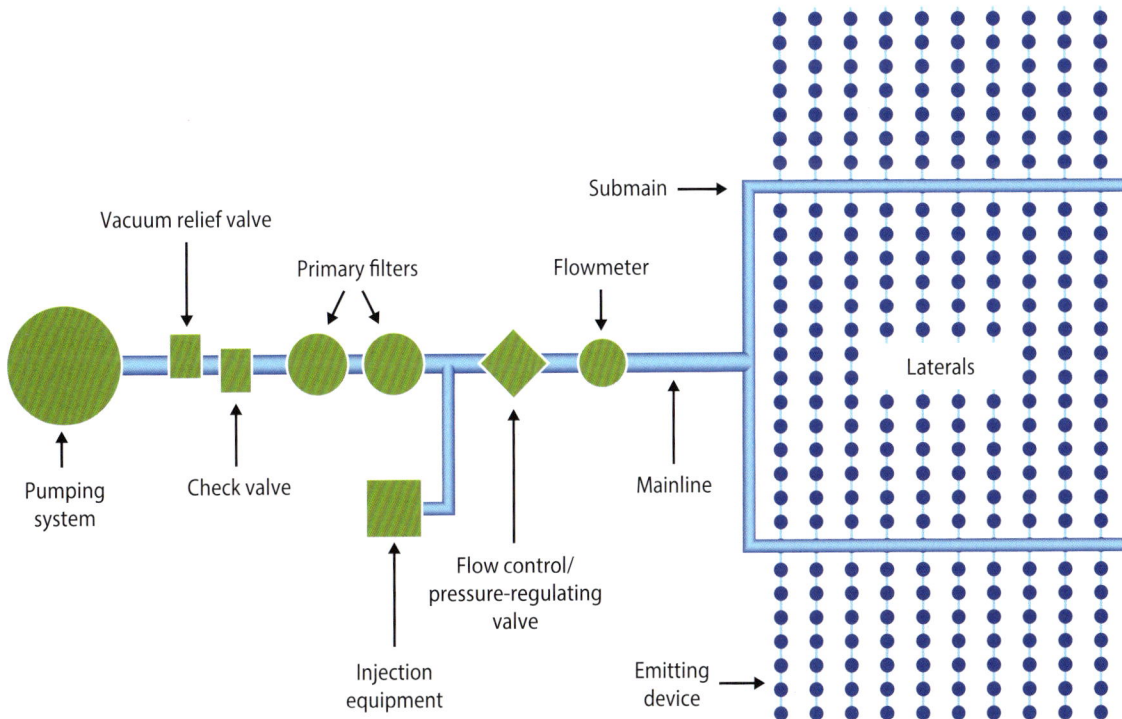

Figure 6.1

Main components of a microirrigation system.

The pumping system, vacuum relief and check valves, filters, injection equipment, flow control and pressure-regulating valve, and flowmeter compose the control head of the microirrigation system. The control head's components are necessary to deliver water at sufficient pressure and appropriate flow rate to the pipe system downstream.

The control head is usually located at a site with convenient access for regular inspection, checkup, maintenance, troubleshooting, and operations. The site must be carefully prepared to ensure drainage of excess water due to storms or leaks from the control head area and also provide reliable access under adverse weather conditions. The control head should be installed on a level concrete pad of sufficient strength and size to mount the pump system, valves, filters, and other control and regulation equipment. The pad should provide a stable foundation to which the equipment can be bolted to reduce vibration, avoid structural stresses, and facilitate inspection and maintenance. Inlet and outlet pipelines should be secured to the concrete pad, and all the equipment must be protected from mechanical damage by agricultural machinery. When possible, the control head should be fenced to keep animals and unauthorized people from damaging the components or being exposed to various hazards.

The selection of microirrigation components may differ from one irrigation system to another based on emitter type, emitter spacing, and discharge rate of the emitter. Higher-discharge emission devices, such as microsprinklers and microsprayers, generally require larger components than those necessary for drip emitters. Emitter discharge rate should be based on the trees' peak water demand and irrigation frequency, while emitter spacing is based on tree spacing and number of emitters per tree. Emitting devices must be capable of supplying each mature tree with sufficient water during the peak water use periods to match the tree's maximum evapotranspiration rate, that is, the maximum amount of water the tree can consumptively use since the last irrigation or rainfall event.

Pumping system

The pumping system pressurizes water so that it flows from the source to the emitting devices and the trees. It includes the pump, pump motor/engine, pump head, suction pipe, and discharge pipe. The pump consists of the housing, or bowl, and inside it an impeller, or impellors, connected by a shaft to the power source, which is an electric motor or fuel-powered engine. The pump capacity should be sufficient to supply the necessary flow rate and the adequate pressure to operate the irrigation system (which depends on the system flow and pressure load) at the maximum possible efficiency.

The system designer usually determines the optimal operating discharge and pressure that the pump must deliver to match the irrigation system's peak demand load. The pump manufacturer and supplier use this information to recommend the most efficient assembly of pump and power unit for a given system and preferred operational setup. The main operating characteristics of pumping units are usually detailed in the technical specifications provided by pump manufacturers and suppliers (technical catalogs of pumps and power units) and include the relationships between discharge, pumping head, rotational speed, and power required. Three types of pumps are commonly used for microirrigation systems:

- **Deep well turbines,** mostly used for pumping groundwater, are installed inside the well casing below water level. They often consist of multiple impeller and casing combinations (or pump stages), to provide sufficient pump output.
- **Centrifugal pumps** (also known as booster pumps), often used to pressurize surface water from ponds or irrigation canals, are placed on the ground above the water level. The impeller casing is spiral-shaped (volute), with the diameter increasing in the flow direction to accommodate increasing discharge as water approaches the outlet.
- **Propeller pumps** (often referred to as axial-flow pumps) are used to pressurize water from ponds, irrigation canals, or other sources with low pressure head and high flow rates.

Selecting the most suitable pumping system will depend on the water source, soil hydraulic properties, water demand of the olive orchard, and climate. The designer should indicate and suggest the pumping system with sufficient capacity and operational flexibility to accommodate changes in operating conditions that may occur over the course of a season and during the orchard's life (such as the reduced water requirements of young trees, possible decline in groundwater levels, pressure changes resulting from expansion of the irrigated acreage, change of emitting devices, or wear of the pumping system's components). Operating efficiency is greatly affected not only by proper selection of pump, power unit, and energy source but also by routine maintenance, testing, and adjustments of the system.

Vacuum relief valve and other control valves

Microirrigation systems use various valves as control, monitoring, and protection mechanisms. These valves include the following:

- **Control valves are** used to turn the irrigation system on and off.
- **Air and vacuum relief valves or vents** allow air to escape when the system is turned on and to enter the system when it is shut off.
- **Check valves** prevent undesirable flow reversal to protect the pump and accessory components and prevent the backflow of chemicals toward the pumping system.
- **Pressure relief valves** open when the system pressure exceeds a certain set level, protecting the various components of the irrigation system against pressure surges that may damage pipes and casings of supplemental parts.
- **Flow control valves** allow water to flow through the pipelines at a constant rate, regardless of changes in operating pressure.
- **Pressure-regulating and pressure-sustaining valves** maintain adequate and constant operating pressure throughout the system.

Maintaining adequate pressure and avoiding pressure fluctuations and variations in the system are crucial for discharging uniform water volume throughout the orchard. The purpose of pressure-regulating valves in mains, submains, and laterals, in conjunction with pressure-compensating emitters, is mainly to overcome pressure differences and consequent variations in emitter discharge rates. These inconsistencies can occur because of friction losses, ground elevation differences across the orchard, and variability in emitter discharge rate.

Primary filters

The filtration system is the best defense against emitter clogging, a very common problem in microirrigation systems. Clogging may be caused by organic suspended material (for example, algae, diatoms, larvae, seed, bacteria, slimes) and inorganic materials (for example, sand, silt, clay, chemical precipitates, iron, lime). Clogging causes poor water distribution along the laterals and among trees in an orchard, with some trees receiving less or no water and nutrients if emitters are clogged for extended periods. Clogged emitters can be difficult to detect and are labor intensive, and therefore expensive, to clean or replace. In addition to filtration, periodic or continuous chemical injections may be necessary to dissolve mineral precipitates and prevent growth of slimes and other organic material inside the pipe system.

Adequate filtration requires processing all the water that enters the microirrigation system. The microirrigation system designer should select appropriate filtering devices based on water quality factors (the type and size of materials suspended in the water) and the water quality required for satisfactory operation of the selected emitters. Any suspended material with particles larger than about one-tenth the emitter flow passage must be filtered and removed from the water before it enters the microirrigation system. This is necessary because several smaller particles may group together and plug the passageway and outlet orifice of emitters.

Adequate filtration is needed not only to prevent emitter clogging but also to avoid deposits of materials in areas of the system where the velocity of water flow is slow. Filtration is also necessary to avoid the growth of organic materials and bacterial slimes. The type and size of the filtering device should be based on the most economical equipment with the lowest friction losses.

Coarse suspended material and particulates can be removed with vortex filters, sand separators, screen filters, disk filters, or sand media filters. Algae, snails, larvae, slime, and other organic materials can be removed using screen filters, disk filters, or sand media filters. Sand media filters are usually preferred for filtering surface water that may contain algae and slime, as such organic matter can easily clog screen and disk filters and is difficult to flush.

All filters must be backwashed periodically to restore their maximum filtering capacity and functionality. Under maximum-capacity, clean operating conditions, each filtering device has a specific pressure drop that depends on its filtering capacity and the flow rate passing through it. Higher pressure drops across the filter relative to the specific pressure drop indicate when backwashing is required. In general, filters should be flushed or cleaned when the pressure drop across the filter is 4 to 10 pounds per square inch (psi) or greater, or as specified by the filter manufacturer. Backwashing can be accomplished either manually or automatically on a predefined schedule or when the filter senses an excessive pressure drop. The water resulting from backwashing must be flushed out and discharged from the system. Selecting filtering devices with automatic backwashing capability is important to reduce labor costs when water containing organic materials or sediments is used as the main irrigation source.

Injection equipment

Injection equipment is needed for the distribution of water-soluble chemicals and fertilizers through a microirrigation system. These compounds are delivered to the root zone to control weeds and pests and to provide nutrients for tree growth and fruit production. The microirrigation system designer usually selects the most suitable injection devices based on cost, types of chemicals applied, precision of injection needed, and injection rate required.

Fertilizers, insecticides, herbicides, fungicides, nematicides, and growth regulators are applied through injection equipment in olive orchards. Amendments, such as gypsum and humic/fulvic acids, can also be applied through injection equipment to condition the soil. Chemicals are also periodically injected into the system for maintenance purposes, that is, to prevent or reduce emitter clogging due to suspended materials, scale, and bacteria.

Microirrigation systems are well suited to chemigation and fertigation due to their high distribution uniformity, ease of operation, adaptability to automation, and lower cost compared with conventional methods of applying chemicals and fertilizers. Operators must follow regulatory requirements and instructions on fertilizer and chemical labels. A major concern with the injection of fertilizers and chemicals is the protection of groundwater aquifers, field personnel, soil, and the environment. A common mistake that prevents uniform distribution is conducting the injection at an excessive rate or for an insufficient duration and then shutting off the microirrigation system right after the injection is finished. Improper timing and duration can lead to excessive or nonuniform application of injected materials, even when the microirrigation system delivers water uniformly. Injecting for too short a duration results in trees hydraulically closer to the injection point receiving more chemicals or fertilizers than trees farther away. Injection devices must be periodically recalibrated to ensure optimal distribution of chemicals and fertilizers.

Flowmeter

Flow measurement and recording devices are crucial components of microirrigation systems, as they enable the operator to determine the flow rate delivered over the course of the irrigation and the total volume of water applied over the orchard block in each irrigation event. Adequate orchard irrigation management entails applying the proper amount of water at the right time and with uniform distribution to meet the trees' evapotranspiration needs. To execute an irrigation schedule, orchard managers must know when to irrigate, how much water to apply, and how much water was actually applied during irrigation events.

A flowmeter can also function as a valuable tool to detect problems and determine troubleshooting and maintenance needs, as it enables orchard managers and irrigators to identify clogging, plugging, leaks, and breaks in the microirrigation system. Clogging and obstruction of pipe flow result in a lower flow rate and possibly higher pressure than expected, whereas leaks and breaks lead to a higher flow rate and lower pressure than normal.

Microirrigation systems need devices that can measure pipe flow and its variations, which are affected by pressure variations along the tubing. Various types of flow measurement and recording devices are available; in general, the devices to consider should be easy to install, maintain, inspect, and read. The most commonly used flowmeter for microirrigation systems is the propeller meter. For detailed information on flowmeters, see Hanson and Schwankl 2009.

Mainline and submains

The largest-diameter pipeline, the mainline must be capable of conveying the flow of the entire microirrigation system or of an individually operated portion of it (sector or block, and sub-block) under favorable hydraulic conditions of flow rate and sufficient pressure head. The submains are smaller-diameter pipelines branching out from the mainline to distribute water to the various orchard blocks and sub-blocks. In modern microirrigation systems for orchards, mainline and submain pipes are usually made of polyvinyl chloride (PVC) or black high-density polyethylene (HDPE). These pipes convey water from the pump to the lateral lines and emitters.

Generally, the irrigation designer sizes the mainline and submains very carefully to provide the system with sufficient capacity and operation flexibility while enabling the system to operate with the lowest possible energy usage and cost. In other words, the designer balances the cost for the different pipe diameters with the cost of energy usage to offset pressure losses caused by elevation differences across the orchard and friction losses as water moves through the pipes and tubing network. Larger-diameter pipelines cost more but, for the same flow rate, require comparatively less energy to sustain a consistent pressure through the system because there is less friction loss and less loss of pressure if water has to be moved forward.

Lateral lines

Laterals, or irrigating lines, are the smallest-diameter pipelines in the microirrigation system; they are usually made of low-density polyethylene (LDPE). Laterals are fitted to the submains at regular distances, laid along the tree rows, and equipped with regularly spaced emitting devices that deliver water to individual trees. The system designer determines the spacing of laterals along the submains based on tree row spacing and number of drip lines per tree row, and the spacing between emitters based on distance between trees along the row and number of emitters per tree.

The designer also selects the length and diameter of the laterals, again balancing the investment costs of larger-diameter laterals with operational energy costs, and also balancing the cost of more submains with the risk of not being able to maintain a consistent pressure along laterals if they are very long. If lateral lines are too small in diameter or too long, or both, emitters may discharge water to trees at different rates as a consequence of relatively high friction losses, resulting in nonuniform applications of irrigation and injected materials (nutrients and chemicals) to trees across the orchard.

Emitting devices

A microirrigation system wets only a small fraction of the soil surface with frequent and small discharges of water through emitting devices that deliver water to precise locations. Microirrigation systems for modern olive orchards use emitting devices that are designed to dissipate pressure and discharge individual, small, and uniform streams of water (point-source emitters), at a constant rate, and avoid flow variations due to pressure fluctuations.

The point-source emitters most frequently used in microirrigated oil olive orchards are in-line drip emitters (see below). On-line drip emitters, microsprinklers, microsprayers, and bubblers can also be used. Multiple emitters per tree are often necessary to meet the tree's water requirements and to wet an adequate volume of soil to allow adequate root development. The irrigation system designer selects the most suitable type, or types, of emitting device on the basis of hydraulic and operational considerations as well as the orchard manager's preference. After selecting the best-suited emitting device, the designer chooses whether to suggest non-pressure-compensating or pressure-compensating emitters, based on orchard topography, shape, dimensions, soil characteristics and variations across the orchard, water quality, and other considerations related to the system's hydraulic balancing.

Drip emitters

Drip emitters, or drippers as they are commonly called, can be coextruded integrally with the drip tubing (in-line drippers) at predetermined spacing during the manufacturing process or punched into the drip line (on-line drippers). With on-line drippers, one emitter per tree can be punched into the drip line for the first couple of years of orchard establishment and additional emitters punched in as the trees grow. The small orifice diameter of both in-line and on-line drippers, around

0.025 to 0.060 inch (0.6 to 1.5 mm), makes them susceptible to clogging, which can be reduced or prevented by periodic chlorine and/or acid injections.

The nominal discharge rates of drip emitters range from 0.5 to 4 gallons (1.9 to 15 L) per hour at specific pressure, typically around 8 to 12 psi. If the emitters are not pressure compensating, changes in the system operating pressure will affect the emitter discharge rate, with the flow rate increasing as the pressure increases and decreasing as the pressure decreases. Pressure-compensating emitters are designed to maintain a constant flow rate within certain ranges of operating pressure.

Single or dual drip lines can be used to irrigate olive orchards, depending on tree planting density, soil infiltration characteristics, targeted wet area, water delivery schedule to the orchard, labor and work shifts, and power availability and tariffs. Single and dual drip lines are both commonly used in HD and SHD orchards, depending on the orchard water requirements and on the most suitable application rate for the specific type of soil. Dual drip lines are commonly used to achieve a larger wetted area and volume in light soils, which are often characterized by poor lateral distribution of water. Despite the higher investment cost, a dual line system can require significantly fewer operating hours to deliver the amount of water necessary to meet the trees' water requirements. Shorter operations may allow orchard personnel to conduct other practices when the system is not operating, and thus reduce the need for long labor shifts, and enable wider use of off-peak utility rates.

Subsurface drip emitters

Subsurface drip irrigation (SDI) systems for orchards usually consist of LDPE drip tubing equipped with in-line drippers, which is durable and easy to install below ground. The in-line drippers for SDI systems have additional features to prevent ingestion of soil particles and root intrusion; these consist of physical barriers or herbicide-impregnated material that creates inhospitable conditions for root growth near the tubing and emitters. An SDI system can consist of single or dual drip lines. The lines are typically installed 10 to 24 inches (25 to 61 cm) deep, depending on soil characteristics and soil profile.

Deeper drip lines usually minimize surface wetting but make installation, troubleshooting, and repairs more difficult and expensive.

Among the advantages of SDI systems are less weed growth, as the soil surface stays dry most of the time, and reduced damage to irrigation equipment during harvest operations or other orchard practices, along with the ability to irrigate during these times. The main disadvantages of SDI systems are the difficulty in detecting emitter clogging and leaks along the drip lines, as well as the risk of roots pinching the tubing and intruding into and clogging emitters. Flushing operations are also more complicated with buried drip lines.

Microsprinklers

Microsprinklers have higher discharge rates, ranging from 4 to 30 gallons (15 to 114 L) per hour at operating pressures from 20 to 40 psi, and wet a larger surface area than drip emitters. They are frequently used for tree crops as they can deliver sufficient amounts of water to meet the orchard's requirement in fewer operating hours than drip emitters. In addition, microsprinklers are often preferred to drip emitters for spreading water more widely in orchards established on soils with poor water infiltration, a high infiltration rate, or very limited lateral subbing.

The wider areas wetted by microsprinklers may also entail higher rates of soil evaporation and more weed growth, especially with young trees. These drawbacks lead to lower application efficiency than is achievable with drip emitters. Additionally, insects may lay eggs in the orifices of microsprinklers, causing clogging.

The flow passageways of microsprinklers range from 0.030 to 0.080 inch (0.08 to 0.20 mm), similar in size to or slightly larger than those of drip emitters. However, the flow velocity of microsprinklers is greater than that of drip emitters, which makes microsprinklers less susceptible to clogging. Larger flow passage orifices usually result in higher discharge rates, which further reduces the risk of clogging. Also, microsprinklers are easier to inspect and address for clogs than are drip emitters.

Bubblers

Bubblers are small, low-volume emitters designed to deliver water in bubbles or streams to small areas. Typically, the flow rate of bubblers ranges from 30 to 65 gallons (114 to 246 L) per hour at operating pressures of 15 to 45 psi. Bubblers are commonly installed on small plastic wedges inserted into the ground and connected to LDPE laterals of various length and diameter. Normally, bubblers are placed in tree basins, one or two per tree; the basin is needed to contain the water as the bubbler discharge usually exceeds the soil intake rate.

Designing efficient microirrigation

Microirrigation systems for tree crops are usually designed and managed to discharge frequent, light applications of water close to each tree while wetting only a portion of the orchard's soil surface. Travel of water over the soil surface or through the air are of minimal relevance. The application uniformity and efficiency of a microirrigation system depend on the rate and uniformity of water discharge from the emitting devices. The rate of water discharge per unit of wet area must be compatible with the soil infiltration rate, and the amount of water applied during an irrigation event must be within the soil storage capacity of the root zone or able to meet the tree water use during the period between water applications, for different irrigation frequencies.

The design strategy for microirrigation systems is centered on achieving the desired application uniformity while providing sufficient flexibility to the system to accommodate various operating conditions (combination of flow rates and pressures) that may be required during the life cycle of the system and the orchard. The process for designing an efficient microirrigation system follows a logical sequence of steps (fig. 6.2).

Design parameters

Typically, microirrigation systems are designed to deliver sufficient water for meeting the full crop evapotranspiration during the peak water demand period. However, the design of resource-efficient irrigation systems must also consider the hydraulic properties of the soil where the trees grow so that the maximum irrigation application does not exceed the capacity of the soil to allow water infiltration and storage, which is critical to minimize inefficiencies and water losses. Several basic parameters are calculated as part of designing resource-efficient microirrigation systems for olive orchards.

1. Maximum gross depth of water per irrigation, *DG*

The maximum amount of irrigation water to apply during an irrigation (gross depth of water per irrigation) is calculated based on soil and tree characteristics, as well as on target application efficiency of the microirrigation system. The basic input information needed for calculating the maximum gross depth of water per irrigation, D_G (in) is as follows:

- maximum allowable depletion for no water stress, *MAD* (%)
- soil water holding capacity, *WA* (in/ft)
- percentage of wetted area by microirrigation emitters, P_W (%)
- tree root depth, *Z* (ft)
- target application efficiency of the microirrigation system, *EffA* (%)

$$D_G = (MAD \times WA \times P_w \times Z) \div Eff_A$$

D_G represents the maximum depth of applied water that the soil can store and make available for tree use. It should not be exceeded when scheduling irrigations. Any water application greater than the maximum may result in deep percolation past the root zone, waterlogging and runoff, and excessive soil evaporation, all of which make irrigation less efficient. Designing a microirrigation system with the capacity to apply water in amounts equal to D_G during peak demand periods is highly recommended for efficient irrigation.

The D_G calculation for a drip-irrigated orchard grown on a silty clay loam with a maximum allowable depletion (*MAD*) of 60 percent, soil water holding capacity (*WA*) of 2.2 inches per foot, percentage of wetted area (*PW*) of 25 percent, root depth (*Z*) of 5 feet, and target application efficiency (*EffA*) of 0.9 percent is as follows:

$$(0.6 \times 2.2 \text{ in/ft} \times 0.25 \times 5 \text{ ft}) \div 0.9 = 1.83 \text{ in}$$

Any water application greater than 1.83 inches may result in water loss. For average values of the water holding capacity of different soil textures, see table 6.1.

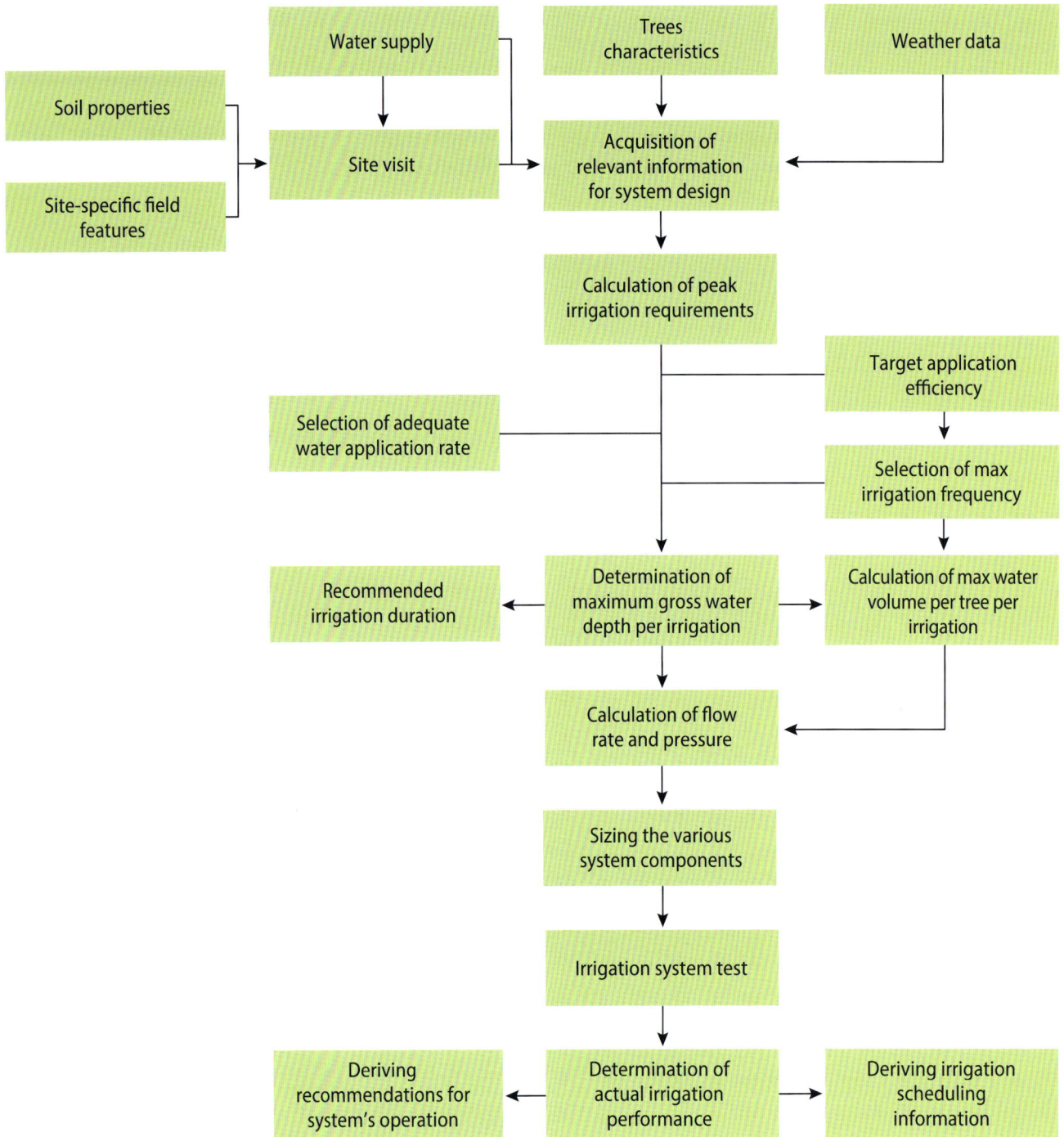

Figure 6.2

Designing a microirrigation system is a process that follows several necessary sequential steps.

2. Maximum volume of water per tree to apply per irrigation, *Vw*

The maximum volume of water to apply per tree per irrigation, *Vw* (gal/tree), depends on the maximum gross depth of water per irrigation, D_G, and the tree density per acre (number of trees per acre), which in turn depends on tree spacing. A conversion factor of 0.623 is used in the calculation:

$$V_w \text{ (gal/tree)} = D_G \times \text{Tree spacing} \times 0.623$$

The *Vw* calculation for a super-high-density olive orchard grown on a silty clay loam, with 726 trees per acre planted at a spacing of 12 feet between rows and 5 feet between trees is, as follows:

Table 6.1

Average available water holding capacity (*WA*) for various soil textures

Soil		W_a (inches of water per foot of soil)
General description	**Texture class**	
light, sandy	coarse sand	0.5
	fine sand	0.9
	sandy loam	1.2
medium, loamy	fine, sandy loam	1.5
	loam	1.8
	silt loam	2.0
heavy, clay	clay loam	2.2
	clay	2.4
	peat/muck	6.0

Source: Modified from U.S. Bureau of Reclamation, Agrimet Irrigation Guide website, (https://www.usbr.gov/pn/agrimet/irrigation.html).

$$1.83 \text{ in} \times 12 \text{ ft} \times 5 \text{ ft} \times 0.623$$
$$= 68.4 \text{ gal/tree}$$

Any water application greater than 68.4 gallons per tree may result in water loss and irrigation inefficiency.

3. Recommended irrigation duration, IT_{REC}

The recommended irrigation duration, IT_{REC} (hr), represents the irrigation duration compatible with the capacity of the soil to infiltrate water and is essential for avoiding waterlogging and runoff. IT_{REC} is calculated on the basis of the maximum gross depth of water to apply per irrigation (D_G) and the basic soil infiltration rate, IR_B (in/hr):

$$IT_{REC} = D_G \div IR_B$$

As an example, the IT_{REC} for a drip-irrigated olive orchard grown on a silty clay loam with a basic soil infiltration rate (IR_B) of 0.2 inch per hour is as follows:

$$1.83 \text{ in} \div 0.2 \text{ in/hr} = 9.15 \text{ hrs}$$

4. Adequate irrigation application rate, *AR*

Efficient microirrigation management entails applying water at a rate compatible with the basic soil infiltration rate and for a duration compatible with the available labor to oversee irrigation

and troubleshoot problems. At the planning and design stage, it is important for the microirrigation system designer to select the proper number, spacing, and configuration of emitting devices per tree and a discharge rate resulting in an adequate irrigation application rate, *AR* (in/hr), lower than or equal to the basic soil infiltration rate, IR_B (in/hr). This ensures that the entire amount of water applied by the irrigation system infiltrates the soil root zone, avoiding waterlogging and runoff.

The duration of irrigation compatible with available labor to troubleshoot any problems and make repairs depends on the individual operation. However, in most orchards, the irrigation system should not be operated for more than 16 hours per day to allow time for field personnel to work on the system if necessary. For typical maximum irrigation water application rates compatible with the basic soil infiltration rate of different soil textures and field slopes, see table 6.2.

5. Adequate irrigation frequency, *IF*

Generally, the orchard manager indicates the irrigation frequency (days) that would most suit the availability of labor and water supply. Field labor is necessary to operate the microirrigation system and oversee the correct execution of irrigation schedules. Typically, orchard personnel work 6

Table 6.2

Maximum irrigation water application rates for various soil textures and slopes

Soil type	Maximum application rate (in/hr) at slope		
	0–5%	5–8%	8–12%
Coarse sandy soil	1.5–2.0	1.0–1.5	0.75–1.0
Light sandy soil	0.75–1.0	0.5–0.8	0.4–0.6
Silt loam	0.3–0.5	0.25–0.4	0.15–0.3
Clay loam, clay	0.15	0.10	0.08

Source: Adapted from USDA NRCS 1997.

days a week, so the weekly irrigation schedule must be completed within that time.

At the irrigation planning stage, the system designer collects the information about available water supply and labor from the orchard manager and determines an adequate irrigation frequency for the peak water demand period. Adequate irrigation frequency is an important design parameter as it affects the determination of the total system's capacity (maximum flow rate and resulting pressure requirements). Typically, microirrigation systems for tree crops are designed with the capacity to apply the amount of water necessary to meet the weekly tree water requirements of the peak period in one irrigation per week with a maximum duration of 16 to 20 hours. However, it is also important to make the system flexible enough so that the orchard can be irrigated beyond the design parameters for special purposes. These include watering more deeply to prevent damage during a heat wave, leaching the soil, refilling the soil profile after a drought, and building deep soil water storage to prevent water stress if irrigation is shut off during harvest or other practices.

Irrigation system evaluation

The main objective of evaluating any irrigation system is to identify deficiencies in the system capacity, operation, or management practices that can be addressed to improve water use efficiency and system efficiency. Evaluating the performance of a microirrigation system is usually simpler than evaluating other irrigation systems, such as surface irrigation or sprinkler irrigation.

Several performance measures can be used, such as application efficiency (AE) and distribution uniformity (DU). AE represents the percentage of the applied irrigation water that is stored in the soil root zone and utilized for crop evapotranspiration. AE drops, for example, if irrigation water percolates below the root zone or runs off or evaporates from the soil. Distribution uniformity (DU) refers to the ability of the irrigation system to distribute the applied water evenly throughout the field. For additional information about DU, see ccuh.ucdavis.edu/measuring-DU-run-time.

In relatively large olive operations with open channels, which are susceptible to evaporation losses, or a buried delivery system, which may be susceptible to seepage losses, water distribution efficiency (WDE) may be measured. WDE is a good indicator of the efficiency of the on-farm irrigation system in relatively large farms. The losses associated with the system are mainly seepage and evaporation and minimizing these losses increases the efficiency of the distribution system. The formula is as follows:

$$\text{Water distribution efficiency (WDE)} = 100 \times Wd \div Wi$$

Wd is water delivered by a distribution system and *Wi* is water introduced into the distribution system.

Maintenance considerations

Microirrigation systems are efficient at providing water, chemicals, and fertilizers but must be maintained for maximum efficiency. The most common problems are emitter plugging, major leaks in tubing caused by rodents, and physical damage caused by farming equipment. One or more of these issues may cause a severe decline in system uniformity and efficiency. Emitter plugging can be reduced with a good filtration system to remove sediment and other particulates before they enter the irrigation system. Emitter plugging is generally caused by one or more of these materials:

- sediments in irrigation water
- organic matter, such as algae or bacteria, in irrigation water
- chemical precipitates due to the chemical composition and the acidity or alkalinity (pH) of irrigation water; an example is calcium carbonate, which as a salt precipitates at high pH—therefore, acid injection is needed to lower the pH and minimize the risk of clogging

Chlorine or acid injection is used in microirrigation systems to prevent clogging caused by biological growths (algae and bacterial slimes) and chemical precipitates (particularly calcium carbonate).

A water quality analysis for chemical composition, hardness hazard, and turbidity should be performed to identify potential issues related to the presence of sediment or chemicals of concern. Then an effective water filtration and chemical mediation measures, such as lowering the pH of irrigation water, can be developed. Surface water typically contains sediments and biological matters, whereas groundwater typically has a higher salinity level and the potential for chemical precipitates and consequent plugging of emitters.

Fertigation overview

Many different substances, including chlorine, acid, fertilizers, herbicides, micronutrients, nematicides, and fungicides, can be injected through irrigation systems. Of these, fertilizers are most commonly injected. Microirrigation systems can be easily utilized for this purpose. The application rate and frequency can be tailored to match the olive orchard's nutrient requirements. The following are the main advantages of fertigation using a microirrigation system:

- higher application efficiency results in more

efficient fertilizer use because the fertilizer is soluble in water
- flexibility in the timing and application rate of fertilizers to meet the trees' demands
- reduced labor cost associated with fertilizer applications
- reduced potential fertilizer losses below the root zone and possible contamination of groundwater
- improved safety and less potential human contact with harmful chemicals

The main disadvantages are as follows:

- potential mixing of incompatible fertilizers and other chemicals
- potential breaks in mixing devices or leaks from storage tanks
- backflow prevention and additional safety measures necessary

Fertilizer solubility

To be injected directly into the irrigation system, liquid fertilizers must be soluble in water. Dry fertilizers must be mixed with water before injection. They must be mixed first in a tank and sufficient water applied to completely dissolve the fertilizer and produce the desired concentration for injection. The solubility of a fertilizer in the irrigation water depends on several factors, such as the physical and chemical properties of the fertilizer, water temperature, and water pH.

Fertilizer application rates

Fertilizer application rates indicate the actual amount of the nutrient required by the crop, in pounds per acre, not the amount of the product that should be applied per acre. To determine the amount of a liquid fertilizer required per acre, it's first necessary to calculate the weight of actual nitrogen delivered per gallon. This is done by multiplying the weight of the product per gallon by the percentage of actual nitrogen in the product. For example, using UAN-32, which weighs 11.1 pounds per gallon (product weight may vary slightly with temperature) and has an N:P:K ratio of 32-0-0 (that is, 32% actual nitrogen), the calculation is as follows:

$$11.1 \times 0.32 = 3.55 \text{ lb actual N/gal}$$

If the recommended fertilizer application rate is

80 pounds of actual nitrogen per acre, the volume of UAN-32 needed per acre is as follows:

$$80 \div 3.55 = 22.5 \text{ gal}$$

To calculate the flow rate (gallons per hour) of liquid fertilizer added to irrigation water, divide the total gallons required by the length of time needed for the irrigation. As an example, to irrigate 35 acres for 6 hours with 4 gallons of product per acre, the calculations are as follows:

$$4 \text{ gal/ac} \times 35 \text{ ac} = 140 \text{ gal needed}$$
$$\text{to irrigate 35 ac}$$

$$140 \text{ gal} \div 6 \text{ hr} = 23.3 \text{ gal/hr}$$

Injection devices

Irrigators wishing to inject chemicals can choose from a variety of injection equipment, including differential pressure tanks, venturi injectors, positive displacement pumps, small centrifugal pumps, and solutionizer machines.

Differential pressure tanks

Differential pressure tanks, also called batch tanks, are the simplest injection devices. The inlet of the tank is connected to the irrigation system at a point of pressure higher than that of the outlet connection. This pressure differential causes irrigation water to flow through the tank, which contains the chemical to be injected. Some of the chemical goes into solution and passes out of the tank into the downstream irrigation system. Because the tank is connected to the irrigation system, it must be capable of withstanding the operating pressure of the irrigation system.

While relatively inexpensive and simple to use, differential pressure tanks have a disadvantage: as irrigation continues, the chemical mixture in the tank becomes increasingly diluted, decreasing the concentration in the irrigation water. If a set amount of a material is to be injected and the concentration during injection is not critical, then a differential pressure tank may be appropriate. However, if the concentration must be kept relatively constant during injection, this system is not appropriate.

Venturi injectors

A venturi injector is installed in the irrigation mainline at the point of a constriction that creates a negative pressure, or suction, at the throat of the constriction, which draws chemical or fertilizer solution from the injector tank into the irrigation stream. It is often referred to as a Mazzei injector, although Mazzei is a brand name; venturi injectors are also available from other manufacturers.

The constriction in the pipeline causes friction that reduces the water pressure by 10 to 30 percent. This means that the pressure at the inlet port must be 10 to 30 percent higher than at the outlet port. Because of this significant pressure loss, the injector should be installed parallel to the pipeline so that a valve can turn off the flow when injection is not occurring. The injection rate is determined by the size of the venturi (constriction) and the pressure differential between inlet and outlet ports. Injection rates as high as 700 gallons (2,650 L) per hour are possible with large venturi injectors. An injector can be installed with a small centrifugal pump that draws water from the irrigation system, increases its pressure while moving water through the venturi, and then returns the water and chemical or fertilizer back into the irrigation system.

Positive displacement pumps

Positive displacement pumps are piston or diaphragm pumps that inject at precise rates. They are powered by electricity or gasoline, or are driven by water; the water-driven pumps can be installed in locations that lack electrical power. Positive displacement pumps are the most expensive injection device. However, this type of device is preferable when a constant and precise injection concentration is needed

Centrifugal pumps

Frequently used for injecting fertilizers, centrifugal pumps have a greater flow rate than positive displacement pumps and most venturi injectors, making them appropriate for higher-injection-rate applications. Centrifugal pumps can be driven by either electricity or gas engines. Using a centrifugal pump in conjunction with a flowmeter can be helpful in controlling the injection rate.

Solutionizer machines

Developed to inject materials that are not readily soluble, solutionizer machines are most commonly used for injecting finely ground gypsum but are also used for injecting fertilizer products, such as potassium sulfate. Solutionizer machines inject a slurry of material into the irrigation line, where it mixes and goes into solution. In microirrigation systems, it is important to inject materials upstream of the system filters to ensure that insoluble elements are filtered out and do not end up clogging the emitters. For example, gypsum materials, which are 95 percent pure, may still contain up to 5 percent insoluble materials. This means that for every 100 pounds (45 kg) of gypsum material injected, 5 pounds (2 kg) of insoluble material might be present. Dry fertilizers may also contain significant insoluble material that could contribute to emitter clogging.

Injection point

The injection point should be properly located so that the injected fertilizer or chemical and the irrigation water can become thoroughly mixed well upstream of any branching of the flow. Because of concerns over fertilizers and chemicals being flushed out when the microirrigation system filters are backwashed, the injection point should be downstream of the filters in all injection systems except the solutionizer system. To ensure that no contaminants are injected into the microirrigation system, a good-quality screen or disk filter should be installed on the line between the chemical or fertilizer supply tank and the injector. The system should be allowed to fill and come up to full pressure before injection begins. Following injection, the system should be operated to flush the chemical or fertilizer from the lines. Leaving residual material in the line may encourage clogging from chemical precipitates or organic sources, such as bacterial slimes.

Injecting materials uniformly

Once injection begins, the injected material does not immediately reach the emitters because of the time it takes for the solution to travel through the microirrigation system. Measurements made in commercial orchards indicate that travel time ranges from 30 minutes to well over an hour,

depending on the microirrigation system design. To ensure that the application of injected material is as uniform as the water application, take the following steps:

1. Determine the travel time of the injected material to the farthest hydraulic point in the microirrigation system. This is a one-time determination and can be done by injecting chlorine into the system (also a good maintenance procedure) and tracing its movement through the system by testing the water for chlorine with a pool or spa test kit.
2. Schedule injection periods at least as long as it took the injected material to reach the end of the last lateral line (determined in step 1). A longer injection period is usually preferable.
3. Once injection stops, continue the irrigation for as long as it took the injected material to reach the end of the farthest lateral (step 1). A longer, post-injection irrigation period is usually preferable.

Especially with injected materials that easily travel with water (for example, nitrate materials), avoid overirrigation because moving the water and injected material through the root zone wastes water and the injected material, and may lead to groundwater contamination.

Safety measures

Appropriate care should be exercised when handling all injected materials, and the safety of personnel should be the highest priority (fig. 6.3). Environmental safety associated with fertigation should also be a priority. Fertigation regulations vary from state to state. In California, chemigation safety regulations apply only to labeled chemicals, not to fertilizers. While there are no California state regulations concerning fertigation, there may be local regulations—and these should be checked and followed. The same safety equipment required for injection of labeled chemicals is also useful for environmental protection when fertigating. The safety devices include the following:

- **Chemigation check valve,** located between the water source and the injection point, prevents injected materials from moving back

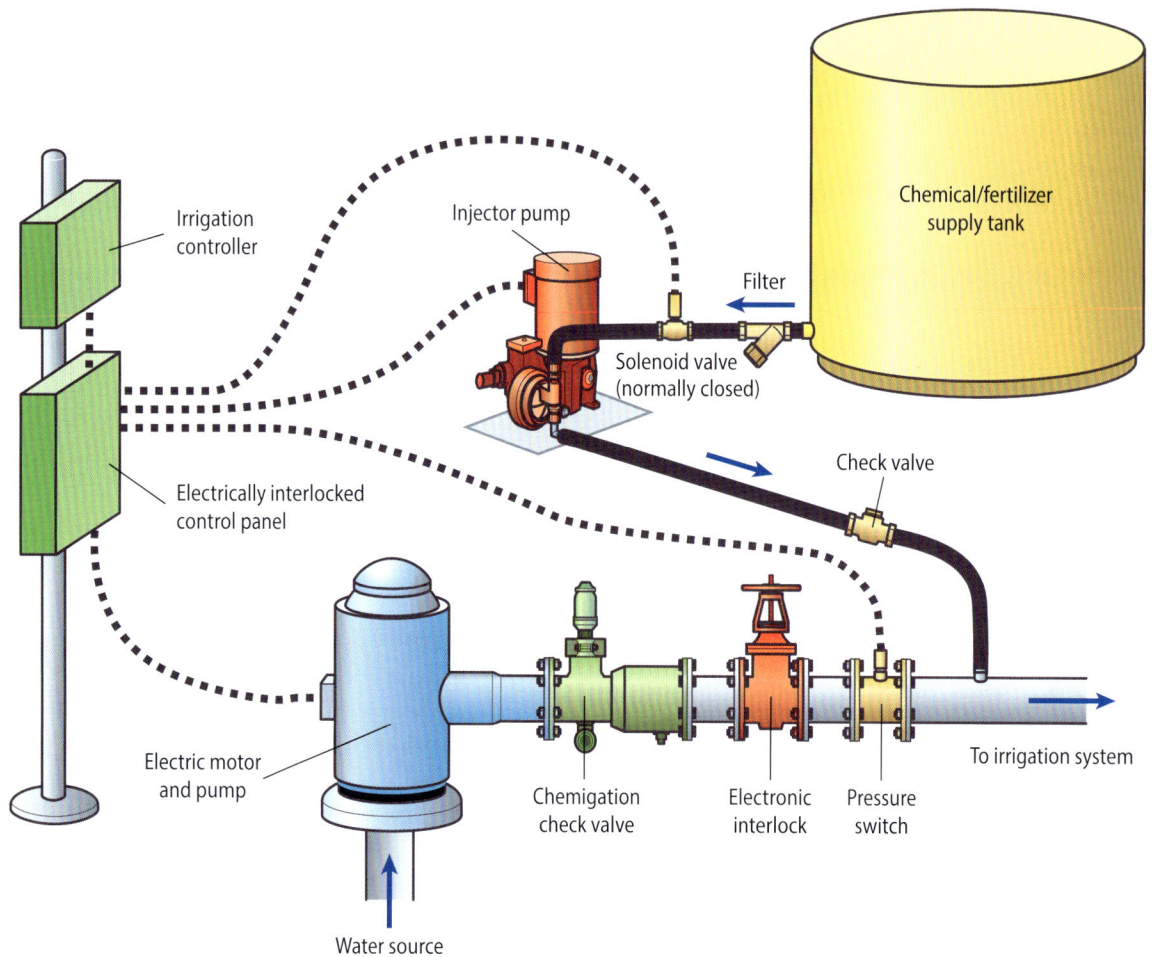

Figure 6.3

A typical injection layout includes many safety devices.

to the water source. The check valve has a one-way, spring-loaded flap inside that allows water to pass only downstream. It also has an air vent and vacuum relief valve and a low-pressure drain upstream of the one-way flap closure. The vacuum relief valve prevents a vacuum, which could draw injected material through the closed check valve. If some does leak past the closed check valve, it will flow out the low-pressure drain, which is open when the irrigation system is shut down but closes when the irrigation system is pressurized.

Even if only fertilizers, no labeled chemicals, are expected to be injected through the irrigation system, installation of a chemigation check valve is prudent (fig. 6.4). Backflow of fertilizer to a well or other water source can result in surface and groundwater contamination. At a minimum, leaving room in the

layout to install a chemigation check valve at a later date is wise.

- **Electronic interlock** between the water pump and the injector pump prevents operation of the injector when water is not being pumped.
- **Check valve** in the line from the injector to the irrigation system prevents water from flowing back through the injector and overflowing the chemical/fertilizer storage tank.
- **Solenoid valve,** normally closed (or a hydraulically operated valve, normally closed), between the chemical/fertilizer supply tank and the injector keeps material in the tank from flowing into the irrigation system when the system is not operating.
- **Pressure switch** in the irrigation system, interlocked to the pump, shuts down the irrigation and injection systems if there is a drop in operating pressure caused, for example, by a break in a pipeline.

Figure 6.4

A double chemigation check valve protects a well against backflow. *Photo:* Lawrence Schwankl.

We'd like to thank L. Schwankl, whose published writing provided the starting point for this chapter on microirrigation systems.

References

Hanson, B., and L. Schwankl. 2009. Measuring irrigation water flow rates. Oakland: UC Agriculture and Natural Resources Publication 21644.

Steduto, P., T. C. Hsiao, E. Fereres, and D. Raes (eds.). 2012. Crop yield response to water. FAO Irrigation and Drainage Paper 66. Rome: Food and Agriculture Organization of the United Nations.

United States Department of Agriculture Natural Resources Conservation Service. 1997. National engineering handbook: Irrigation guide.

Suggested reading

International Olive Council. 2015. International olive oil production costs study—2015. Madrid: IOC.

Lightle, D., D. A. Sumner, and J. Murdock. 2016. Sample costs for olive oil: Establish a super-high density olive orchard and produce olives for oil: Arbequina variety—drip irrigation: Sacramento Valley—2016. UC Agriculture and Natural Resources Cooperative Extension, Agricultural Issues Center.

Schwankl, L. 2016. Microirrigation and fertigation systems. In L. Ferguson, and D. R. Haviland, eds., Pistachio production manual. Oakland: UC Agriculture and Natural Resources Publication 3545.

Schwankl, L. J., and T. Prichard. 2001. Chemigation in tree and vine microirrigation systems. Oakland: UC Agriculture and Natural Resources Publication 21599.

Schwankl, L., B. Hanson, and T. Prichard. 1995. Micro-irrigation of trees and vines. Oakland: UC Agriculture and Natural Resources Publication 3378.

———. 1998. Micro-irrigation of trees and vines. Oakland: UC Agriculture and Natural Resources Publication 3378.

———. 2008. Maintaining microirrigation systems. Oakland: UC Agriculture and Natural Resources Publication 21637.

Sibbett, G. S., and L. Ferguson, eds. 2005. Olive production manual. 2nd edition. Oakland: UC Agriculture and Natural Resources Publication 3353.

Irrigation Management

Giulia Marino,
UC Cooperative
Extension (UCCE)
Specialist in Orchard
Systems, Department
of Plant Sciences, UC
Davis

Luke Milliron, UCCE
Orchard Systems
Advisor, Butte,
Tehama, and Glenn
Counties

Highlights

- Although olive can survive with very little water, precise irrigation optimizes oil yield and quality.

- The goal of irrigation management is to provide the right amount of water at the right time to achieve the grower's yield, quality, and orchard health goals.

- Olive is very sensitive to water stress early in the growing season, when vegetative growth, flowering, fertilization, fruit set, and rapid fruit growth occur.

- Olive seasonal water usage (crop evapotranspiration, ETc) from April to November is about 25 inches (630 mm), one of the lowest water usage levels of all California crops.

- Irrigating to match crop evapotranspiration needs allows growers to obtain maximum fruit yield, but not necessarily maximum oil yield and quality.

- Orchard water use is site- and time-specific, hence information on soil, environment, crop, and irrigation system must be integrated for a more precise assessment of irrigation needs.

- Soil monitoring can show moisture trend and depth, but soil variability and the placement, cost, and complexity of monitoring systems can limit the reliability of soil monitoring as the only irrigation management tool.

- Stem water potential readings with a pressure chamber are the gold standard for determining when to irrigate, but measurements are labor intensive and limited to recording a single point in time.

- Continuous monitoring of plant water status can facilitate irrigation management. However, interpretation of the information can be complex.

- Sustained deficit irrigation can save water and sustain yields if water stress early in the growing season is avoided.

- Regulated deficit irrigation during stress-tolerant stages can reduce excessive vegetative growth and improve oil yield and oil quality, resulting in a more profitable crop.

areful irrigation management can help oil olive growers achieve uniform tree structure, robust tree health, consistent and high yields, and excellent oil quality. By instituting best irrigation practices in the orchard, growers can improve their prospects for an efficient operation and a high return on investment.

New concept of irrigation management

Olive is a drought-tolerant species that historically has been cultivated in arid and semiarid Mediterranean regions characterized by limited water availability and high evaporative demand. In these areas, traditional olive groves were (and still partially are) rain-fed and planted in very low densities to maximize the availability of stored soil water per tree. However, with the aim to increase yield per surface unit, many traditional groves have been converted to irrigation. Three other significant changes have influenced and deeply transformed the concept of irrigation management in olive:

- A new growing system based on low-vigor olive varieties planted at super-high densities was developed in Spain and has been rapidly adopted by several countries. Trees in these orchards have a limited root zone but a greater exposed leaf surface per canopy volume and, therefore, are more susceptible to water stress.

- With increased attention on healthy foods, consumers are seeking high-quality olive oil.

- Climate change and the associated water scarcity has challenged irrigated agriculture around the world to focus on environmental sustainability, water footprint, and water productivity.

In response to these global developments, the concept of irrigation management in olive orchards has gained attention but also complexity. A change toward more precise and efficient irrigation is key for the economic sustainability of future olive systems. This change will be strongly supported by the next generation of technological advancement and by the capability of growers, researchers, and industry stakeholders to move together quickly in this direction.

Water requirements for olive

Olive has some of the lowest water usage of all the California crops, and this makes olive a relatively more sustainable crop in terms of future climatic and regulatory water restrictions. However, optimal management of the water is essential not only for industry sustainability but also for good yields. An important first step to achieve optimal water management is to estimate the orchard water requirements.

Water and olive performance

Olive is considered a drought-tolerant species and can survive and produce with very little water. Its small leaves are thick and leathery with a waxy cuticle and hairs mainly on the lower surface, which are characteristics that limit water loss. The stomata are located on the lower surface in depressions, which also reduces water loss. Although these adaptations help olive trees to survive dry conditions, the crop does not achieve optimal oil yield without sufficient irrigation. Conversely, prolonged wet conditions lead to a decline in tree health.

Bloom, shoot growth, and alternate bearing

As an evergreen, olive is affected by excess and deficit water availability throughout the year. Before bloom, in late winter and early spring as olive flower buds form, water stress can affect the number of blossoms, the timing of bloom, and even the percentage of blossoms that are self-pollinating (Hartmann and Panetsos 1961).

Water stress after bloom can reduce fertilization and fruit set (Grattan et al. 2006).

If water is not limited, shoot growth will commence in late winter or early spring and continue into early fall. New shoot growth is critical because olive bears on 1-year-old shoots. Water stress can stop shoot growth and reduce flowering and fruit production potential for the next year (Grattan et al. 2006). Conversely, excessive irrigation can lead to excessive vegetative growth (Grattan et al. 2006). Although excessive vegetative growth sets up more bloom and crop potential for the following year, it also sets up poor return bloom and crop in the second year, therefore setting up a cycle of alternate bearing (Grattan et al. 2006). Excessive vegetative growth can also shade out lower fruit wood, raise pruning costs, and increase limb breakage and wounding during mechanical harvest (Lightle and Connell 2018).

The balance between current season fruit growth and current season shoot growth, which will provide the bloom for next year's crop, is critical. Because fruit is produced on last season's shoots, olive should be irrigated with a 2-year horizon in mind. To moderate alternate bearing, increased irrigation may be needed in heavy cropping years to promote shoot growth; in light cropping years, water savings can likely be achieved with still plenty of shoot growth. In other words, irrigation can be used to help achieve more consistent year-to-year yields.

Fruit size, total yield, and quality

Although fruit size and yield are optimized at high rates of irrigation, a high moisture level in the fruit at harvest reduces oil extractability (Grattan et al. 2006). Hence, oil yield can be optimized with a deficit irrigation strategy that limits water application prior to harvest.

Irrigation management can have a strong impact on the concentration of phenols in the oil. Phenols are chemical compounds that increase shelf life and beneficial health effects; if not excessive, they also improve the oil's sensory properties (Servili et al. 2009). Generally, moderate irrigation optimizes oil flavor, with bitter oils produced by very low rates of irrigation and bland oils by very high rates (Berenguer et al. 2006), but the specific periods when irrigation is reduced determine the quality response (García et al. 2020).

Tree health

Although olive orchards are drought tolerant, previously irrigated orchards that lose access to water for an extended period will have yield losses and can eventually suffer tree decline and death (L. Milliron, personal observation). However, overwatering is a more common problem in California's irrigated orchards (Joseph Connell, UCCE farm advisor emeritus, personal communication). Poorly drained or layered soils can become waterlogged, resulting in poor aeration and root deterioration. More common in winter and early spring, waterlogging leads to poor shoot growth, yellow foliage, and tree loss. Soil that becomes saturated after fruit set contributes to fruit shrivel. Olive trees suffering from root damage are predisposed to winter cold damage, presumably because they have lower amounts of stored carbohydrates. Phytophthora root rot (fig. 7.1) and Verticillium wilt are also greater problems in orchards not properly irrigated (see chapter 12, Diseases of Olive).

The concept of evapotranspiration

Trees take up water through the roots and lose it through the leaves, mainly through stomata, small pores located on the lower surface of the leaves. Such water loss is called transpiration. If this water is not replaced and soil water availability decreases, plants close stomata to limit further

Figure 7.1

Yellowing foliage and canopy decline afflicted this California olive orchard in February. This portion of the orchard contains a heavy clay soil series, and a root sample from the area tested positive for *Phytophthora*. *Photo:* Luke Milliron.

water loss and excessive tissue dehydration. In addition to being transpired by the plants, water evaporates directly from the soil surface, can run off the surface, or percolate below the root zone. The sum of the water that is transpired by the plants and that evaporates from the soil is called evapotranspiration (ET).

Evapotranspiration and yield

In addition to being the way out for water, stomata are the way in for carbon dioxide, a gas used in the photosynthetic process to synthesize the sugars needed to support the development of vegetative and reproductive organs and to build stored resources. Hence, irrigation should aim to replace water lost by evapotranspiration to keep stomata open, maximize carbon assimilation, and obtain maximum fruit production while reducing water loss due to run off and percolation. To replace water lost by evapotranspiration, a crop evapotranspiration (ETc) is calculated, which is the amount of water that the crop would use in a defined location and time frame under no limitation of water availability in the soil.

Water application lower or higher than ETc may impair olive fruit and oil yield (fig. 7.2). In an olive oil production and irrigation study (Grattan et al. 2006), applying 60 percent of ETc resulted in 80 percent of maximum fruit yield and 90 percent of maximum oil yield. Applying 80 percent of ETc resulted in 94 percent of maximum fruit yield. Oil yield was maximized at water applications lower than 100 percent of ETc

and was negatively impacted by overirrigation. Maximum oil yield was obtained at about 75 percent ETc; however, multiple factors (for example, rain before harvest and time of harvest) can alter the relation between irrigation and oil yield.

Estimating evapotranspiration

Estimating crop evapotranspiration for oil olive requires calculations that can be accomplished in the following five steps.

Step 1: Identify reference evapotranspiration

As both evaporation and transpiration are processes associated with the vaporization of water, environmental parameters such as temperature, relative humidity, wind, and radiation have a strong influence on water use in the orchard. For example, plants lose more water on a dry, hot day than on a humid, cold day. To take into account local environmental conditions, a calculation of ETc uses reference evapotranspiration (ETo), which represents the measured water use of a healthy close-cut grass growing under no limitation of water and nutrients.

Real-time daily, weekly, and monthly values of ETo are accessible at the website of the California Irrigation Management Information System (CIMIS), a network of automated weather stations located throughout California and operated by the California Department of Water Resources. Another resource for growers, although a less precise guide to irrigation

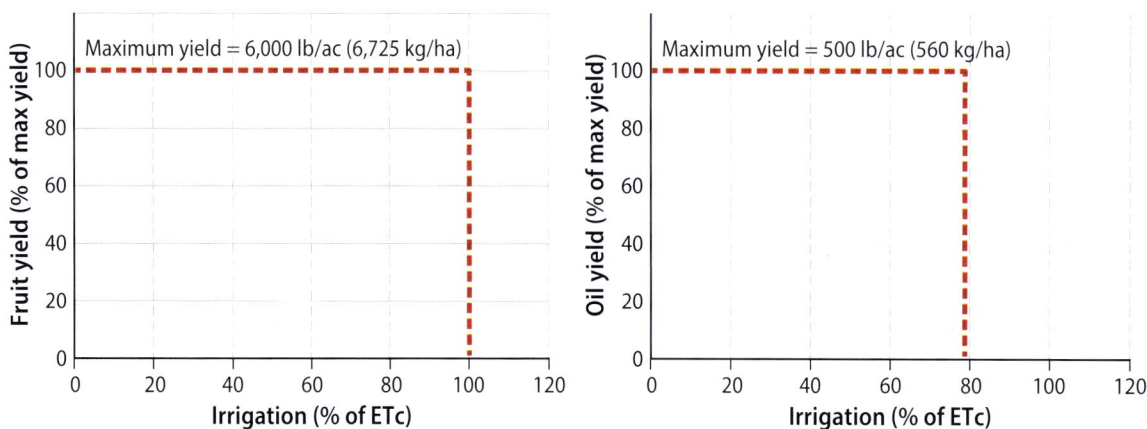

Figure 7.2

The relationship between productivity (yield of fruit and oil) and water application is shown for an Arbequina super-high-density (SHD) orchard. *Source:* Adapted from Grattan et al. 2006.

scheduling than real-time data, is long-term historical average monthly ETo data compiled for all locations in California by Snyder et al. (1995).

Step 2: Select the crop coefficient

The crop coefficient (Kc) represents the amount of water that a nonstressed crop uses compared with the ETo. For example, a Kc of 0.5 means that the crop water use is 50 percent of the water used by close-cut grass, and a Kc of 1.1 means the crop uses 10 percent more water.

The average seasonal Kc suggested for a California clean-cultivated Manzanillo table olive orchard in which 60 percent or more of the ground is shaded is 0.75 (Goldhamer et al. 1994). Beede and Goldhamer (2005) suggested a slightly lower value of 0.65 for oil olive orchards. Studies conducted by Orgaz and Fereres (2004) in Spain on oil olive Kc suggest a slightly higher Kc value in winter and spring (between 0.65 and 0.75) and a lower value in fall (0.60 to 0.70) and summer (0.50 to 0.55).

Reported Kc values have been developed and tested mainly for traditional and intensive olive orchards with round-canopied trees; less information is available for hedgerow systems used in super-high-density (SHD) and high-density (HD) oil olive orchards. Based on a study in northern Spain (Martínez-Cob and Faci 2010), the Kc reported for round-canopied olive trees could potentially be adopted for the SHD hedgerow system.

Kc values are determined experimentally by researchers directly measuring water use of an olive orchard. As Kc is affected by several orchard-specific characteristics (tree shape, soil type, crop load, and general orchard management practices), the reported values can be slightly variable from orchard to orchard. Growers should use direct monitoring of soil and plant water status to adjust the Kc to their site-specific conditions (see the sections "Soil moisture monitoring" and "Plant-based irrigation management," below).

Step 3: Estimate crop evapotranspiration

ETc can be estimated as the factor of the parameters identified in steps 1 and 2: the reference evapotranspiration from the CIMIS station closest to the orchard (or from historical data) and the reported crop coefficient (Doorenbos and Pruitt 1977):

$$ETc = ETo \times Kc$$

Table 7.1 shows the estimated monthly ETc of a traditional olive orchard in Willows, Sacramento Valley, California, with round-canopied trees shading 60 percent of the ground at midday. For example, the ETc estimate for July is 5.6 inches: the product of 8.7 inches (the historical ETo for July in that geographical area) and 0.65 (the Kc of olive).

Step 4: Modify for smaller canopy size

Reported Kc values apply to mature orchards with 60 percent or more of the ground shaded by tree canopies at midday. The total leaf area intercepting solar radiation is the most significant plant factor affecting ETc. Thus, a smaller canopy, with less ground shaded, will reduce ETc. To account for this effect, a reduction coefficient (Kr) is calculated for young orchards with canopies shading less than 60 percent of the ground. The research by Fereres (1981) on almond has been

Table 7.1

Average estimated crop evapotranspiration (ETc) for a Sacramento Valley clean-cultivated, round-canopied olive orchard shading 60 percent of the ground at midday

Month	ETo* (in)	Kc[†]	ETc (in)	ETc (in/day)
Jan	1.6	0.65	1.0	0.03
Feb	2.2	0.65	1.4	0.05
Mar	3.7	0.65	2.4	0.08
Apr	5.1	0.65	3.3	0.11
May	6.8	0.65	4.4	0.14
Jun	7.8	0.65	5.1	0.17
Jul	8.7	0.65	5.6	0.18
Aug	7.8	0.65	5.1	0.16
Sep	5.7	0.65	3.7	0.12
Oct	4.0	0.65	2.6	0.08
Nov	2.1	0.65	1.4	0.05
Dec	1.6	0.65	1.0	0.03
Total	57.0		37.0	

*Zone 14, Snyder et al. 1995.

[†]Beede and Goldhamer 2005.

widely used to calculate the *Kr* of orchards containing trees with spherical or conical canopies:

$$Kr = 2 \times fc$$

The fraction of ground shaded by olive trees at midday, or *fc*, which is known to correlate well with the canopy area exposed to sunlight, is calculated as follows:

$$fc = (\pi \times D2 \times N) \div 174,240$$

D is the average diameter of the canopy in feet; *N* is the number of trees per acre; and 174,240 is the factor of 43,560, the number of square feet per acre, and 4, the denominator of the formula of the circumference area.

For hedgerow systems, in which tree canopies form a continuous hedge wall with a squared geometry, *fc* can be estimated as the rectangular base of the hedgerow:

$$fc = (N \times d \times r) \div 43,560$$

N is the number of trees per acre, *d* is the average width of the canopy in feet, *r* is the distance between trees in feet, and 43,560 is the number of square feet per acre.

Step 5: Modify if needed for site-specific conditions

Site-specific conditions can affect the orchard water use. Two conditions in particular warrant modifying the ETc calculation:

- **Row orientation:** Orientation of a hedgerow orchard has a big effect on percentage of ground shaded, with north–south hedgerows intercepting up to 9 percent more radiation annually than east–west hedgerows. In addition, the seasonal dynamics of light interception are different, with north–south hedgerows intercepting more radiation in summer and east–west more in winter (Connor et al. 2014; Trentacoste et al. 2015). Considering only the irrigation period, differences in light interception can be as high as 33 percent in summer, increasing water use up to 20 percent at that time of year (Connor et al. 2014).
- **Crop load:** Trees with a high fruit load can transpire as much as 30 percent more than low-yielding or nonbearing trees (Bustan et al. 2016). Water application should be adjusted during years of particularly high or low yields.

Scheduling irrigation

An effective irrigation schedule focuses on two objectives: when to irrigate and how much water to apply. The most common approaches to scheduling irrigation are the water budget method, soil moisture monitoring, plant water status monitoring, and remote sensing. There is no best irrigation scheduling method. Each grower should have a well-defined orchard production target and decide which irrigation scheduling method (or combination of methods) will meet this target in the most efficient way. Growers should be aware of the different options available and their advantages and pitfalls.

Water budget method

The water budget method, which is commonly used with surface and sprinkler irrigation systems, accounts for the water stored in the soil and budgets its depletion through evapotranspiration, deep percolation, and runoff (outputs) and its restoration with irrigation and rainfall (inputs) (fig. 7.3). The following steps are helpful in scheduling irrigation with the water budget method.

Step 1: Calculate available water content

The soil water available for plant use is called the available water content (AWC). It is expressed in inches per foot of soil. It is calculated as the difference between field capacity (FC) and permanent wilting point (PWP):

$$AWC = (FC - PWP)$$

FC is the amount of water a soil can hold after drainage has reached a very slow rate. PWP is the point at which the water remaining in the soil pores is held so tightly that plants cannot extract it and remain wilted when watered.

AWC depends primarily on soil texture and structure. Coarse-textured sands have larger pores but less total pore space than silty clay loams, so they have less water holding capacity. Clay soils possess many small pores and hold the most water. Estimates of AWC for various soil textures are given in table 7.2.

After selecting AWC based on the orchard's soil texture, multiply this value by the depth of the root zone to obtain the total AWC, which

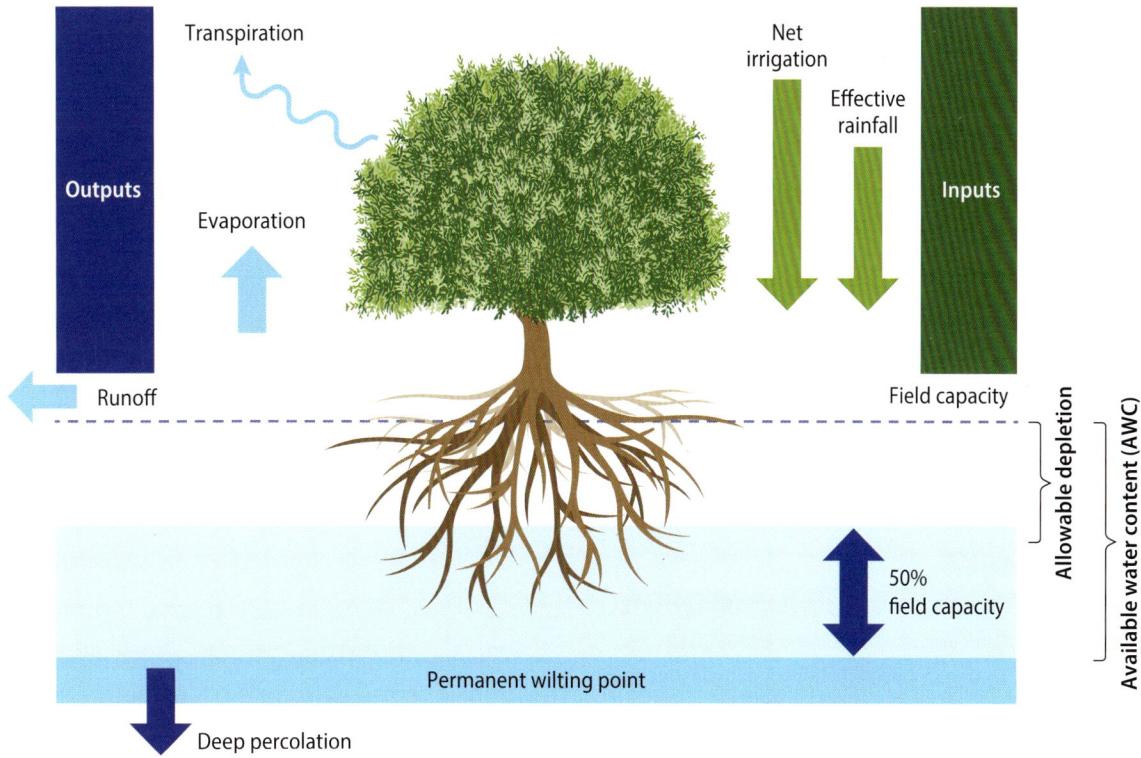

Figure 7.3

The water budget method estimates irrigation needs by calculating inputs and outputs.

Table 7.2

Average values of available water content (AWC) for different soil textures

Soil texture	Available water (in/ft)
Sand	0.7
Sandy loam	1.4
Loam	1.8
Clay loam	1.6
Silty clay	2.4
Clay	2.2

is the amount of water available to the trees. In spaced orchards where trees have round canopies and large crowns, the root zone is commonly 3 to 4 feet (0.9 to 1.2 m) deep (Beede and Goldhamer 2005). In SHD orchards, the root zone is more limited, generally 1.5 to 2 feet (0.5 to 0.6 m) deep (Díaz-Espejo et al. 2012; Fernández et al. 2013). An excellent method for evaluating root depth of an orchard is to backhoe along the tree drip line at three or four sites.

Step 2: Calculate the allowable water depletion

As AWC decreases from field capacity, roots must expend more energy to extract water. This limits water uptake as soils dry out and decreases crop growth before the entire root zone reaches permanent wilting point. Therefore, growers should irrigate before soil water is depleted to a point that restricts growth.

Allowable depletion (AD) is the percentage of AWC that indicates irrigation is needed. No single percentage can be recommended for all situations. Growers typically irrigate when AWC reaches about 50 percent of depletion. In stress-sensitive periods, such as during flower and fruit development, trees should be irrigated at relatively small depletion levels, about 30 percent.

Step 3: Estimate effective rainfall

Effective rainfall (Re) is the amount of rain stored in the soil profile and available to the trees. Part of rainfall evaporates from the soil surface, runs off the surface, or percolates below the root zone. These losses depend on the intensity and duration of the rain event and on soil type. For California orchard systems, 50 to 70 percent of winter rain is considered Re. To estimate Re in a single rain event, subtract 0.25 inch (6 mm) from the rain amount and reduce the remaining amount by 20 percent:

$$\text{Re (in)} = (\text{Rain - 0.25}) \times 0.8$$

A single rainfall of less than 0.25 inch (6 mm) is not included in the estimation of Re since it likely evaporates before entering the soil.

Step 4: Calculate net irrigation requirement

The orchard's net irrigation requirement (NIR) can be calculated as the difference between accumulated ETc and Re, presuming a full soil water profile at the beginning of the calculation period. This is the amount of water needed to refill the soil to field capacity. When NIR is higher than the AD calculated in step 2, water should be applied through irrigation to bring the soil reservoir to 100 percent AWC.

Step 5: Calculate gross irrigation requirement

The total amount of water that should be applied is the orchard's gross irrigation requirement (GIR). When the orchard is irrigated, some losses occur due to runoff, percolation below the root zone, and (with sprinklers) spray evaporation and drift. Application efficiency (Ae) describes the percentage of applied irrigation water that is stored in the root zone and available for crop use. Ae can be used to determine GIR as follows:

$$\text{GIR} = \text{NIR} \div \text{Ae}$$

The type of irrigation system used, soil, weather conditions, and water management practices largely determine Ae. In general, microirrigation systems can be operated with higher efficiencies than surface methods. Because Ae varies, it must be estimated for each orchard. Some general estimates for different systems are shown in table 7.3.

Table 7.3

Application efficiencies (Ae) typical of various irrigation systems

Irrigation system	Ae (%)
Surface	65–80
Microsprinkler	85–90
Drip	90–95

Example

The orchard is a traditional, clean-cultivated orchard of trees with round canopies that shade 60 percent of the ground at midday. It is drip irrigated and located in the Sacramento Valley, California, where the ETc in July is 5.6 inches (see table 7.1). The soil is clay loam. In July, only one rain event of 0.03 inch was recorded during the entire month. The steps for determining the amount of water to deliver with irrigation in July are as follows:

Step 1. The AWC in clay loam is 1.6 inches per foot (see table 7.2), and the root depth of a traditional orchard is 4 feet. Thus, the total AWC is 6.4 inches of water (1.6 × 4).

Step 2. The AD is 3.2 inches (50% of the AWC).

Step 3. Because the rain during July was only 0.03 inch, it is not considered to be effective rain (Re) and is not included in the calculation.

Step 4. The NIR is the difference between ETc, which is 5.6 inches, and Re, which is zero. Since the NIR (5.6 in) is higher than the AD of 3.2 inches (see step 2), irrigation is needed. To understand when water should be applied during July, the water budget is recalculated on a weekly basis: each week the orchard uses 1.4 inches of water, and after 3 weeks, 4.2 inches of water have been depleted, which is above the maximum AD, so irrigation should be applied at the end of the second week of July, before the maximum AD is reached.

Step 5. Assuming the drip irrigation system has an Ae of 90 percent (see table 7.3), the amount of water that must be delivered during the month to achieve the GIR is

$$\text{5.6 in} \div \text{0.9} = \text{6.2 in}$$

This amount can be split between two irrigation events: 3.1 inches at the end of the second week of July and 3.1 inches at the end of July. Or it can be split into four weekly irrigation events of 1.55 inches.

Soil moisture monitoring

Technology developed to measure soil moisture is based on an understanding that managing soil water levels is also managing plant water status. Specifically, less water in the soil means that soil particles will hold that water at a greater tension, making uptake more difficult for roots. The objective of soil-based irrigation scheduling is to keep soil in an acceptable moisture range that does not adversely stress the trees (Shackel et al. 2012).

Strengths and weaknesses of soil moisture monitoring

Evaluating actual soil moisture, that is, the soil water bank in the orchard at a point in time, can be used as a standalone irrigation management approach or as a way to check and refine the water budget method. Soil moisture monitoring indicates where water is in the root zone, how quickly it is being depleted, and how well moisture levels are recharged by irrigation. Knowing how deep water is infiltrating during irrigation helps inform irrigation run times. Longer-term trends can help indicate if irrigation run time should be increased or decreased as the growing season progresses. A potential weakness of soil moisture monitoring is that although moisture levels in parts of the root zone can be monitored, it is impossible to know the moisture levels in the complete extent of the root system beyond the monitored area.

Soil moisture by feel

A simple and still-relevant method of monitoring soil moisture is by feel. This technique involves extracting a soil sample at a depth of at least 3 feet (0.9 m) and squeezing the soil by hand, estimating the level of moisture depletion by sensation and appearance (fig. 7.4). A soil auger or a probe is more suitable than a shovel for extracting a soil sample at depth, and many types are available. Choose equipment that makes soil sampling to a required depth easy and rapid.

Figure 7.4

Differences can be learned in the feel (evident in the amount of soil crumbling or moisture on the hand) and appearance of a soil at wet, medium, and dry moisture levels. *Photo:* Jack Kelly Clark.

Monitoring soil moisture by feel is inexpensive and available for quickly monitoring multiple orchard areas. A disadvantage is the time required to take samples and to gain experience in assessing soil moisture accurately. Monitoring soil moisture by feel can complement other monitoring technologies, requiring only an investment in the skill set and the inclusion of a soil probe or auger in the toolbox, but shouldn't be used as a stand-alone technique to manage irrigation. An interpretive guide is available online from USDA NRCS (1998).

Soil moisture monitoring technology

Whichever technology is used, the information from soil moisture monitors is limited by the potentially wide variability of soil texture and structure in even a small area. They should be installed in an area of soil that is as representative as possible of the orchard. They are usually positioned about 4 feet (1.2 m) from the microsprinkler and 3 to 5 feet (0.9 to 1.5 m) from the tree, where most of the soil moisture uptake occurs. A common reason for using soil moisture monitors is to evaluate the average water holding capacity or to assess an area with less water holding capacity (because of coarse-textured soil or a shallow hardpan) to prevent trees from experiencing severe stress.

Some soil moisture monitoring technology is either very expensive or so complex that it requires professional consultation. The high cost can often limit monitoring to just a single site in the orchard, which can be problematic because soil may vary significantly across the orchard. Close contact between sensors and the soil is often critical, with the drying out and cracking of heavy soils causing some technologies to fail. Some technologies provide one-time snapshots of soil moisture, while others continuously log moisture levels, creating a more complete picture.

Tensiometer

Soil water is held under increasing tension as moisture is depleted. A tensiometer, which consists of tubing with a vacuum gauge at one end and a porous ceramic tip at the other, provides an indirect way of measuring this tension (fig. 7.5). It is filled with distilled water and air is vacuumed out. As soil moisture is depleted, soil tension and tension at the interface of the porous tip increase. Tension draws water through the porous tip of the tensiometer, expressing tension in centibars on the gauge. Tensiometers come with charts that correlate gauge reading with percentage of soil moisture depletion based on soil texture (Beede and Goldhamer 2005).

Some tensiometers can be integrated with automated irrigation systems. A major limitation of tensiometers is that they break suction at high tension levels in excessively dry soils and even moderately dry soils with significant clay content. They may not be appropriate for many loam and clay loam soils because of the high clay content (Shackel et al. 2012). Another limitation is that tensiometers measure tension from a very small volume of soil, only the soil in immediate contact with the porous ceramic cup.

Electrical resistance block

Electrical resistance can be used to monitor soil moisture. The resistance between two electrodes separated by a small gypsum or ceramic-sand block increases as soil moisture is depleted and water flows out of the block; there is less resistance when the soil is wetted and water enters the block. Electrical resistance blocks have wires that attach to a panel above the soil surface. Readings are logged continuously and can be downloaded either at the site or remotely via a cell phone signal; they can be displayed as tables or graphs.

As with tensiometers, electrical resistance blocks indicate when irrigation is needed but not how much irrigation is required (Shackel et al. 2012). Like tensiometers, soil texture must be considered; however, unlike tensiometers, electrical resistance blocks perform best under drier soil conditions and are therefore more appropriate for loam and clay soils (Shackel et al. 2012). One weakness of electrical resistance blocks is that they show false high moisture levels in salt-affected soils, which affects their reliability under these conditions (Beede and Goldhamer 2005). Benefits of electrical resistance blocks include their relatively low cost, which allows for the installation of multiple sensors at the desired depths, and the capacity for remote downloading of soil moisture graphs.

At monitoring sites, blocks should be placed at least in the top one-third and bottom one-third of the root zone (Shackel et al. 2012). The root zone is commonly 3 to 4 feet (0.9 to 1.2 m) deep in traditional orchards with large round-shaped

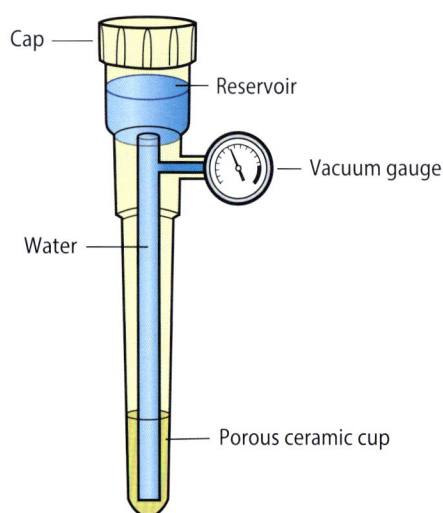

Figure 7.5

A tensiometer's vacuum gauge reading corresponds to percentage of soil moisture depletion.

canopies (Beede and Goldhamer 2005) and 1.5 to 2 feet (0.5 to 0.6 m) deep in SHD orchards (Díaz-Espejo et al. 2012; Fernández et al. 2013). Consultants who install soil moisture monitoring sites using resistance blocks often place blocks at 1-foot intervals throughout the root zone (Shackel et al. 2012).

Neutron probe

Radiation provides a way to measure soil moisture. The neutron probe emits low-level radiation of high-energy neutrons that collide with the hydrogen atoms in the soil water, which slows down the neutrons. Increased soil moisture results in more of these collisions and the detected return of more slowed neutrons to the radiation source (Beede and Goldhamer 2005). A calibration curve correlates these slowed neutron readings with soil moisture volume.

The probe consists of a display unit and a shielded radiation source tethered to a cable (fig. 7.6). The display unit is placed on top of a previously buried PVC or metal access tube, and the radiation source is lowered into the access tube to take readings at different depths (for example, at 1, 2, 3, 4, and 5 ft for an access tube buried 5.5 ft deep). Once the readings are complete, the trained technician transports

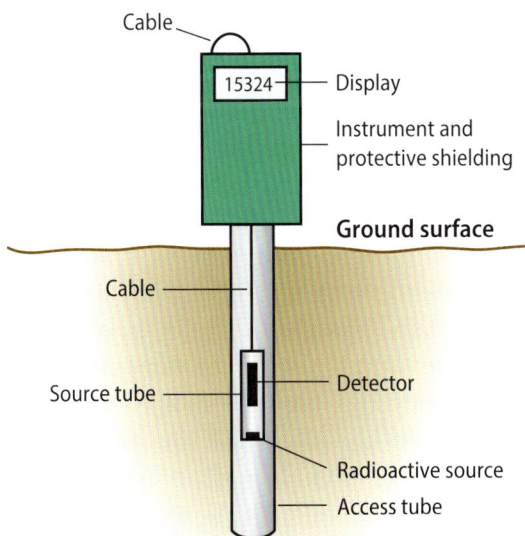

Figure 7.6

A neutron probe uses radiation to record soil moisture levels.

the probe to the next access tube, either in the same or a different field.

A properly calibrated neutron probe is considered one of the most direct and accurate soil moisture monitoring techniques. It measures an approximately basketball-size sphere of soil, which is a much larger soil volume than other moisture monitors. Additionally, because a neutron probe measures inches of water per foot of soil, it is immediately clear how many inches of irrigation water are required when the allowable depletion is reached, with no calibration chart needed for the probe readings.

A single operator with a neutron probe can easily cover the irrigation monitoring for several hundred acres per week, assuming there are the suggested two access tubes per 40 acres (16 ha). However, the strict regulations surrounding the operation and storage of neutron probes means they are typically restricted to use by irrigation consultants and large operations (Beede and Goldhamer 2005). Another significant drawback is that readings are episodic and in-person, without the capability of continuous remote monitoring.

Dielectric constant soil moisture devices

A material's ability to establish an electric field, or dielectric constant, can be used to monitor soil moisture. The dielectric constant of water is 20 times higher than that of soil. This constant can be measured either by time domain reflectometry (TDR), which uses electromagnetic wave technology, or by frequency domain reflectometry (FDR), also known as capacitance, which utilizes radio wave technology (Beede and Goldhamer 2005). TDR sensors typically come in the form of paired stainless steel rods, buried at varying depths, while FDR sensors are available as a permanently installed multisensor tube, or as a mobile unit, like the neutron probe, lowered to various depths into installed access tubes (Shackel et al. 2012). As with neutron probe readings, calibration curves have been developed to correlate dielectric constant sensor readings with soil moisture content.

The huge advantage of TDR and FDR sensors is they provide continuous logging and remote downloading of graphical soil moisture trends, a benefit offered also by electrical resistance blocks. As with the neutron probe, using a professional service may ease the complexity of installation and interpretation. Dielectric sensors also have disadvantages. As is the case with tensiometers, excellent contact between soil and sensor must be maintained, and only a very small volume of soil is measured. Installation is challenging and may be particularly difficult in soils that are gravelly or have a high clay content. Also, the cost of these sensors is high.

Plant water status monitoring

Past irrigation management practices were commonly based on soil water status or environmental indices. Severe and recurrent drought events and adoption of microirrigation have led to more efficient plant-based irrigation management practices that measure the water stress experienced by the plant. Plant water status is considered the most precise indicator of water need as it integrates the physiological response of plants to both the soil and the environment.

A simple way to think about plant water status is the visual cue provided by a wilted plant. Wilting indicates severe water stress, but the water stress began before the plant showed any obvious visual cues. Plant water status assesses water stress before the plant wilts.

Stem water potential

The most used method of assessing plant water status is taking readings of stem water potential (SWP), which is a measure of water tension in the plant. Just as soil water is under tension, so is plant water. Soil water tension, measured with a tensiometer, gives an estimate of soil water content. Similarly, plant water tension, measured with a pressure chamber, also known as a pressure bomb, gives an estimate of plant water status.

SWP readings are commonly taken between noon and four o'clock (midday SWP), when olive trees experience peak and relatively stable water stress. A mylar plastic bag is sealed around the tip of a shaded terminal shoot, low in the canopy and close to the trunk. The bag is left for at least 10 minutes, and then the stem is cut with a razor blade. The bagged shoot is placed inside the pressure chamber and pressurized, with only the cut end of the shoot exposed. When the positive

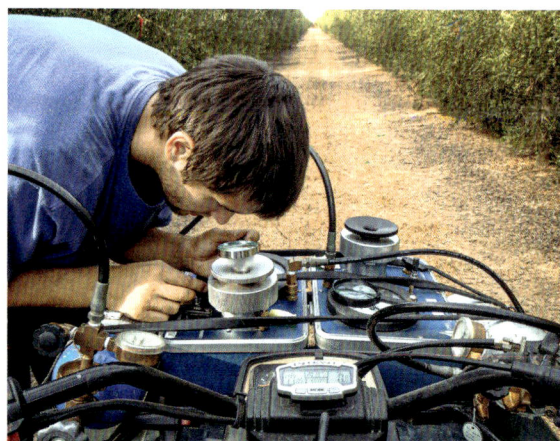

Figure 7.7

To take an SWP reading, an operator pressurizes a bagged terminal shoot tip and when water appears on the cut stem surface, the operator records the pressure gauge reading. *Photo: Luke Milliron.*

pressure inside the chamber exceeds the shoot's water tension, water becomes visible on the stem surface and a gauge records the pressure in negative bars or megapascals (fig. 7.7).

SWP readings are affected by weather, and olive response to water stress changes through the growing season, so the SWP threshold that triggers irrigation will be different on a cool, high-humidity spring day than on a hot, dry summer day. To use SWP readings to decide when to irrigate, the operator records the temperature and relative humidity in the orchard and consults a baseline chart that shows SWP readings for a fully watered orchard at different temperatures and relative humidity.

The SWP baseline developed by Shackel et al. (2021) for SHD olive orchards is reported in table 7.4. Using the baseline, an operator would know that an average SWP reading of -11.1 bars (-1.11 MPa) at midday on a 95°F (35°C) day with 20 percent relative humidity indicates the orchard is at the fully watered baseline. However, that SWP reading on a 77°F (25°C) day with 60 percent relative humidity would be 6.4 bars below the fully watered baseline of -4.7 bars (-0.47 MPa), indicating that the trees are experiencing some water stress.

Managing irrigation by comparing field SWP readings with baseline values is a best practice in almond, prune, and walnut production (Fulton et al. 2014). A lesson learned in those crops is to avoid keeping trees at the fully watered baseline

Table 7.4

Baseline midday stem water potential (negative bars) for fully irrigated SHD olive orchard across a range of temperature and relative humidity

Air temperature (°F)	Air relative humidity (%)					
	10	**20**	**30**	**40**	**50**	**60**
41	-3.0	-2.7	-2.3	-1.9	-1.6	-1.1
50	-4.1	-3.7	-3.3	-2.8	-2.3	-1.8
59	-5.4	-5.0	-4.4	-3.9	-3.3	-2.6
68	-6.9	-6.3	-5.7	-5.1	-4.4	-3.6
77	-8.4	-7.8	-7.1	-6.4	-5.6	-4.7
86	-10.0	-9.3	-8.6	-7.8	-6.9	-5.9
95	-11.9	-11.1	-10.2	-9.3	-8.3	-7.2
104	-14.0	-13.0	-12.0	-11.0	-9.8	-8.6

Source: Adapted from Shackel et al. 2021.

for an extended period, as this can lead to root damage from waterlogging. For example, the recommendation for walnut growers is to wait until SWP drops 2 to 3 bars below baseline before starting irrigation. In olive, the recommended adjustment to the baseline shifts throughout the season based on the sensitivity of the phenological stage of development (see the "Sensitive periods" section, below).

SWP is a gold standard for accurately measuring plant water stress, and newer technologies are evaluated against it. However, SWP measurements are labor and time intensive, and they provide only a snapshot, not a continuous log.

Continuous proximate monitoring sensors

Recent developments in low-cost technology have led to many plant-based continuous monitoring sensors that allow for automated irrigation. Generally, most of these sensors do not measure water status directly but rather evaluate processes at the stem, fruit, and leaf levels affected by water status. Irrigation management protocols that

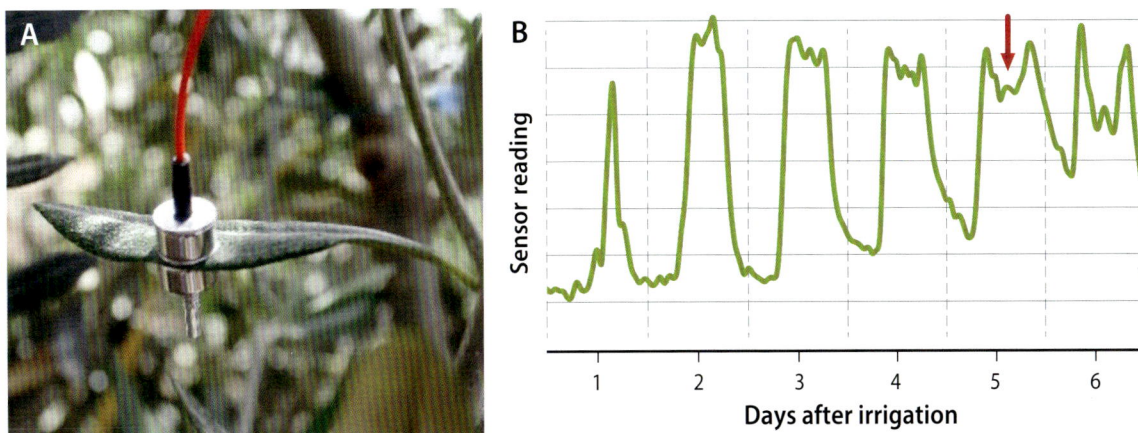

Figure 7.8

A leaf patch clamp pressure probe (A) is installed on an olive leaf. Sensor output (B) shows water stress increasing in the days after irrigation. The red arrow shows a midday depression corresponding to stress onset (dotted vertical line indicates midnight). *Source:* Marino et al. 2021. *Photo:* Giulia Marino.

compare these indirect indicators with SWP values have been developed. Because olive varieties differ in their response to drought, it is important that these irrigation protocols are developed or adjusted for the main varieties grown locally. Two continuous measuring sensors studied in recent years show promising results for irrigation management of oil olive orchards.

Leaf patch clamp pressure probe

A leaf patch clamp pressure (LPCP) probe has two small magnets that are clamped to a leaf (fig. 7.8). As leaves lose water during the day, their thickness decreases, the two magnets get closer, and the sensor reading increases accordingly. Under optimal soil water conditions, water lost from leaves is replaced by water taken up by the roots, and the shape of the daytime reading from the sensor is a typical bell-shaped curve, as depicted for days 2 and 3 after irrigation in figure 7.8. For Arbequina, this corresponds to a condition of no stress and a SWP above -12 bars (-1.2 MPa) (Fernández et al. 2011). When water lost by transpiration is not replaced by water uptake from the roots, leaves lose water and the daily shape of the sensor reading changes, showing a midday depression, as visible on day 5 after irrigation in figure 7.8. The onset of the midday depression in Arbequina corresponds to a SWP between -12 and -17 bars (-1.2 and -1.7 MPa)

and can be used by growers as a visual indicator of the need to irrigate (Padilla-Díaz et al. 2016).

Dendrometer and fruit gauge

A dendrometer is a gauge that measures shrinkage in a tree trunk or branch as an indicator of water stress. A fruit gauge measures the shrinkage in a fruit. The daytime transpiration process depletes water from plant tissues; it is recovered during the night, when stomata close and transpiration ceases. This daily variation in tissue water content is reflected in changes in trunk, branch, and fruit volume. Volume decreases during the day and increases during the night, and the cycle of shrinkage and swelling is more pronounced as soil water availability decreases.

Trunk diameter changes monitored with dendrometers have been widely studied in olive; researchers have proposed approaches to interpreting the data that take into account climatic conditions (Corell et al. 2019) and could potentially be used in automating olive orchard irrigation (Fernández 2014, 2017). Researchers are also investigating the potential for using fruit shrinkage as an indicator of water stress (fig. 7.9).

Remote sensing

Remote sensing uses drones, aircraft, or satellites to collect information about an orchard without physical contact with the trees. It provides information on surface temperature and reflectance

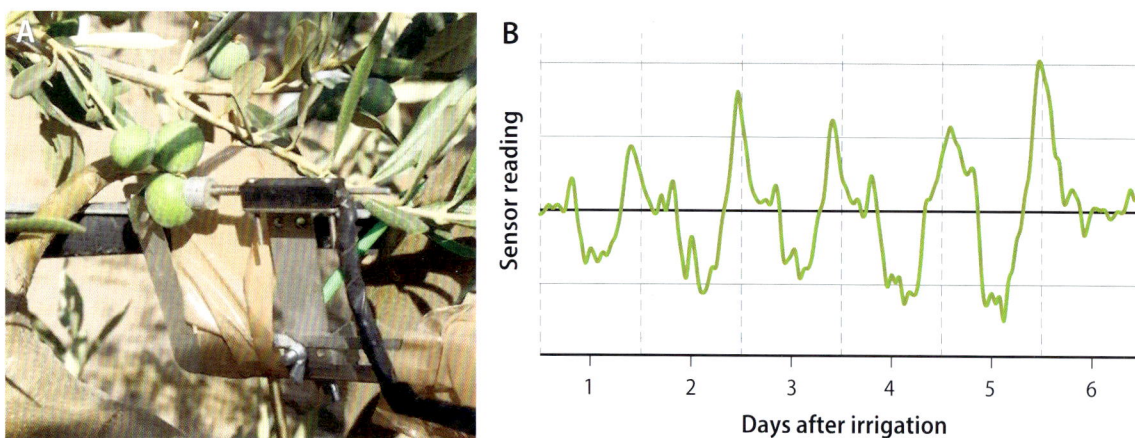

Figure 7.9

Researchers use a fruit gauge to measure an olive fruit (A). Sensor output (B) shows water stress increasing in the days after irrigation. Values above the black horizontal line show increases in fruit diameter happening during the night; values below the horizontal line show reductions in fruit diameter happening during the day (dotted vertical line indicates midnight). *Source:* Marino et al. 2021. *Photo:* Giulia Marino.

that are highly correlated with plant water stress and actual orchard evapotranspiration. This technology has the advantage of covering wider areas than soil or proximate plant-based monitoring systems. Research has developed indices to calibrate remote data with ground-based information for irrigation management, but more investigation is needed to provide guidance for growers.

Deficit irrigation strategies

In oil olive orchards, optimal yield and quality can be achieved by providing less than full ETc (Grattan et al. 2006; Rosecrance et al. 2015). In addition, in years with limited water supply it may not be possible to replace full water loss. There are two main strategies for supplying less water than full ETc and still achieving yield and quality goals: sustained deficit irrigation and regulated deficit irrigation.

Sustained deficit irrigation

Sustained deficit irrigation (SDI), in which irrigation is reduced proportionally throughout the growing season, is the simpler deficit strategy to manage and can achieve substantial water savings while preserving oil yield and quality (Grattan et al. 2006; Moriana et al. 2012). The strategy can be implemented as an ETc-based percentage deficit, as a soil moisture depletion–based deficit, or as a plant water status–based deficit.

Although SDI reduces fruit yield, oil yield can be maintained or even optimized under mild SDI (Grattan et al. 2006). This is possible because of the positive impact on oil extractability of reduced water application just before harvest. Growers using SDI should seek a balance between fruit yield reduction and oil extractability increase that both saves water and optimizes oil yield. Under severe SDI, increased oil extractability does not compensate for the severe reduction in fruit yield, and total oil yield will be reduced.

Two years of SDI research on an SHD orchard of 30-month-old trees in Butte County, California, found that total oil yield per tree for these young trees was maximized when 16 inches (408 mm) of irrigation water was applied through the season, from April-May to October, which, in the calculations used by the researchers, equated to 75 percent ETc (Grattan et al. 2006).

Irrigation at 33 to 40 percent ETc produced oils with more balanced organoleptic profiles—with higher but not excessive levels of bitterness and pungency, and higher fruitiness and other appealing flavors—and may be appropriate if oil quality is the grower's main objective (Berenguer et al. 2006). However, the water stress caused by the 33 to 40 percent ETc treatment reduced shoot growth by about a third and the fruit yield by about half compared with trees irrigated at full ETc. Such low irrigation levels are outside of the range associated with optimum oil extraction (40 to 89% ETc). Growers should consider for their individual operation where loss of yield intersects with increased income from higher-quality oils.

More research is needed to develop clear protocols for SDI in California oil olive orchards using soil- or plant-based irrigation management. Results to date suggest that SWP readings of -13 to -17 bars (-1.3 to -1.7 MPa) after the spring vegetative growth and flower development period do not affect oil yield. A reading of -20 bars (-2 MPa) has been correlated with stomatal closure and photosynthesis reduction in olive trees (Marino et al. 2018), and if that level of water stress occurs during a sensitive period, it could impact productivity. Sustaining a moderate deficit, such as the 75 percent ETc in the Butte County study, is likely a successful SDI strategy.

One reason to be cautious of the SDI strategy is the consequences of water stress early in the growing season on yield. Since the irrigation requirement in early spring is generally very low, reducing water application in this period may challenge orchard productivity without saving much water.

Regulated deficit irrigation

A regulated deficit irrigation (RDI) strategy applies less water during specific periods when the trees are more resistant to water stress, so yield is not impacted, and in some cases the water stress has a positive impact. This strategy should allow growers to maximize marketable yield per unit of water used by the crop or to achieve higher net income than they would with full irrigation. To successfully manage an RDI strategy, growers should be able to identify the sensitive and tolerant stages of the olive orchard.

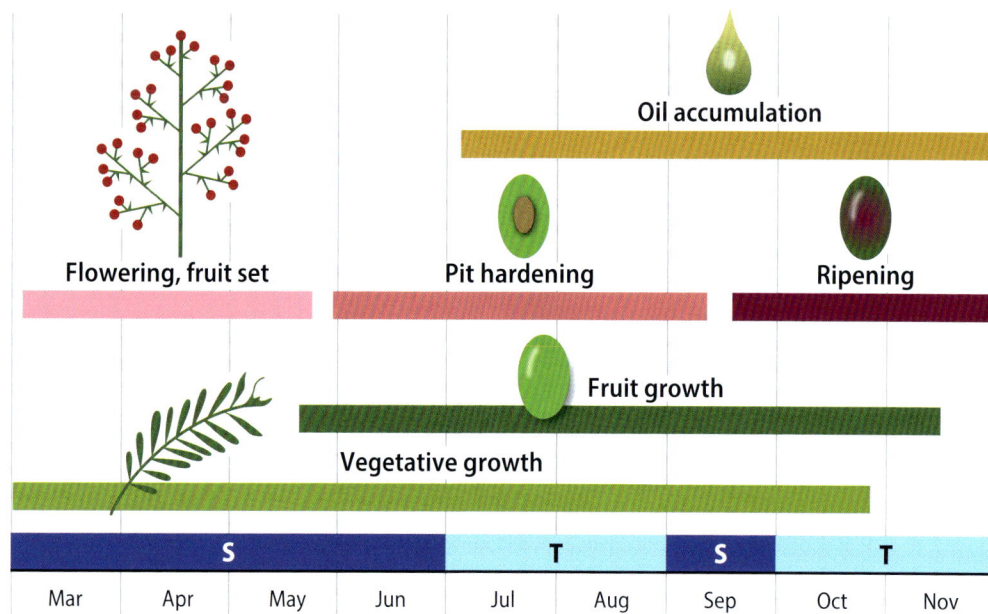

Figure 7.10

During the growing season, there are two sensitive (S) periods, when RDI should not be implemented, and two tolerant (T) periods, when RDI is appropriate. The exact dates are affected by site- and year-specific factors such as variety, management practices, and soil and climate conditions. *Sources:* Fernández 2017; Fernández et al. 2013; Fernández et al. 2018; Hernandez-Santana et al. 2017.

Sensitive periods

Growers should apply the orchard's full water requirement and maintain SWP values close to the baseline thresholds during two sensitive periods. The first is at the beginning of the growing season, in California approximately from the beginning to middle of March to the middle to end of June (fig. 7.10). This sensitive period covers shoot growth, flowering, fertilization, fruit set, and rapid fruit cell division (Moriana et al. 2003; Rapoport et al. 2012). Water stress during this period can affect fruit set and size (Rapoport et al. 2004) and reduce shoot growth (Moriana et al. 2012), and thus affect the current year's and next year's yield.

The second sensitive period lasts from the end of August to the end of September (see fig. 7.10), when a marked increase in oil accumulation occurs (Lavee and Wodner 1991; Moriana et al. 2003). Stress during oil synthesis can impact yield and decrease oil phenols (García et al. 2020). These phenols are natural antioxidants linked to oil sensory and health properties highly appreciated by consumers.

Tolerant periods

The period most tolerant of water stress for California olive orchards is from the end of June to the end of August, during pit hardening but after rapid fruit cell division has ended (see fig. 7.10). Water stress during this period reduces vegetative growth rather than fruit growth. Water applications can be reduced up to 50 percent during July and August to control canopy growth, a key objective for SHD orchards, without reducing yield (Rosecrance et al. 2015). Water stress at pit hardening increases oil phenol content (Gómez del Campo and García 2013; Gucci et al. 2019). During this tolerant period, the midday SWP values should range between -20 and -25 bars (-2.0 and -2.5 MPa) (Girón et al. 2015; Iniesta et al. 2009; Rosecrance et al. 2015). Values below -30 bars (-3 MPa) would affect yield (Moriana et al. 2003).

RDI is implemented at the beginning of the pit hardening phenological stage, which can be estimated in the field with weekly measurements of the longitudinal growth of the fruit. A constant increase in longitudinal diameter of about 0.08 inch (2 mm) per week suggests that massive pit hardening is still not taking place; this value is only indicative and can vary with several parameters, including crop load, variety, and temperature (Gijón et al. 2010). No increase in fruit longitudinal diameter compared with the previous week's measurement means that the massive pit hardening stage has started (Rapoport et al. 2013) and irrigation can be reduced.

Irrigation can also be limited in the last stage of fruit ripening, in October and November, when most of the oil has already accumulated in the fruit. At this stage, water stress reduces fruit water content, improves oil extractability, and increases oil yield per unit of milled fruit. However, suggesting a threshold for irrigation for this period, when rains are generally expected, is difficult due to the many factors that determinine oil quality. In studies in Spain, reducing irrigation during this period to just 25 percent of ETc had no impact on oil yield and increased oil polyphenol content (Motilva et al. 2000). Fernandez et al. 2013 suggest applying 40 percent of the net irrigation requirement (NIR) during this period.

References

Beede, R. H., and D. A. Goldhamer. 2005. Olive irrigation management. In G. S. Sibbett and L. Ferguson, eds., Olive production manual. 2nd edition. Oakland: UC Agriculture and Natural Resources Publication 3353. 61–69.

Berenguer, M. J., P. M. Vossen, S. R. Grattan, J. H. Connell, and V. S. Polito. 2006. Tree irrigation levels for optimum chemical and sensory properties of olive oil. HortScience 41:427–432. https://doi.org/10.21273/HORTSCI.41.2.427

Bustan, A., A. Dag, U. Yermiyahu, R. Erel, E. Presnov, N. Agam, D. Kool, J. Iwema, I. Zipori, and A. Ben-Gal. 2016. Fruit load governs transpiration of olive trees. Tree Physiology 36(3):380–391. https://doi.org/10.1093/treephys/tpv138

Connor, D. J., M. Gómez-del-Campo, M. C. Rousseaux, and P. S. Searles. 2014. Structure, management and productivity of hedgerow olive orchards: A review. Scientia Horticulturae 169:71–93. https://doi.org/10.1016/j.scienta.2014.02.010

Corell, M., M. J. Martín-Palomo, I. Girón, L. Andreu, E. Trigo, Y. E. López-Moreno, A. Torrecillas, A. Centeno, D. Pérez-López, and A. Moriana. 2019. Approach using trunk growth rate data to identify water stress conditions in olive trees. Agricultural Water Management 222:12–20. https://doi.org/10.1016/j.agwat.2019.05.029

Díaz-Espejo, A., T. N. Buckley, J. S. Sperry, M. V. Cuevas, A. de Cires, S. Elsayed-Farag, M. J. Martin-Palomo, et al. 2012. Steps toward an improvement in process-based models of water use by fruit trees: A case study in olive. Agricultural Water Management 114:37–49. http://dx.doi.org/10.1016/j.agwat.2012.06.027

Doorenbos, J., and W. O. Pruitt. 1977. Crop water requirements. FAO Irrigation and Drainage Paper 24. Rome: Food and Agriculture Organization of the United Nations.

Fereres, E., tech. ed. 1981. Drip irrigation management. Oakland: Division of Agricultural Sciences, University of California, Publication Leaflet 21259.

Fernández, J. E. 2014. Plant-based sensing to monitor water stress: Applicability to commercial orchards. Agricultural Water Management 142:99–109. https://doi.org/10.1016/j.agwat.2014.04.017

—— 2017. Plant-based methods for irrigation scheduling of woody crops. Horticulturae 3(2):35. https://doi.org/10.3390/horticulturae3020035

Fernández, J. E., A. Diaz-Espejo, R. Romero, V. Hernandez-Santana, J. M. García, C. M. Padilla-Díaz, and M. V. Cuevas. 2018. Precision irrigation in olive (*Olea europaea* L.) tree orchards. In I. F. García Tejero and V. H. Durán Zuazo, eds., Water scarcity and sustainable agriculture in semiarid environment. Cambridge: Academic Press. 179–217.

Fernández, J. E., A. Perez-Martin, J. M. Torres-Ruiz, M. V. Cuevas, C. M. Rodriguez-Dominguez, S. Elsayed-Farag, A. Morales-Sillero, J. M. García, V. Hernandez-Santana, and A. Diaz-Espejo. 2013. A regulated deficit irrigation strategy for hedgerow olive orchards with high plant density. Plant Soil 372:279–295. https://doi.org/10.1007/s11104-013-1704-2

Fernández, J. E., C. M. Rodriguez-Dominguez, A. Perez-Martin, U. Zimmermann, S. Rüger, M. J. Martín-Palomo, J. M. Torres-Ruiz, et al. 2011. Online-monitoring of tree water stress in a hedgerow olive orchard using the leaf patch clamp pressure probe. Agricultural Water Management 100:25–35. https://doi.org/10.1016/j.agwat.2011.08.015

Fulton, A. E., J. A. Grant, R. P. Buchner, and J. H. Connell. 2014. Using the pressure chamber for irrigation management in walnut, almond, and prune. Davis, CA: UC Agriculture and Natural Resources Publication 8503.

García, J. M., A. Hueso, and M. Gómez-del-Campo. 2020. Deficit irrigation during the oil synthesis period affects olive oil quality in high-density orchards (cv. Arbequina). Agricultural Water Management 230:105858. https://doi.org/10.1016/j.agwat.2019.105858

Gijón, M. C., D. Pérez-López, J. Guerrero, J. F. Couceiro, and A. Moriana. 2010. Riego deficitario controlado en olivo y pistachero. Agricultura 930:458–462.

Girón, I. F., M. Corell, M. J. Martín-Palomo, A. Galindo, A. Torrecillas, F. Moreno, and A. Moriana. 2015. Feasibility of trunk diameter fluctuations in the scheduling of regulated deficit irrigation for table olive trees without reference trees. Agricultural Water Management 161:114–126. https://doi.org/10.1016/j.agwat.2015.07.014

Goldhamer, P. A., J. Dunai, and L. Ferguson. 1994. Irrigation requirements of olive trees and responses to sustained deficit irrigation. Acta Horticulturae 356:172–176. https://doi.org/10.17660/ActaHortic.1994.356.36

Grattan, S. R., M. J. Berenguer, J. H. Connell, V. S. Polito, and P. M. Vossen. 2006. Olive oil production as influenced by different quantities of applied water. Agricultural Water Management 85:133–140. https://doi.org/10.1016/j.agwat.2006.04.001

Gucci, R., G. Caruso, C. Gennai, S. Esposto, S. Urbani, and M. Servili. 2019. Fruit growth, yield and oil quality changes induced by deficit irrigation at different stages of olive fruit development. Agricultural Water Management 212:88–98. https://doi.org/10.1016/j.agwat.2018.08.022

Hartmann, H. T., and C. Panetsos. 1961. Effects of soil moisture deficiency during floral development on fruitfulness in the olive. Proceedings of the American Society for Horticultural Science 78:209–217.

Hernandez-Santana, V., J. E. Fernández, M. V. Cuevas, A. Perez-Martin, and A. Diaz-Espejo. 2017. Photosynthetic limitations by water deficit: Effect on fruit and olive oil yield, leaf area and trunk diameter and its potential use to control vegetative growth of super-high density olive orchards. Agricultural Water Management 184:9–18. https://doi.org/10.1016/j.agwat.2016.12.016

Iniesta, F., L. Testi, F. Orgaz, and F. J. Villalobos. 2009. The effects of regulated and continuous deficit irrigation on the water use, growth and yield of olive trees. European Journal of Agronomy 30:258–265. https://doi.org/10.1016/j.eja.2008.12.004

Lavee, S., and M. Wodner. 1991. Factors affecting the nature of oil accumulation in fruit of olive (Olea europaea L.) cultivars. Journal of Horticultural Science 66(5):583–591. https://doi.org/10.1080/00221589.1991.11516187

Lightle, D., and J. H. Connell. 2018. Drought tip: Drought strategies for table and oil olive production. Davis, CA: UC Agriculture and Natural Resources Publication 8538.

Marino, G., T. Caruso, L. Ferguson, and F. P. Marra. 2018. Gas exchanges and stem water potential define stress thresholds for efficient irrigation management in olive (Olea europaea L.). Water 10(3):342. https://doi.org/10.3390/w10030342

Marino, G., A. Scalisi, P. Guzmán-Delgado, T. Caruso, F. P. Marra, and R. L. Bianco. 2021. Detecting mild water stress in olive with multiple plant-based continuous sensors. Plants 10(1):131. https://doi.org/10.3390/plants10010131

Martínez-Cob, A., and J. M. Faci. 2010. Evapotranspiration of an hedge-pruned olive orchard in a semiarid area of NE Spain. Agricultural Water Management 97(3): 410–418. https://doi.org/10.1016/j.agwat.2009.10.013

Moriana, A., F. Orgaz, M. Pastor, and E. Fereres. 2003. Yield responses of a mature olive orchard to water deficits. Journal of the American Society for Horticultural Science 128(3):425–431. https://doi.org/10.21273/JASHS.128.3.0425

Moriana, A., D. Pérez-López, M. H. Prieto, M. Ramírez-Santa-Pau, and J. M. Pérez-Rodriguez. 2012. Midday stem water potential as a useful tool for estimating irrigation requirements in olive trees. Agricultural Water Management 112:43–54. https://doi.org/10.1016/j.agwat.2012.06.003

Motilva, M. J., M. J. Tovar, M. P. Romero, S. Alegre, and J. Girona. 2000. Influence of regulated deficit irrigation strategies applied to olive trees (*Arbequina* cultivar) on oil yield and oil composition during the fruit ripening period. Journal of the Science of Food and Agriculture 80(14):2037–2043. https://doi.org/10.1002/1097-0010(200011)80:14<2037::AID-JSFA733>3.0.CO;2-0

Orgaz, F., and E. Fereres. 2004. Riego. Capítulo 10. El cultivo del olivo. Madrid: Mundi Prensa.

Padilla-Díaz, C. M., C. M. Rodriguez-Dominguez, V. Hernandez-Santana, A. Perez-Martin, and J. E. Fernández. 2016. Scheduling regulated deficit irrigation in a hedgerow olive orchard from leaf turgor pressure related measurements. Agricultural Water Management 164:28–37. https://doi.org/10.1016/j.agwat.2015.08.002

Rapoport, H. F., G. G. Costagli, and R. Gucci. 2004. The effect of water deficit during early fruit development on olive fruit morphogenesis. Journal of the American Society for Horticultural Science 129:121–127. https://doi.org/10.21273/JASHS.129.1.0121

Rapoport, H. F., S. B. M. Hammami, P. Martins, O. Pérez-Priego, and F. Orgaz. 2012. Influence of water deficits at different times during olive tree inflorescence and flower development. Environmental and Experimental Botany 77:227–233. https://doi.org/10.1016/j.envexpbot.2011.11.021

Rapoport, H. F., D. Pérez-López, S. B. M. Hammami, J. Agüera, and A. Moriana. 2013. Fruit pit hardening: Physical measurement during olive fruit growth. Annals of Applied Biology 163(2):200–208. https://doi.org/10.1111/aab.12046

Rosecrance, R. C., W. H. Krueger, L. Milliron, J. Bloese, C. Garcia, and B. Mori. 2015. Moderate regulated deficit irrigation can increase olive oil yields and decrease tree growth in super high density 'Arbequina' olive orchards. Scientia Horticulturae 190:75–82. https://doi.org/10.1016/j.scienta.2015.03.045

Servili, M., S. Esposto, R. Fabiani, S. Urbani, A. Taticchi, F. Mariucci, R. Selvaggini, and G. F. Montedoro. 2009. Phenolic compounds in olive oil: Antioxidant, health and organoleptic activities according to their chemical structure. Inflammopharmacology 17(2): 76–84. https://doi.org/10.1007/s10787-008-8014-y

Shackel, K., A. Moriana, G. Marino, M. Corell, D. Pérez-López, M. J. Martin-Palomo, T. Caruso, et al. 2021. Establishing a reference baseline for midday stem water potential in olive and its use for plant-based irrigation management. Frontiers in Plant Science 12:791711. https://doi.org/10.3389/fpls.2021.791711

Shackel, K. A., T. L. Prichard, and L. J. Lawrence. 2012. Irrigation scheduling and tree stress. In R. P. Buchner, ed., Prune production manual. Oakland: UC Agriculture and Natural Resources Publication 3507. 115–122.

Snyder, R. L., W. O. Pruitt, and D. A. Shaw. 1995. Determining daily reference evapotranspiration (ETo). Oakland: UC Agriculture and Natural Resources Publication 21426.

Trentacoste, E. R., D. J. Connor, and M. Gómez-del-Campo. 2015. Row orientation: Applications to productivity and design of hedgerows in horticultural and olive orchards. Scientia Horticulturae 187:15–29. https://doi.org/10.1016/j.scienta.2015.02.032

[USDA NRCS] US Department of Agriculture Natural Resources Conservation Service. 1998. Estimating soil moisture by feel and appearance.

8

Soil and Nutrient Management

Patrick H. Brown,
Distinguished Professor, Department of Plant Sciences, UC Davis

Saiful Muhammad,
Project Scientist, Department of Plant Sciences, UC Davis

Sat Darshan S. Khalsa, Assistant Professional Researcher, Department of Plant Sciences, UC Davis

Highlights

- Soil nutrients exist in the mineral pool, the exchangeable pool, and the soluble pool. Nutrients in the mineral pool are largely unavailable. Soluble pool and exchangeable pool nutrients are governed by the soil's cation exchange capacity.

- The three methods of monitoring nutrient supply and diagnosing problems are soil analysis, plant tissue analysis, and visual diagnosis of plants.

- Soil analysis determines nutrient availability and potential toxic elements and describes chemical and physical properties that may contribute to deficiency or toxicity.

- Plant tissue analysis measures nutrient status in the plant. Concentrations can be compared with established critical values.

- Identification of nutritional disorders by symptoms is the quickest method of deficiency diagnosis, but accurate assessment of symptoms requires years of experience.

- Nitrogen (N) and potassium (K) are required by olive in the largest quantities and have the greatest impact on productivity and oil quality.

- California regulations mandate the management of N to minimize its loss, which requires an understanding of the N cycle in the soil and tree.

- The 4R approach to nutrient management—right source, right rate, right time, and right place—provides an ideal basis for managing nutrients.

- Identifying nutrient deficiencies as soon as possible is important so that corrective action can be taken to avoid yield loss.

- Salinity damage is largely a consequence of the quality of the irrigation water.

To produce high-yielding orchards of high-quality olives in an environmentally sound fashion, growers need a basic knowledge of soil and plant nutrition, including soil nutrient availability and plant nutrient uptake. Understanding how soil and water interact to impact plant nutrient uptake enables growers to make informed decisions. Balanced, effective, and efficient management is critical if the olive orchard is to attain early and consistently high yields while maintaining high fruit and oil quality and protecting the environment.

Olive nutrition needs

Olive requires seventeen elements for growth and development; if any of these nutrients are deficient, growth and production are impaired. These seventeen essential elements are carbon (C), hydrogen (H), oxygen (O), nitrogen (N), phosphorus (P), potassium (K), sulfur (S), calcium (Ca), magnesium (Mg), boron (B), zinc (Zn), copper (Cu), iron (Fe), manganese (Mn), chlorine (Cl), molybdenum (Mo), and nickel (Ni). C, H, and O come from air and water, while the other nutrients are present in the soil or supplied by fertilizers and soil amendments. N, P, and K are primary macroelements, which are required in the largest quantities. S, Ca, and Mg are secondary macroelements, needed in smaller quantities. B, Zn, Cu, Fe, Mn, Cl, Mo, and Ni are microelements, required in much smaller quantities.

Soil nutrient availability

Soil nutrients exist in three pools: the mineral pool, the exchangeable pool, and the soluble pool (fig. 8.1). The bulk of nutrients in the mineral pool are locked up in the structural framework of primary minerals, organic matter, clay, and humus. The mineral pool is determined by the geology, minerology, and age of the soil. Nutrients in the mineral pool are largely insoluble and hence unavailable to the plant. Under certain conditions, such as high or low soil pH, or in the presence of specific K-fixing clay minerals, applied nutrients from fertilizer or amendments can become fixed in the mineral pool of certain soils and become unavailable for plant uptake. However, these nutrients can become available over time. Growers can influence soil fertility with tillage, deep ripping, or drainage, or with amendments, such as gypsum, manure, or compost.

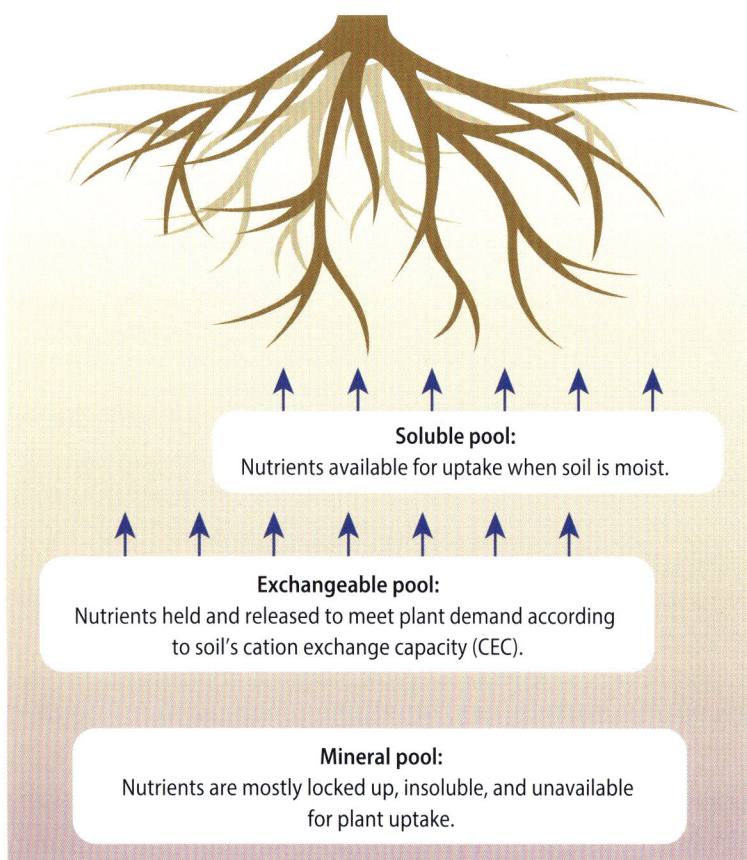

Soluble pool:
Nutrients available for uptake when soil is moist.

Exchangeable pool:
Nutrients held and released to meet plant demand according to soil's cation exchange capacity (CEC).

Mineral pool:
Nutrients are mostly locked up, insoluble, and unavailable for plant uptake.

Figure 8.1

The three pools of nutrients in soils play different roles in soil nutrient availability and plant nutrition.

Nutrients in the exchangeable pool are lightly held by negative charges on the functional groups (COO-) of soil minerals and organic matter. Negative soil charges attract positively charged nutrients known as cations, including N in the form of NH_4^+ (ammonium), Ca^{2+}, Mg^{2+}, K^+, Mn^{2+}, Fe^{2+}, Cu^{2+}, and Zn^{2+}. The number of exchange sites available for holding cations is defined as the soil's cation exchange capacity (CEC), which plays a role in the soil's native fertility. Higher CEC increases the potential for soil to hold and release nutrients to meet plant demand. Organic matter and clay have the greatest CEC, while sand has the lowest CEC (fig. 8.2). Under conditions of high salinity, sodium ions (Na^+) can dominate soil CEC and displace Ca^{2+} and Mg^{2+}, leading to a loss of soil structure. Low-CEC soils hold nutrients poorly and require more intensive management and monitoring.

Nutrients in the soluble pool are available for uptake by plant roots, but only when the soil contains moisture. Drought conditions reduce soil water availability and nutrient uptake. Soluble pool and exchangeable pool nutrients exist in dynamic equilibrium governed by CEC: nutrients are released and move at different rates from the exchangeable pool into the soluble pool as plants deplete the soluble pool through uptake (see fig. 8.1). Nutrients added to soil with fertilizer either move into solution for immediate plant uptake or enter the exchangeable pool. In K-fixing soils, K^+ may undergo fixation in the mineral pool. Nutrients in the soluble pool, particularly negatively charged anions that are repelled by soil CEC, such as nitrate (NO_{3-}), may leach below the root zone with excess irrigation or precipitation.

Nutrients in soil differ in mobility depending on their form and tendency to form compounds with other cations and anions in the soil solution. Mobile nutrients, such as NO_{3-}, Mg, S, B, Ca, and K, move freely with water via a process called mass flow. Immobile nutrients, such as Fe, Zn, P, Cu, Mn, and K in K-fixing soils, move only very short distances via diffusion and require roots to grow within close proximity for uptake. Nutrient mobility and solubility of fertilizer or soil amendments affect how growers must manage applications. Nutrient sources with low solubility, such as some K and P fertilizers, and immobile nutrients require placement near roots and adequate soil moisture to achieve plant uptake.

Soil pH strongly influences nutrient solubility and availability for uptake. Strongly acid or moderate alkaline soil pH can produce nutrient deficiencies or toxicities by reducing nutrient solubility and causing fixation of nutrients in the mineral pool (fig. 8.3). Olive orchards in California are most commonly planted in soils with neutral to alkaline or slightly acidic soil pH. As soil pH increases above 7.5, Zn, Mn, Cu, and Fe become less available, and plant deficiencies can occur.

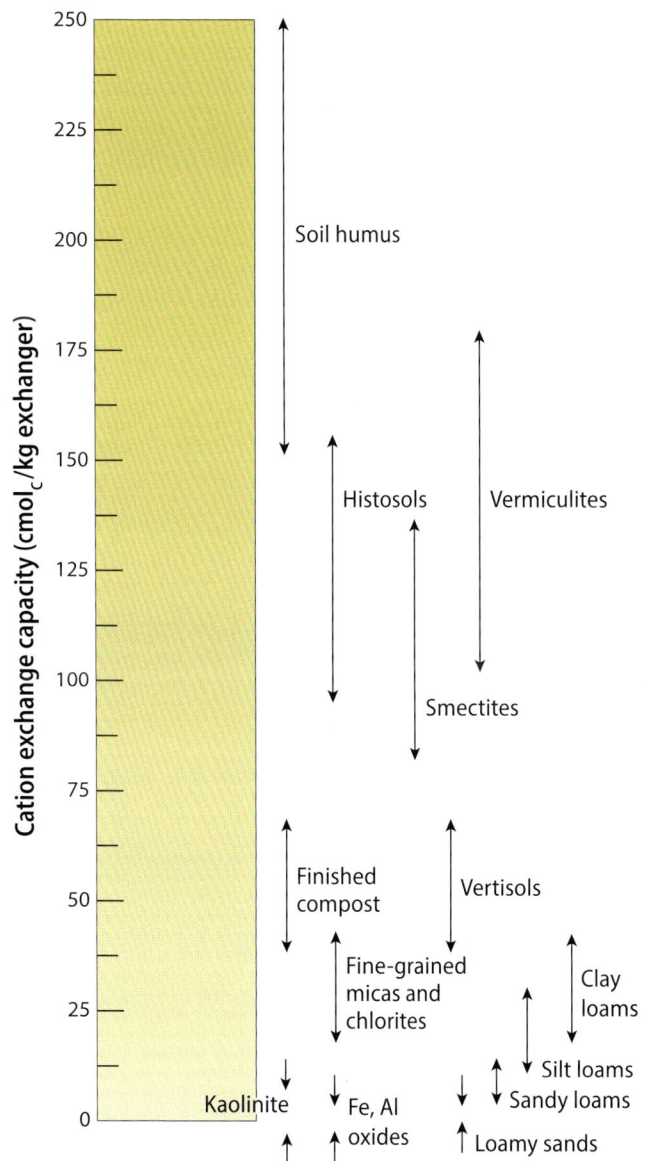

Figure 8.2

Ranges of cation exchange capacities at pH 7 are shown for a variety of soils and soilless materials. *Source:* Brady and Weil 2017.

Plant nutrient uptake and mobility

First and foremost, plant nutrient uptake is dependent on root health and root distribution in the soil profile. Soil properties or environmental conditions that limit root growth, such as soil compaction, hardpan, or waterlogging, reduce root health and growth, which, in turn, reduces plant nutrient uptake. Improving soil health by ensuring good aeration, increasing soil organic matter, maintaining adequate but not excessive moisture, and building good soil structure will improve nutrient uptake, plant productivity, and fruit quality.

After uptake, nutrients are transported to growing shoots, flowers, new roots, and fruit. Some nutrients remain mobile within the plant, while others quickly become immobile. The immobile nutrients Fe, Mn, Cu, and Ca only move upward with water in the unidirectional transpiration stream known as xylem. The mobile elements N, P, K, S, B, Mg, and Cl move in both the xylem and the bidirectional phloem. Zn is intermediate in its mobility; a small fraction of tissue Zn is mobile in spring and fall, but the majority is immobile. The mobility of an element influences its management.

Immobile nutrients accumulate at sites of high transpiration, mainly the leaves; once deposited, they cannot be remobilized to other parts of the plant. Therefore, to avoid deficiencies, immobile nutrients must be continually supplied to the plant from the soil or by foliar fertilizers. Deficiencies of immobile elements always occur in the youngest parts of the tree, while toxicities are expressed in the oldest leaves.

Mobile nutrients can move freely within the plant and, when present in mature leaves and tree organs, can be remobilized to supply new growth as needed. Remobilization will continue until the nutrient is depleted from mature leaves. Thus, deficiencies of mobile elements are expressed in the oldest leaves first.

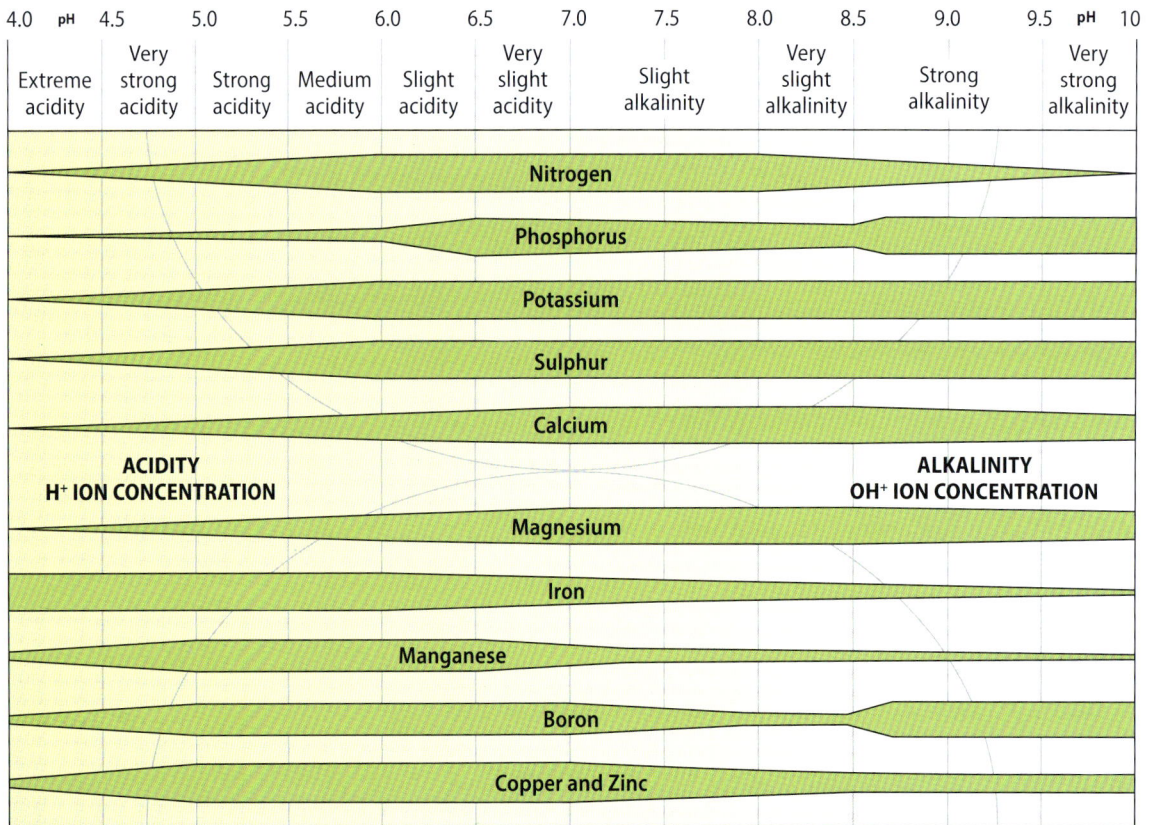

Figure 8.3

Soil pH influences nutrient availability. The thicker the green bar, the more available a nutrient is for plant uptake at a given pH.

Diagnostic methods

Nutrient management should focus on ensuring optimal productivity while minimizing biological, environmental, and financial costs from excess fertilization. Supplying and maintaining balanced plant nutrition requires growers to predict soil nutrient availability and monitor plant nutrient status during the growing season. The three methods of monitoring nutrient supply and diagnosing problems are soil analysis, plant tissue analysis, and visual diagnosis of plants for deficiency or toxicity symptoms.

Soil analysis

Soil analysis is used as a tool for preplant planning and for monitoring performance of established orchards. It determines nutrient availability and potential toxic elements and describes chemical and physical properties that may contribute to deficiency or toxicity. The soil chemical parameters frequently measured include pH, CEC, sodium adsorption ratio (SAR), and salinity (ECe). The soil physical parameters measured include soil texture and bulk density; they help identify impaired soil layers, essential in guiding preplant soil modifications and designing irrigation systems.

In established orchards, soil analysis is useful for monitoring the efficacy of fertilizer use and soil amendments, as well as for identifying potential toxicity from an overabundance of Na, Cl, or B or imbalances of Ca, Mg, and Na, which both may impede water infiltration. Soil analysis is most valuable when conducted in combination with plant tissue analysis and visual diagnosis of plants for deficiency or toxicity symptoms.

Soil sampling procedures

To be useful in identifying plant nutrition problems, soil samples should be collected within the area of active root growth. In regions with little in-season rainfall, the active root zone is largely determined by irrigation water distribution. While olive roots can reach depths of greater than 6 feet (1.8 m), the vast majority of root uptake occurs in the irrigated root zone, which is restricted to the top 2 to 20 inches (5 to 51 cm) of wetted soil during most of the year (P. H. Brown and A. Olivos, unpublished data). In flood-irrigated orchards, rain-fed orchards, and orchards under high rainfall conditions, roots occupy the entire soil volume. To identify toxicities, sampling up to 4 feet (1.2 m) deep may be necessary if roots occasionally access contaminated soils at depth. Analysis of pH, CEC, SAR, ECe, and toxic elements can be useful in identifying the cause of problems.

In uniformly flood-irrigated orchards, soil samples can be collected from anywhere under the tree canopy, 3 to 6 feet (0.9 to 1.8 m) from the trunk. Under microirrigation, the wetted zone is smaller, and samples should be restricted to this zone. A total of three to ten samples, each about 1 cup in volume, should be collected from the wetted root zone in the top 2 to 20 inches (5 to 51 cm) at different positions around a single tree or from trees growing in similar soil. If soil type varies by depth, separate samples should be taken from each depth and labeled and analyzed separately. Samples can be composited by depth if the soil is uniform. Orchards under microirrigation require sampling approximately halfway between the emitter source and the edge of the wetted area. If salinity is an issue, samples should be collected from the middle of the wetted zone and the margins, and analyzed separately, because of different accumulations of nutrients and toxic elements.

Collection of representative soil samples in an orchard can be a challenge given the heterogeneity of orchard soils by area and depth. Nutrient deficiencies in a part of an orchard can be due to a change in soil type; an old riverbed, corral, or pasture; differences in topography; or cuts and fills from land leveling. Depending on the soil heterogeneity, samples should be collected for each soil type and zone under different management and analyzed separately.

Methods and soil analysis interpretation

Samples should be submitted to a laboratory for analysis as soon as possible. The purpose of laboratory tests is to predict the fraction of soil nutrients that the olive trees have access to in the exchangeable nutrient pool. Various extraction methods have been developed to provide the most accurate results based on the region where the crop is grown. In areas where olive is not widely grown, care must be taken to ensure that the test recommended by the laboratory is appropriate for olive.

The analysis of soil physical and chemical properties is an important step to predict the performance of olive in given soil conditions and to guide the use of fertilizers and soil amendments. For example, a heavy clay soil may restrict root growth or water movement, a situation that can be improved by incorporating manures and organic matter. A high pH soil may reduce the availability of Zn, Fe, Mn, and Cu, but steps can be taken to lower soil pH. Soil analysis can assist in identifying toxicities caused by excess Na, Cl, and B. However, the value of soil analysis to guide fertilization practices is limited by the difficulty in predicting the relationship between soil nutrient concentrations and plant nutrient uptake.

Soil pH

Soil pH is a measure of H+ concentration in the soil. It is expressed on a logarithmic scale ranging from 0 to 14, with 0 to 4 strongly acidic and 9 to 14 strongly alkaline, or basic; 7 to 8 is ideal for olive production. Excellent olive production can be achieved in native soils outside this ideal range, such as pH 5.5 to 6.5 and pH 7.5 to 8.5. However, outside the desired range, more intensive management is required to avoid nutrient imbalances.

Because soil pH strongly affects nutrient availability, it is used to predict deficiencies and toxicities, verifiable with tissue sampling. Most nutrients are ideally available for plants at pH 6.5 to 7.5, while nutrient availability is affected at higher or lower pH values (see fig. 8.3). Fortunately, olive is well adapted to high pH soils of California and deficiencies are uncommon. The availability of P, Mg, and Ca may become compromised at pH values below 6; at very low pH levels (below 4.5), Al (aluminum), Fe, and Mn availability becomes excessive and toxicity may occur. Soils with such low pH are very rare in California, although low levels can be induced with acidifying ammonium-based fertilizers.

Soil salinity

Soil and irrigation water salinity due to high levels of Na and Cl negatively impact plant growth in three critical ways. In the short term, within days to weeks, high salt concentrations (> 8 dS/m) in the active root zone will inhibit water uptake and induce plant water stress. Over time, and if levels of salinity are moderate (< 5 dS/m), olive can adapt to these levels of salinity by producing its own internal metabolites to enhance water uptake. As the duration of salinity exposure continues, however, Na and Cl can accumulate in the plant, resulting in tissue toxicity and compromised photosynthesis.

High levels of soil Na, common in sodic soils, can also compromise soil structure by dispersing soil aggregates that are essential for good drainage, particularly under alkaline conditions. An accumulation of salts over time in the plant and in the root zone will damage productivity. With the exception of orchards planted on naturally saline or sodic soils, salt damage in irrigated olive is most commonly a consequence of poor-quality irrigation water along with the lack of good-quality water for leaching. Access to better-quality water for occasional leaching can reverse salinity damage. Amending the soil with gypsum or organic matter under saline conditions can improve soil aeration and root growth, facilitating soil leaching. Leaching is best and most efficiently performed with irrigation and rainfall during the dormant season.

Salinity standards for soil (table 8.1) and irrigation water (table 8.2) have been developed for olive. Soil salinity is determined by collecting soil from the active root zone and measuring electrical conductivity (ECe) with a soil and water paste extract. A saline soil has an ECe of greater than 4 dS/m. In irrigation water, even good-quality irrigation water with an ECw of 1 dS/m can cause soil salinity over time in the absence of leaching. For example, 1 dS/m contributes about 5,100 pounds (2,313 kg) of salt per 3 acre-feet (3,699 m³). Olive is relatively salt tolerant; it exhibits a 10 percent yield decrease with ECe of 4 dS/m, and a 50 percent yield loss at 8 dS/m.

In olive orchards located on well-drained soils and irrigated with good-quality water, winter rainfall may be adequate to leach out the accumulation of annual salts. Leaching the soil with irrigation requires directing water at the accumulated salts in the root zone. Uniform distribution across the orchard floor is required in flood- or sprinkler-irrigated fields. In orchards under microirrigation, salts accumulate at the periphery of the wetted soil volume during the growing season, and leaching must likewise target

Table 8.1

Soil analysis guidelines for olive

Soil problem	Degree of problem		
	Starting	**Increasing**	**Severe**
Salinity (affects crop water availability)			
ECe (dS/m)	4	5	8
Olive yield decrease (%)	10	25	50
Sodium (ESP)			
Soil permeability and plant toxicity problems stunt growth (%)	4.0–8.0	8.0–12.0	>12.0
Boron (ppm)	2		
pH			
Range for most crops: 5.5–8.4			
Optimal range for most crops: 6.5–7.5			
Known harmful in olive: > 8.5			

Source: Freeman et al. 2005.

Table 8.2

Irrigation water quality guidelines for olive

Irrigation problem	Degree of problem		
	Starting	**Increasing**	**Severe**
Salinity (affects crop water availability)			
ECw (dS/m)	< 2.0	2.5–4.0	> 5.5
Permeability (affects soil infiltration rate)			
ECw (dS/m)			
If SAR = 0–3	> 0.7	0.7–0.2	< 0.2
If SAR = 3–6	> 1.2	1.2–0.3	< 0.3
Specific ion toxicity			
Boron (ppm or mg/L)	0–1.0	> 1.0	

Source: Freeman et al. 2005.

this zone. Leaching should be used with caution on poorly drained soils as olive is sensitive to soil saturation and root diseases. The water required for leaching salts, known as the leaching fraction, can be calculated for specific soil types, existing soil salinity, and the salinity of the available water. In general, 1 acre-foot (1,233 m³) of excess water reduces salinity of the upper 1 foot (30 cm) of soil by 70 to 80 percent.

Sodic (alkali) soils

Soils that contain excessive amounts of exchangeable Na relative to Ca and Mg are termed sodic, or alkali, soils. Sodic soils are characterized by a dispersion of soil particles, which reduces soil permeability to both water and air. By definition, a sodic soil has an exchangeable sodium percentage (ESP) greater than 15, which indicates that 15 percent of the soil CEC is associated with Na and the remainder with Ca, Mg, and other cations, such as K. Olive trees are negatively affected when ESP levels reach 20 to 40 percent (see table 8.1).

When a sodic condition is identified, a laboratory analysis can be performed to determine the amount of gypsum required to displace the excessive Na. After a Ca-containing amendment is applied, the displaced Na must be leached below the root zone. Leaching with irrigation water during dormancy minimizes root damage caused by poor aeration.

Sodium hazard in water. Irrigation water with a high level of Na relative to Ca and Mg produces a sodic soil. The Na hazard in water is indicated by the sodium adsorption ratio (SAR). An adjusted SAR (SAR_{adj}) is commonly used when carbonate and bicarbonate salts are present in soil or water. Acidification of irrigation water can dissolve bicarbonates and carbonates. Water with a low SAR_{adj} dissolves lime from the soil and increases the exchangeable Ca. However, water with a high SAR_{adj} precipitates Ca, decreasing the exchangeable Ca and leading to sodic conditions.

Irrigation water should ideally have an SAR_{adj} less than 6, with hazards when greater than 9. Gypsum should be used to replace precipitated Ca in irrigation water with a high SAR_{adj} level. Because gypsum must be applied in wetted soil, placement in orchards under microirrigation can be a challenge. An application of 2 to 4 tons per acre (4.5 to 9 t per ha) of gypsum every year or two is common; soil incorporation is not necessary.

Low salinity water. Irrigation with water containing less than 1 dS/m salts may infiltrate slowly in sandy loam or fine-textured soils. This is a common issue in California on the east side of the San Joaquin Valley, where the irrigation water originates as snowmelt from the Sierra Nevada and contains only 0.5 to 1 dS/m. Water infiltration can drop to less than 0.1 inch (2.5 mm) per hour, making it difficult to satisfy orchard water demand.

Gypsum applications of 1 to 2 tons per acre (2.24 to 4.48 t per ha) just before peak evapotranspiration in late spring or early summer can increase the infiltration rate by as much as fivefold. Gypsum is generally beneficial for three to five irrigations. Infiltration can also be improved by mixing the gypsum directly into the irrigation water using specially designed equipment that reduces the possibility of emitter blockage.

Plant tissue analysis

While soil analysis aids in the prediction of nutrient availability and helps identify soil constraints, plant tissue analysis directly measures nutrient status in the plant and integrates multiple factors, including plant development stage, climatic conditions, nutrient availability, root distribution and activity, and irrigation. Olive tissue analysis requires correct leaf sampling and analysis protocols, and comparison of nutrient and toxic element concentrations with established critical values (table 8.3).

Tissue analysis is most useful when the results are integrated with information on crop productivity, current management practices, orchard heterogeneity, and the timing and location of sampling. Multiple years of tissue analysis with good record keeping is a valuable approach for determining how an orchard responds to soil and nutrient management practices as well as changes in environmental conditions.

Critical values

Critical values were developed by correlating nutrient concentrations in plants grown under controlled experimental conditions with leaf symptoms or crop yield. The critical value of a nutrient is the concentration at which the crop yield or growth is 90 to 95 percent of maximum (Ulrich 1952).

The critical value curve is composed of four zones: deficient, marginal, adequate, and toxic (fig. 8.4). In the deficient zone, leaf symptoms are clearly evident, and an increase of tissue nutrient concentration results in an increase in yield and diminished symptoms. The marginal zone is where deficiency symptoms may or may not be present, but yield is compromised. The adequate zone is where growth is maximal and no symptoms are evident; an increase in nutrient concentration has no effect on relative yield, and nutrient additions have little effect on tissue nutrients. In the toxic zone, nutrient additions result in yield loss, disease, or quality loss.

Table 8.3

Olive critical values from leaf tissue sampled in July in the northern hemisphere for nutrient and toxic elements

Element	Deficient	Adequate	Toxic
Nitrogen, N (%)*	1.40 (1.20)	1.5–2.0 (1.3–1.7)	(> 1.7)
Phosphorus, P (%)†	0.05	0.1–0.3	—
Potassium, K (%)	0.40	> 0.8	—
Calcium, Ca (%)	0.30	> 1.0	—
Magnesium, Mg (%)	0.08	> 0.1	—
Manganese, Mn (ppm)	—	> 20.0	—
Zinc, Zn (ppm)	—	> 10.0	—
Copper, Cu (ppm)	—	> 4.0	—
Boron, B (ppm)	14.0	19–150	185
Sodium, Na (%)	—	—	> 0.2
Chloride, Cl (%)	—	—	> 0.5

Source: Adapted from Fernández-Escobar 2018.

*Nitrogen levels proposed by Molina-Soria and Fernández-Escobar (2012) are given in parentheses.

†Toxicity symptoms were observed at 0.21% in young plants (Jiménez-Moreno and Fernández-Escobar 2016).

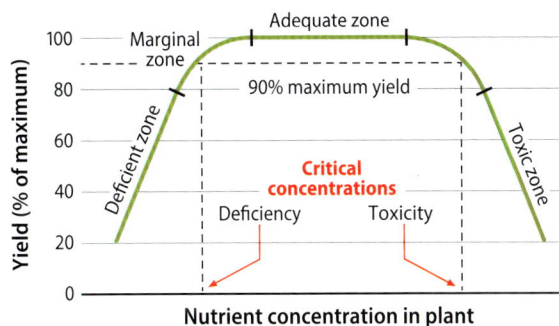

Figure 8.4

The critical value curve for diagnosing nutrient deficiency and toxicity in plants consists of four zones: deficient, marginal, adequate, and toxic. *Source:* Smith and Loneragan 1997.

Tissue sampling procedures and strategies

Leaf nutrient concentrations and dry mass vary by season, leaf age, canopy position, and presence or absence of fruit or sun exposure (fig. 8.5). This variability makes following a protocol essential so that results can be compared with standard critical values. In olive, leaves with petioles are sampled from the mid to basal portion of fruitless shoots in current year growth (fig. 8.6). The position

chosen in the canopy is generally well exposed and 5 to 8 feet (1.5 to 2.4 m) from the ground. Leaves sprayed with foliar micronutrients should not be sampled because they cannot be reliably analyzed for those nutrients due to surface contamination. In the northern hemisphere, the ideal sampling time is usually during the last 2 weeks of July and the first 2 weeks of August, a period that corresponds to initiation of pit hardening.

Sampling strategies vary depending on the objective of the tissue analysis. If investigating poor tree performance due to nutrient deficiency or toxicity in an isolated orchard area, then targeted sample collection from those trees and comparison with samples from healthy trees should suffice. Isolated areas of poor performance are frequently a consequence of soil disturbance or differences in soil type. Making a soil analysis and comparing the results with results from a plant tissue analysis can help identify the cause of poor performance.

In orchards where oil quality is a particular emphasis, separately analyzing samples from areas of varying tree vigor or leaf greenness could identify orchard zones of excessive N uptake and compromised oil quality. With the advent of improved

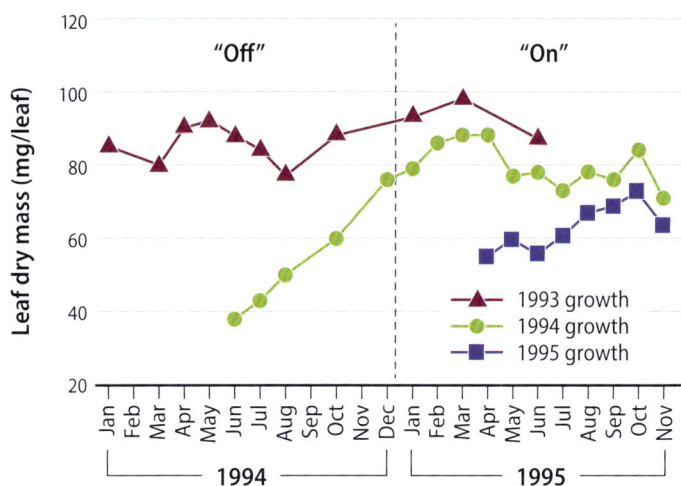

Figure 8.5

Seasonal changes occurred in the dry mass of olive leaves during three growth periods and an alternate-bearing cycle. *Source:* Adapted from Fernández-Escobar et al. 1999.

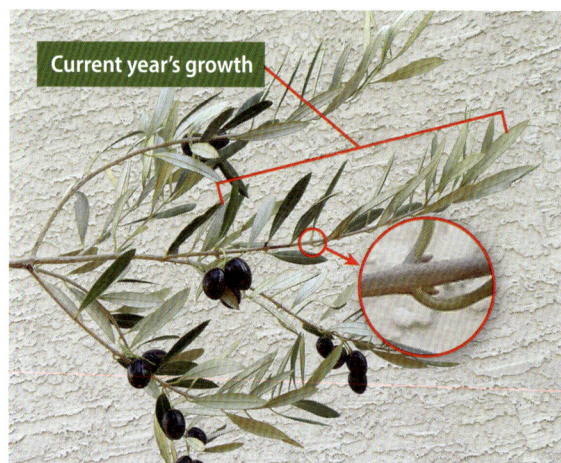

Figure 8.6

Leaves for tissue sampling are taken from the current year's growth. *Photo:* Louise Ferguson.

remote-sensing tools, winegrowers are adopting precision management on a limited scale. Similar technology is under development for olive growing (Gómez-Casero et al. 2007).

If the purpose of nutrient analysis is to determine the nutrient status of an orchard and verify nutrient management practices, then sampling throughout the orchard is needed. Research in nut crops confirms that leaf sampling from five to eight well-exposed shoots from fifteen to thirty trees, with each tree a minimum of 30 yards (27 m) apart, is required. This level of sampling is necessary to account for the variability in trees and to ensure sampling represents the field average. All leaves can be combined into a single sample for submission to a reputable laboratory. In California, many laboratories are evaluated by the state's Environmental Laboratory Accreditation Program.

Sample labeling with the exact location of the samples, orchard conditions, yield, fruit and oil quality, and management practices used in a given year is essential. Careful record keeping and integration with soil analysis will greatly enhance the value of the tissue analysis by allowing comparison across years and after management practice changes.

Sample preparation and laboratory analysis

Leaf tissues degrade if samples are exposed to extreme heat and excessive moisture inside a closed vehicle or prolonged storage in a plastic bag. Collected samples should be kept in cool, dry conditions and submitted promptly. In the laboratory, standard procedures include leaf washing with a weak, phosphate-free detergent mixture, rinsing with distilled water, and drying at 140°F (60°C). Dried samples are ground, and a subsample is analyzed to determine concentrations of nutrients and toxic elements.

Many testing laboratories routinely conduct analysis of all relevant nutrients and toxic elements in the tissue samples. A full nutrient panel including N, P, K, S, Ca, Mg, B, Zn, Cu, Fe, and Mn provides useful information. However, routine analysis of N, P, and K may provide sufficient information if no other deficiency or toxicity is suspected. Leaf Cl is a separate analysis using a different technique at an added cost.

Visual diagnosis of symptoms

Deficiencies or toxicities of essential elements cause metabolic disturbances that eventually result in characteristic visual symptoms, including reduced growth and specific foliage disorders. Identification of nutritional disorders by symptoms can be extremely useful and, when performed by a professional, is the quickest method of deficiency diagnosis.

However, visual diagnosis has important disadvantages and limitations. By the time visual symptoms appear, tree productivity has already been negatively affected. Accurate visual assessment of foliar symptoms requires years of experience, and symptoms vary with time and environment. Diagnosis can be complicated by the occurrence of multiple deficiencies and environmental conditions, and marginal deficiencies are particularly difficult to identify. The effects of low temperatures, waterlogging, root or crown diseases, toxic elements, and mechanical and spray damage can all be misinterpreted as nutrient deficiencies.

Often, the pattern of symptom development, such as the site of deficiency within the tree, can be used to identify a nutrient deficiency. Because the mobility of different elements varies within the tree, symptoms of various nutrient deficiencies appear first in different portions of the canopy. In general, immobile element deficiencies appear first in the youngest tissues, whereas mobile element deficiencies appear first in the oldest leaves.

Using tissue analysis in fertilizer decisions

Tissue analysis is excellent at identifying nutrient deficiencies and monitoring the effect of management practices in the orchard. However, it does not provide sufficient information to be the sole guide in nutrient management. For example, in all but deficient trees, tissue analysis does not always show significant changes after changing fertilizer management practices. Furthermore, analysis results may lack recommendations for how much nutrient to apply. Finally, tissue analysis does not identify why a nutrient is deficient. Nutrient deficiencies may be the result of inefficient fertilizer use, incorrect nutrient sources, high soil pH, nutrient imbalances or toxicity, poor root growth, or saturation or other soil conditions that reduce nutrient availability.

For example, a tissue analysis that returns an N value within the adequacy range of 1.5 to 1.7 percent does not prove that the nutrient has been used efficiently and does not provide sufficient information to guide fertilization rates. Similarly, determining how much N is needed to bring a marginally deficient tissue result of 1.2 percent to a sufficient concentration of 1.5 percent is difficult to determine. Furthermore, overfertilization may generate ideal tissue N values.

Nitrogen and potassium management

All seventeen essential nutrients are required for the production of olive, and a deficiency of any element can impair growth and production. However, N and K are required by olive in the largest quantities and have been demonstrated to have the greatest impact on productivity and oil quality. N is an integral constituent of proteins, nucleic acids, chlorophyll, coenzymes, phytohormones, and secondary metabolites. K has an important role in stomatal opening, plant growth and cell expansion, and fruit growth and development; it is also required for phloem loading and transport of carbohydrates.

N deficiency reduces photosynthesis and plant growth. Deficiency symptoms appear first in older leaves as a result of N remobilization from older to younger leaves. The impact of N depletion over time in trees with sufficient N nutrition may have little impact on yield for a couple of years. N fertilization in excess of tree demand is poorly utilized and may be lost to leaching or result in excessive vegetative growth and greater disease susceptibility. Overfertilization with N fertilizers is a leading cause of groundwater contamination. Efficient N management to ensure full productivity while reducing N losses is essential to the ongoing success of the olive industry.

Large amounts of K are present in olive fruit; K removed in harvested fruit generally exceeds N removal. K deficiency reduces photosynthesis and carbohydrate transport, and accelerates premature leaf senescence and abscission. A deficiency may take several years to appear because fruit and nut trees can store significant amounts of K (Reidel et al. 2004). Rain-fed olive orchards are more susceptible to K deficiency due to limited K mobility and, as a result, reduced uptake. Irrigation appears to increase soil K availability as well as uptake (Zipori et al. 2015). While some studies found no response in vegetative growth or yield due to K fertilization, other research showed significantly higher yield in K-fertilized trees because of more intense flowering and more fruit per tree (Zipori et al. 2020).

The total fruit demand for N, P, and K were calculated for an intensively managed olive orchard (Rosecrance and Krueger 2020), with removal numbers normalized per 1,000 pounds (454 kg) of fruit (table 8.4). Total nutrients found in prunings for different commercial orchard systems are found in table 8.5.

Changes in olive leaf by season and leaf age

Changes in leaf N concentration and content per leaf are shown in figure 8.7 for current year, prior year, and 2-year-old leaves across two seasons from a dryland, low-yielding, and alternate-bearing 12-year-old orchard of Picual in Spain. In current year and 1-year-old leaves in both the "on" and "off" years, N concentration decreased slowly from January through June, then decreased

Table 8.4

Nutrient amounts removed in fruit from high-density fertilized olive orchards in California*

Variety	N	P	P$_2$O$_5$	K	K$_2$O
	lb/ac				
Arbosana 5 t/ac	32.00	6.81	15.60	81.40	98.00
Arbequina 5 t/ac	34.10	7.57	17.40	83.60	101.00
Koroneiki 5 t/ac	37.30	5.43	12.50	61.20	73.60
Manzanillo 5 t/ac	40.30	4.03	9.24	55.20	66.50
	lb/1,000 lb fruit				
Arbosana	3.20	0.68	1.56	8.14	9.80
Arbequina	3.41	0.76	1.74	8.36	10.10
Koroneiki	3.73	0.54	1.25	6.11	7.36
Manzanillo	4.03	0.40	0.92	5.52	6.65

Source: Rosecrance and Krueger 2020.

*The top panel assumes a fresh weight yield of 5 tons per acre at a fruit moisture content of 58 percent. The bottom panel normalizes the nutrient amount removed per 1,000 pounds of fruit.

Table 8.5

Nutrient amounts in prunings from high-density fertilized olive orchards

Variety	Tree density	Prunings fresh weight	N	P	P$_2$O$_5$	K	K$_2$O
	trees/ac		lb/ac				
Arbequina*	532	1,821	4.51	0.49	1.12	4.28	5.16
Barnea†	145	16,319	40.40	4.39	10.10	38.40	46.20
			lb/1,000 lb fresh prunings				
			2.48	0.27	0.62	2.35	2.83

Note: The top panel shows tree densities and expected prunings fresh weight for each orchard type with a prunings moisture content of 58 percent. The bottom panel normalizes the nutrient amount removed per 1,000 pounds of fresh prunings.

*Hedge-style trellised orchard in California, with trees spaced at 13 by 5 feet. *Source:* Rosecrance and Krueger 2020.

†Orchard in Israel, 8 to 11 years old, with trees spaced at 23 by 13 feet. *Source:* Zipori et al. 2020.

rapidly until late August, before increasing to a maximum in January or February. In 2-year-old leaves, a rapid decline in tissue N concentration occurred from January until June.

In the dryland orchards used in these experiments, K behaved differently than N (fig. 8.8). In current year and 1-year-old leaves in both the "on" and "off" years, K concentration decreased rapidly and uniformly from January through August, indicating high K demand. Leaf K concentration remained relatively constant through February until a large increment in both K content and concentration were observed. Leaf K concentrations were below the leaf K critical value of 0.8 percent; the trees may have been K-limited and overly responsive to

rainfall-induced changes in soil K availability. It is uncertain if the same patterns of change in K would occur in irrigated orchards with higher yields.

Olive oil quality

In oil olive orchards, the complex relationship between tissue N concentrations and vegetative growth, yield, and oil quality makes nutrient management critical. This relationship is shown in figure 8.9 for irrigated intensive olive orchards. When tissue N is in the deficient zone (see fig. 8.4), an application of N will have a strong positive effect on vegetative growth and yield. At N levels at or less than 1.5 percent, the impact on current and future yield will be significant.

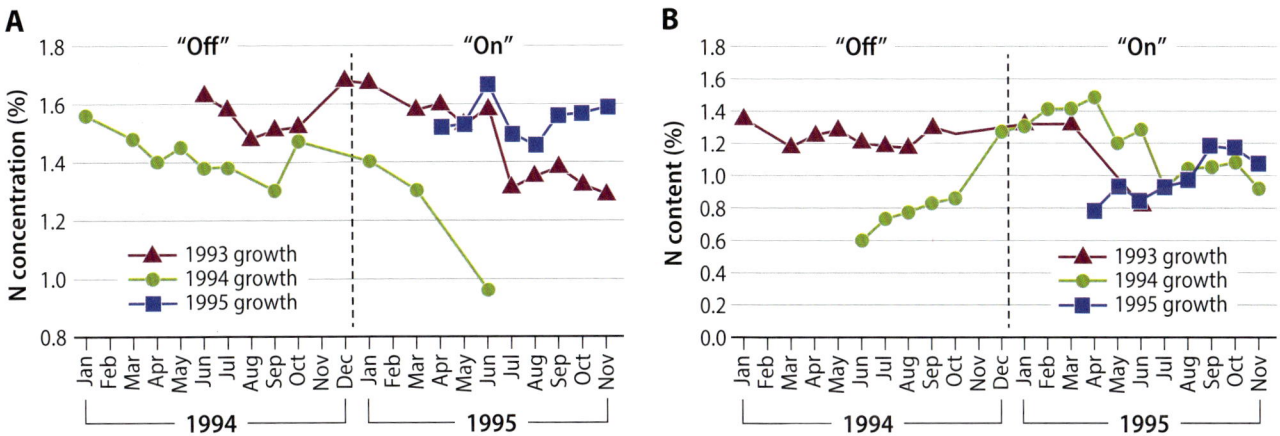

Figure 8.7

Seasonal trends of N concentration (A) and N content (B) in olive leaves occurred during three growth periods and an alternate-bearing cycle. *Source:* Adapted from Fernández-Escobar et al. 1999.

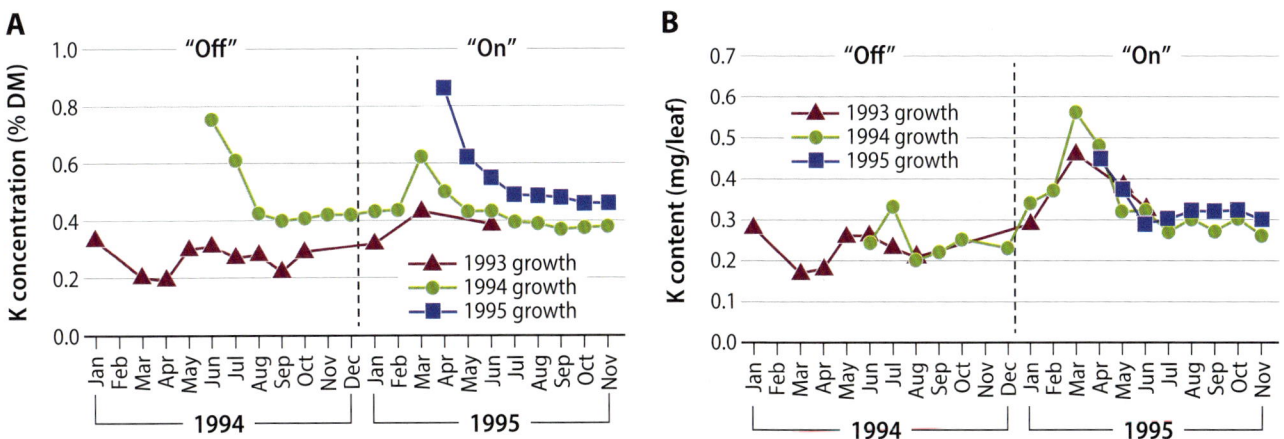

Figure 8.8

Seasonal trends of K concentration (A) and K content (B) in olive leaves that occurred during three growth periods and an alternate-bearing cycle are shown in "on" and "off" years. *Source:* Adapted from Fernández-Escobar et al. 1999.

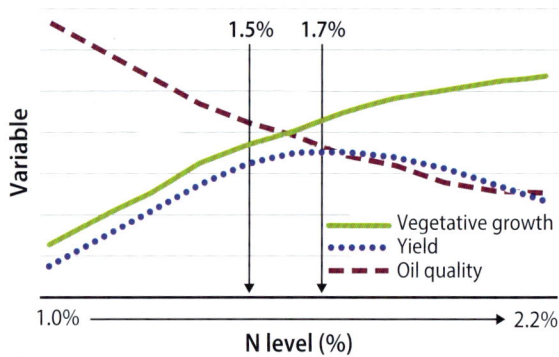

Figure 8.9

Vegetative growth, yield, and oil quality respond to olive leaf N levels. *Source:* Adapted from Zipori et al. 2020.

However, as tissue N increases to well above 2 percent, oil quality can be impaired as free fatty acids increase and polyphenols decrease in the oil (Dag et al. 2009; Erel et al. 2013). Overfertilization with N can also cause excess growth and increased susceptibility to disease, potentially reducing tree health and yield.

The ideal N application rate depends on the grower's goals. The critical values for olive leaf tissue in midsummer (mid-July and early August in the northern hemisphere) are presented in table 8.3. These values are suitable for both table olive and oil olive orchards, although research indicates that the higher levels of N concentrations (> 1.7%) should be avoided if the highest quality oil is desired. Recent experimentation with limited rates of applied K suggest that the current critical value of 0.8 percent may be too low for high-yield irrigated olive orchards. A critical value of greater than 1 percent may be beneficial (Haberman et al. 2019).

Nitrogen cycle

To protect groundwater resources, California regulations mandate the management of N to minimize its loss below the active root zone. Careful management requires an understanding of the N cycle in the soil and plant. Over time, microbes decompose organic residue, converting it into mineral N in a process called mineralization. N from mineralization of soil organic matter is available as ammonium (NH_4^+) in the spring; the process of nitrification rapidly converts the NH_4^+ to nitrate (NO_3^-) as the soil warms. Urea and NH_4^+-based fertilizers also undergo rapid

nitrification, often in a matter of days as soil temperatures rise.

Because negatively charged NO_3^- is repelled by the negatively charged soil particles, it is easily leached from the root zone by irrigation water or rainfall and is readily lost to groundwater. Under waterlogged conditions, such as during flood irrigation, soil microbes convert NO_3^- into nitrogen gases, like dinitrogen (N2) gas, in a process called denitrification. During this process, nitric oxide (NO), a precursor to smog, or nitrous oxide (N2O), a potent greenhouse gas, can also be released to the environment.

Application of N fertilizers or organic matter amendments (OMA) can also result in N losses via volatilization. If urea fertilizer is broadcast, part of it may convert into NH_4^+ and volatilize. This loss can be reduced by ensuring that soils are moist or by incorporating the urea into the soil. High concentrations of NH_4^+-based fertilizers from banding as well as soils with alkaline pH can also favor volatilization losses.

NO_3^- leaching is the primary way that N is lost to groundwater. Excess NO_3^- supplied by N fertilizer, irrigation water, or nitrification of NH_4^+ has the potential to move below the root zone. This excess amount beyond plant uptake, combined with movement of water past the root zone, results in leaching. Use of fertigation to deliver N fertilizer in split applications reduces N losses by ensuring N is maintained in the root zone.

Potassium cycle

The majority of native soil K is present in the mineral form, with only a very small amount of soil K available for plant uptake. Soil K can be divided into three groups:
- K contained in minerals (90 to 98% of total soil K) and nonexchangeable K (1 to 10% of total soil K), released as soil minerals, are weathered at such a slow rate as to be unavailable to plants.
- K fixed or trapped between layers of certain soil clay minerals may become slowly available (1 to 2% of total soil K).
- Readily available K includes K present in the soil solution and exchangeable K in organic matter and clay (0.1 to 0.2% of total soil K).

The main K additions to an olive orchard are in the forms of fertilizers, crop residues, and OMA. The main K removal from an olive orchard is in the harvested crop.

Nitrogen and potassium uptake

The primary forms of N in fertilizers are NH_4^+, NO_3^-, and urea, all of which are taken up readily by plant roots. The uptake of NH_4^+ results in the efflux of H^+ from the root, resulting in soil acidification (Meyer 1996). N assimilation is the process in which organic N forms from NH_4^+ and NO_3^- and is transported from roots to shoots and leaves through the xylem along with water. All the steps of N assimilation require large amounts of energy supplied by carbon substrates from photosynthesis. As a consequence of the high cost to the plant of N uptake and assimilation, N uptake rates are highly selective and closely match demand.

Uptake of K is also highly selective and closely matches demand. K is highly mobile in plants, including within cells and tissues, and via xylem and phloem. It plays an important role in plant-water relations and balances the charge of anions in the plant.

An important principle of olive nutrient management is that the tree cannot be forced to utilize more nutrients than the current crop demands. Poorly growing trees will have compromised nutrient uptake that can only be remediated by correcting the factors limiting growth. The 4R approach to nutrient management provides growers with the means to optimize productivity without wasting resources.

4R approach to nutrient management

The 4R approach to nutrient management, which integrates the right source, right rate, right time, and right place, has gained widespread acceptance and provides an ideal basis for managing nutrients in crops (IPNI 2012). In the 4R approach, the choice of the correct fertilizer or nutrient (right source) applied in proportion to the demand of the crop (right rate), timed with periods of nutrient uptake (right time), and placed within the active root zone or to the plant foliage (right place) is essential to satisfy plant demand and avoid nutrient application in excess of plant capacity for uptake. Together, the 4Rs maximize nutrient use efficiency and ensure optimum yield and quality.

Fertilizers and soil amendments: Right source

N and K are the two nutrients needed most by olive. Many N and K fertilizers are available in granular and liquid forms. They are popular due to their widespread availability and effectiveness; however, cost, convenience, potential soil acidification, and susceptibility to loss will vary. Soil amendments, cover crops, and prunings also play a role in nutrient management.

Nitrogen fertilizers

Urea and NH_4^+-based fertilizers transform into NO_3^- and are susceptible to leaching. Urea hydrolyzes to NH_4^+, which converts to NO_3^-. Neutrally charged urea and NO_3^- may leach below the root zone under heavy rainfall or excessive irrigation.

A soil pH between 6.5 and 7.5 provides optimum nutrient availability for olive. However, many olive orchard soils in California have a higher pH, and growers may choose a fertilizer source to help reduce it. NH4+-based fertilizers frequently have an acidifying effect on soil: ammonium sulfate, (NH4)2SO4, is very acidifying; urea, ammonium nitrate, NH4NO3, and urea ammonium nitrate solution, UAN32, are moderately acidifying. In contrast, calcium nitrate, Ca(NO3)2, creates an alkaline reaction in the soil.

Potassium fertilizers

The appropriate selection of K fertilizer sources is important in olive production. Considerations include price, salt content, solubility, and fertilization method. The units of K fertilizer are reported as potassium oxide (K_2O). Values of K must be multiplied by 1.0247 in order to equal K_2O equivalents in fertilizer. If the soil or irrigation water has a high chloride (Cl) content, avoid using potassium chloride (KCl) as it may increase Cl toxicity in the trees. In most California olive orchards, potassium sulfate (SOP or K_2SO_4) is the major K source. It is frequently applied in winter so that rain can assist with dissolving and moving it into the root zone. SOP can also be applied followed by irrigation or solubilized with a dedicated SOP injection system. In the case of surface application of SOP in orchards under microirrigation, application should be made below or adjacent to the drip emitters.

If soil or water salinity is not an issue and the orchard is located in a high-rainfall area, KCl can be used as long as tissue Cl levels are monitored. Liquid KCl is also available for fertigation. Potassium nitrate (KNO_3) is an excellent source of both N and K; it is highly soluble and suitable for fertigation systems. Excellent results have also been obtained using a combination of low-cost granular SOP with soluble KNO3 or potassium thiosulphate (KTS).

Nitrogen in irrigation water

NO_3- in irrigation water is a free and effective N fertilizer that should be accounted for in any N budget (see the section "Calculation of an annual nutrient budget," below). A water report may present NO_3- in irrigation water as either nitrate or nitrate-N, that is, NO_3- (ppm) or NO_3-N (ppm); be sure to check which form the laboratory has used. To calculate how much N could be contributed by irrigation water, use the appropriate formula:

$$\text{lb N/ac} = NO_3\text{- concentration (ppm)} \times \text{inches of irrigation water} \times 0.052$$

$$\text{lb N/ac} = NO_3\text{-N concentration (ppm)} \times \text{inches of irrigation water} \times 0.23$$

For example, applied irrigation water of 30 inches at a concentration of 45 ppm NO_3. (10 ppm NO_3-N) contributes about 70 lb N per acre (78 kg per ha) annually.

Organic matter amendments

OMA, which include compost, manure, and green waste, such as olive mill waste, can be safely utilized during orchard planting by either backfilling or incorporating in the tree berm, or spreading and incorporating in the alleyway. Different OMA sources possess different nutrient profiles, and laboratory analysis is recommended (table 8.6).

The majority of OMA in bearing orchards are spread on the tree berm and left on the soil surface as a mulch (Khalsa and Brown 2017). For sources of OMA posing a potential food safety risk, placement in bearing orchards should be limited to up to 120 days before harvest to allow adequate time for decomposition. If the amendment has a high C:N ratio, greater than 17:1 to 20:1, placement in the irrigated zone may alter the availability of fertilizers delivered through the irrigation system due to N immobilization.

In addition to their N content, OMA include salts, such as Na and Cl. If allowed to accumulate in the soil, these salts can contribute to toxicities. Different sources of OMA can have different levels of Na and Cl salt loading. Thus, the rate of OMA application should consider total salt load and available irrigation water to leach excess salts. If well managed, OMA represent a valuable

Table 8.6

Chemical characteristics of composted manure and green waste compost prior to application

	pH	ECe	TC	TN	C:N	NH$_{4+}$-N	NO$_3$-N
		dS/m	%	%		ppm	ppm
Composted manure	7.93	29.1	28.3	2.35	12:1	351	223.00
Green waste compost	4.69	22.5	34.8	1.87	19:1	1,250	6.80
	Olsen-P	X-K	Ca	Mg	Na	Cl	B
	ppm	ppm	meq/l	meq/l	meq/l	meq/l	mg/l
Composted manure	2,655	26,895	6.26	9.43	94.7	145.0	7.96
Green waste compost	271	7,410	63.00	87.40	69.7	96.3	5.43

KEY

ECe = electroconductivity of saturated paste extract

TC = total carbon

TN = total nitrogen

X-K = exchangeable potassium

source of N, P, and K for olive; improve soil structure and health; and increase the capacity of soils to retain nutrients and water.

Cover crops

Like OMA, cover crops increase soil organic matter, which improves soil health through reduced nutrient losses and improved soil water and nutrient retention. The C:N ratio of cover crop residues impacts the balance of N mineralization and immobilization. In general, leguminous cover crops have lower C:N ratios than grasses, and legume-grass mixes have become more common. Cover crop management includes sowing seed in the orchard alleyway while maintaining the tree berm free of vegetation. Termination of annual cover crops can occur in the spring, and perennial cover crops or conservation cover are maintained by mowing.

Prunings

The typical practice in many tree crops is to mulch tree prunings on the soil surface by flail mowing. This is a best management practice to recycle nutrients and organic matter back into the soil. Nutrient contents of tree prunings are shown in table 8.5.

Fertilizer application methods: Right placement

Proper placement of fertilizers is important to optimize the availability of nutrients to roots. The goal is to apply fertilizers to the active root zone to maximize uptake and avoid losses. Under microirrigation systems in environments with limited rainfall, the majority of active roots are restricted to the soil zone wetted during an irrigation cycle. In regions with substantial winter and spring rainfall, roots are active to a greater depth. In almond studies conducted in California's Kern County, a region with minimal winter rain, with irrigation cycles of 5 to 14 days, 90 percent of active roots were found between 5 and 20 inches (13 and 51 cm) deep (Muhammad et al. 2017). A small percentage of roots were present at depths of 5 feet (1.5 m) or greater.

Broadcast N fertilizers are at risk of substantial N loss through leaching or surface runoff, while broadcast K is at greater risk of fixation. Banding K in a strip 12 to 18 inches (30 to 46 cm) wide

or in the herbicide strip has long been used in orchards and is still used in furrow-, flood-, and solid-set sprinkler–irrigated fields. Broadcasting K fertilizer is generally an inefficient approach as tree roots can be restricted to the drip line or the wetted zone of the irrigation system. Banding creates a concentrated zone of K, where fixation will be reduced and plant roots can scavenge the K, provided soil moisture is available.

The availability of liquid N and K fertilizers and microirrigation have made fertigation an efficient option as it encourages root proliferation in the soil wetted zone and facilitates targeted placement of fertilizers. However, poor fertigation or irrigation practices can also result in substantial N losses. As a guideline, fertilizers should be injected into the irrigation system in the final one-third of the irrigation event, and enough water following fertigation should be used to clean out the irrigation lines. The uniformity of nutrient distribution will always be limited by the distribution uniformity of the irrigation system.

Fertilizer application methods: Right rate and right time

Traditionally, nutrient management was based on leaf tissue analysis and comparison with standard critical values. Fertilizer practices were adjusted when tissue analysis for a particular element fell below a specified critical value. While this approach has been a useful tool for diagnosis of nutrient deficiency or excess, it does not provide sufficient information to define the rate and timing of fertilizer applications.

The goal of any nutrient management strategy is to ensure that adequate nutrients are available to supply the current demand of the plant. Although all plant processes and organs require N and K, fruit development represents by far the largest sink for N and K use in olive. N and K uptake is demand driven, that is, the size of the crop determines how much N or K is taken up by plant roots and when that uptake occurs. While a shortage of N and K can reduce yield, adding N and K in excess of plant demand will not increase yield and may result in N losses and reduced nutrient use efficiency.

Under California conditions, tree activity during winter months is minimal and thus nutrient uptake is low. A study of high-yield irrigated olive orchards in Israel showed that the total amount of N should be applied from March or April until the end of August or early September (Haberman et al. 2019). The end of this period represents the time of maximum oil accumulation. This timing approach allows for any residual N to be taken up prior to winter rains to reduce risk of N loss, and it reduces tree exposure to high N levels during oil accumulation, which can impair oil quality (Dag et al. 2009).

The concept of a nutrient budget to determine the right fertilizer rates combined with the right timing, placement, and source to guide efficient nutrient management has gained widespread acceptance. To make a nutrient budget, first the total nutrient demand of the orchard for a year is calculated by adding together the estimated nutrient demand of the crop and the estimated nutrient demand for tree growth. Then the necessary supply of nutrients to meet that demand is calculated, taking into account the fertilizer delivery efficiency factor, which ranges from 0.7 to 0.9 (70 to 90%) in many orchards. The right fertilizer rates can then be established by deducting from the necessary supply any nutrient credits from nonfertilizer sources like soil mineralization, leaf litter, OMA, cover crops, and irrigation water. These are the budget equations:

(1) Demand = Crop + Tree

(2) Supply = Demand ÷ Efficiency factor

(3) Rate = Supply - Credits

Calculation of an annual nutrient budget

Because orchards vary in yield due to age, planting density, and growing conditions, all calculations of N and K demand are based on expected yield expressed as N or K per 1,000 pounds of fruit weight. This requires that growers make a reasonable estimate of their yield potential at the beginning of the year, then use that estimate to calculate the upcoming season's demand. This initial estimate can be reassessed throughout the year and fertilizer rates adjusted if the yield appears to be better or worse than predicted.

The following case study shows the calculation of an annual N and K budget for a hypothetical orchard based on the Arbequina orchard referenced in tables 8.4 and 8.5: Leaf tissue analysis

for the previous year showed the orchard to be sufficient in leaf N at 1.6 percent and leaf K at 0.9 percent. The orchard is intensively farmed with irrigation and fertigation. The N and K removed the previous year in fruit is 3.41 and 8.36 pounds per 1,000 pounds of fruit, respectively (from table 8.4); the estimated targeted crop yield for the coming growing season is 10,000 pounds per acre; the N and K removed in prunings the previous year is 2.48 and 2.35 pounds per 1,000 pounds of prunings, respectively (from table 8.5; we assume annual tree growth nutrient demand is equivalent to nutrients reported for tree prunings), and the anticipated annual tree growth for the coming growing season is 5,000 pounds per acre. The efficiency factor for fertilizer delivery in this orchard is 0.7 for N and 0.9 for K. There is only one fertilizer credit, which is 10 pounds of N per acre from nitrate in the irrigation water.

The calculations are as follows:

(1) Demand = Crop + Tree

Crop N = 3.41 lb N ÷ 1,000 lb fruit
× 10,000 lb/ac = 34 lb N/ac

Crop K = 8.36 lb K ÷ 1,000 lb fruit
× 10,000 lb/ac = 84 lb K/ac

Tree growth N = 2.48 lb N ÷ 1,000 lb tree growth
× 5,000 lb/ac = 12 lb N/ac

Tree growth K = 2.35 lb K ÷ 1,000 lb tree growth
× 5,000 lb/ac = 12 lb K/ac

N demand = 34 lb N/ac + 12 lb N/ac = 46 lb N/ac

K demand = 84 lb K/ac + 12 lb K/ac = 96 lb K/ac

(2) Supply = Demand ÷ Efficiency factor

N supply = 46 lb N/ac ÷ 0.7 = 66 lb N/ac

K supply = 96 lb K/ac ÷ 0.9 = 107 lb K/ac

(3) Rate = Supply - Credits

N rate = 66 lb N/ac - 10 lb N/ac from
irrigation nitrate = 56 lb N/ac

K rate = 107 lb K/ac - 0 lb K/ac
in credits = 107 lb K/ac

K_2O rate = 107 lb K/ac
× 1.2047 = 129 lb K_2O/ac

Based on the preceding calculations, the right rate to apply is 56 pounds of N per acre and 107 pounds of K per acre (129 lb K_2O per ac).

Nutrient deficiencies

Nutrient deficiencies impair orchard productivity over time. Identifying nutrient deficiencies as soon as possible is important so that corrective action can be taken to avoid yield loss. Olive orchards with sufficient nutrition will maintain tree health and more easily resist pests, disease, and other abiotic stress. Knowledge and careful documentation of soil conditions, irrigation water, and fertilizers provide insights to help identify and correct specific nutrient deficiencies. Several nutrients are at risk of deficiency in olive orchards.

Nitrogen

In olive leaves, N deficiency is readily identified by smaller-than-usual size and yellowing (fig. 8.10). Olive trees deficient in N are smaller than N-sufficient trees, with sparse foliage and a light crop. Individual shoots may be less than 8 inches (20 cm), with very little shoot growth and dieback of some shoots. The fruit appears normal, but there is less of it. In mature trees, which have limited new growth, symptoms may occur uniformly throughout the tree. If almost all trees in an orchard are N deficient, symptoms may be difficult to discern as there will be few healthy trees for contrast. The deficiency can be corrected by applying N fertilizers.

Phosphorus

P deficiency in olive orchards is very rare. Olive tree uptake of P is very efficient due to an extensive root system and low demand for the nutrient. Soil P mobility is very low, and most uptake occurs when roots grow toward P in soils. Fertigation offers an opportunity to increase the potential for P uptake. No P fertilization is recommended as long as leaf P concentrations are at or above 0.1 percent. Many OMA contain sufficient amounts of P, so there is a limited need to apply P fertilizers.

Potassium

Leaves deficient in K become pale in the same way that leaves deficient in N do, show apical necrosis, and show reduced leaf size and shoot growth (fig. 8.11). Trees deficient in K can resemble a weeping willow, and branches appear to lack strength. Shoots display short internodes with short shoot growth and branchlet defoliation. Fruit color is normal. Olive has a high demand for K; while a deficiency can occur in any region, historically it has been more prevalent in areas with K-fixing soils, such as heavy clays, and sandy soils. Young, vigorously growing trees are particularly susceptible. Correction of K deficiency in trees may take more than a year.

Calcium and magnesium

Ca is an important constituent of plant cell walls. As an activator for many enzymes involved in energy transfer and growth processes, Mg is the central element in the structure of chlorophyll with an essential role in photosynthesis. Deficiencies of Ca and Mg have not been

Figure 8.10

The lower leaves on this olive shoot are paler in color, a sign of N deficiency. *Photo:* Louise Ferguson.

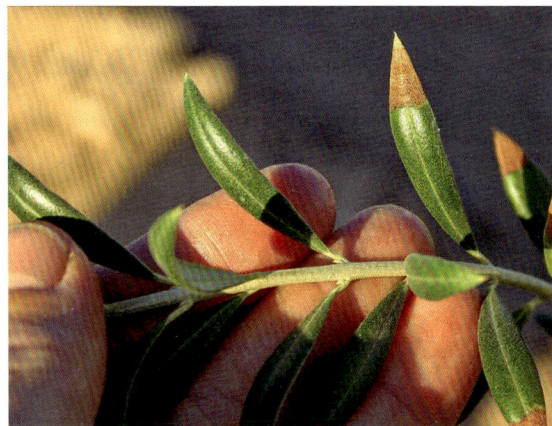

Figure 8.11

Typical symptoms of K deficiency are dead leaf tips or margins, a light green leaf color, and branch dieback. *Photo:* Louise Ferguson.

observed in California olive orchards. Mg is usually found in large quantities in soil solution, so deficiencies are rare. Mg deficiency may be caused by high concentrations of K, Ca, and NH_4^+, which compete for plant uptake in soil solution.

Iron

Fe is involved in the production of chlorophyll and is required for certain enzyme functions. A deficiency of the nutrient, known as Fe chlorosis, can occur in highly calcareous, or chalky, soil with a pH above 7.5 or in heavy, poorly drained soil, especially during a cold, wet spring. Leaf symptoms of chlorosis include yellowing of the leaf blade with green veins, accompanied by a decrease in leaf size, especially in terminal leaves. The chlorosis may disappear as the growing season progresses.

Increasing the availability and uptake of Fe can be achieved by lowering the soil pH with a banded application of elemental S or by injecting sulfuric acid ($H2SO4$) into the irrigation system. This will create a zone of low pH and increase the solubility of Fe, as well as some other nutrients, such as Zn. In soils with a high pH and high residual lime, foliar applications of a soluble Fe source or injection of a soluble Fe chelate directly into the microirrigation system can be effective.

Manganese, zinc, and copper

Because Mn is important in photosynthesis and activates many enzymes, a deficiency can greatly reduce photosynthesis and crop yield. An essential constituent of a large number of enzymes and proteins, Zn plays a key function in flowering and early spring growth, maintenance of membrane integrity, and hormone synthesis. Zn is essential in maintaining the plant hormone auxin, which plays a key role in cell elongation and growth. Additionally, Zn is required for chlorophyll synthesis and function, and is critical in plant defenses against environmental stresses. Cu is a component of many enzymes and plays a role in energy metabolism. Because Cu is immobile in the phloem, its deficiency occurs in the youngest leaves. Some pesticides applied to olive trees, such as copper sulfate fungicides, contain Cu.

Little information is available about the requirement of Mn, Zn, and Cu for olive.

Management practices aimed at lowering pH could make these elements more available. However, foliar applications or direct injection of these elements as sulfates or chelates may be more economical to correct a potential deficiency. However, deficiencies of these elements in olive are uncommon.

Boron

B is required for new cell division and cell wall growth as well as fruit set, and it may have a role in maintaining cell membrane integrity. Reproductive structures have a higher demand for B and are particularly sensitive to a deficiency. Olive is considered to have a high B requirement and is more tolerant to excess levels of B in soil solution than other tree crops. B availability decreases under drought conditions and as soil pH increases, especially in calcareous soils. B deficiency shows as chlorosis on the leaf margins, formation of shoots known as witches'-broom, and fruit malformations. Visual symptoms of B deficiency may be confused with K deficiency (fig. 8.12). Correcting B deficiency is relatively easy to do with fertilizer applied to the soil. Foliar application is preferable in high-pH soils. Foliar application of B immediately prior to flowering has been shown to increase fruit set and yield under conditions in which no leaf deficiency symptoms were apparent (Perica et al. 2001).

Figure 8.12

Symptoms of B deficiency are leaves with a dead tip and a yellow band, but still green at the base. *Photo:* Joseph H. Connell.

Nutrient toxicities

Toxicities of B, Na, and Cl accumulate over time as the trees transpire water. It is ultimately the accumulation of these elements or salts in the plant and in the root zone that decrease productivity. Salinity damage in olive is largely a consequence of the quality of the irrigation water and the duration of irrigation with poor-quality water.

Boron toxicity

B is present in excessive amounts in many olive production areas in the world. Saline- and B-contaminated ground and surface irrigation water are increasingly used for olive irrigation. B-affected soils are reclaimed only slowly and with great difficulty. Irrigation water is the most significant source of B for most olive production regions, and long-term irrigation with water exceeding 1 ppm B can result in the accumulation of B in olive tissue and the development of toxicity symptoms, including premature fruit drop. Boron toxicity symptoms initially appear as leaf tip and marginal chlorosis that becomes necrotic, eventually covering the entire leaf and causing leaf drop.

Sodium and chloride toxicity

Compared to most tree crops, olive is very tolerant of saline and sodic soils, which usually contain high levels of Na and Cl. Leaf analysis is the most useful tool for diagnosing salt injury in trees. Soil and water analyses may also be needed to determine the source of salts or whether a problem exists before planting. Problems with Cl toxicity often occur in orchards planted on saline soils that are not fully reclaimed or when high-Cl irrigation water is used. The soil should be completely reclaimed before planting because it is much more difficult to correct a salinity problem after planting. Cl toxicity sometimes occurs after high concentrations of fertilizers containing Cl, such as KCl, are applied to poorly drained soils. Excesses of both Na and Cl produce leaf tip burn that look similar and progress down the leaf. In severe cases, the necrotic area at the tip increases during the growing season so that by late summer the apical half of the leaf, or more, is necrotic.

References

Brady, N., and R. Weil. 2017. The nature and properties of soils. Columbus, OH: Pearson Education Inc.

Dag, A., E. Ben-David, Z. Kerem, A. Ben-Gal, R. Erel, L. Basheer, and U. Yermiyahu. 2009. Olive oil composition as a function of nitrogen, phosphorus and potassium plant nutrition. Journal of the Science of Food and Agriculture 89:1871–1878. https://doi.org/10.1002/jsfa.3664

Erel, R., Z. Kerem, A. Ben-Gal, A. Dag, A. Schwartz, I. Zipori, L. Basheer, and U. Yermiyahu. 2013. Olive (Olea europaea L.) tree nitrogen status is a key factor for olive oil quality. Journal of Agricultural and Food Chemistry 61:11261–11272. https://doi.org/10.1021/jf4031585

Fernández-Escobar, R. 2010. Fertilisation. In D. Barranco, R. Fernández-Escobar, and L. Rallo, eds., Olive growing. Translated by Susan E. Hovell, and William A. Hovell. Pendle Hill, NSW, Australia: RIRDC. Originally published as El cultivo del olivo. Madrid: Junta de Andalucia, Consejeria de Agricultura y Pesca, and Ediciones Mundi-Prensa, 2004.

———. 2018. Trends in olive nutrition. Acta Horticulturae 1199(35): 215–224. https://doi.org/10.17660/ActaHortic.2018.1199.35

Fernández-Escobar, R., R. Moreno, and M. García-Creus. 1999. Seasonal changes of mineral nutrients in olive leaves during the alternate-bearing cycle. Scientia Horticulturae 82(1–2):25–45. https://doi.org/10.1016/S0304-4238(99)00045-X

Freeman, M., K. Uriu, and H. T. Hartmann. 2005. Diagnosing and correcting nutrient problems. In G. S. Sibbett and L. Ferguson, eds., Olive production manual. 2nd edition. Oakland: UC Agriculture and Natural Resources Publication 3353. 88.

Gómez-Casero, M. T., F. López-Granados, J. M. Peña-Barragán, M. Jurado-Expósito, L. García-Torres, and R. Fernández-Escobar. 2007. Assessing nitrogen and potassium deficiencies in olive orchards through discriminant analysis of hyperspectral data. Journal of the American Society for Horticultural Science 132:611–618. https://doi.org/10.21273/JASHS.132.5.611

Haberman, A., A. Dag, N. Shtern, I. Zipori, R. Erel, A. Ben-Gal, and U. Yermiyahu. 2019. Long-term impact of potassium fertilization on soil and productivity in intensive olive cultivation. Agronomy 9:525. https://doi.org/10.3390/agronomy9090525

[IPNI] International Plant Nutrition Institute. 2012. 4R plant nutrition: A manual for improving the management of plant nutrition. Arlington: The Fertilizer Institute.

Jiménez-Moreno, M. J., and R. Fernández-Escobar. Response of young olive plants (*Olea europaea*) to phosphorus application. HortScience 51(9):1167–1170. https://doi.org/10.21273/HORTSCI11032-16

Khalsa, S. D. S., and P. H. Brown. 2017. Grower analysis of organic matter amendments in California orchards. Journal of Environmental Quality 46:649–658. https://doi.org/10.2134/jeq2016.11.0456

Meyer, R. D. 1996. Nitrogen on drip irrigated almond. Years of discovery (1972–2003). Modesto, CA: Almond Board of California. 291–292.

Molina-Soria, C., and R. Fernández-Escobar. 2012. A proposal of new critical leaf nitrogen concentrations in olive. Acta Horticulturae 949:283–286. https://doi.org/10.17660/ActaHortic.2012.949.41

Muhammad, S., S. Saa, S. D. Khalsa,, S. Weinbaum, and P. H. Brown. 2017. Almond tree nutrition. In R. Socias i Company and T. M. Gradziel, eds., Almonds: Botany, production and uses. Wallingford, UK: CABI.

Perica, S., P. H. Brown, J. H. Connell, A. M. S. Nyomora, C. Dordas, H. Hu, and J. Stangoulis. 2001. Foliar boron application improves flower fertility and fruit set of olive. HortScience 36(4):714–716. https://doi.org/10.21273/HORTSCI.36.4.714

Reidel, E. J., P. H. Brown, R. A. Duncan, R. J. Heerema, and S. A. Weinbaum. 2004. Sensitivity of yield determinants to potassium deficiency in 'Nonpareil' almond (Prunus dulcis (Mill.) D.A. Webb). Journal of Horticultural Science & Biotechnology 79:906–910. https://doi.org/10.1080/14620316.2004.11511864

Rosecrance, R. C., and W. H. Krueger. 2020. Total fruit nutrient removal calculator for olive in California. https://rrosecrance.yourweb.csuchico.edu/Model/OliveCalculator/OliveCalculator.html

Smith, F. W., and J. F. Loneragan. 1997. Interpretation of plant analysis: Concepts and principles. In D. J. Reuter, and J. B. Robinson, eds., Plant analysis: An interpretation manual. Collingwood, VIC 3066 Australia: CSIRO Publishing.

Ulrich, A. 1952. Physiological bases for assessing the nutritional requirements of plants. Annual Review of Plant Physiology 3:207–228. https://doi.org/10.1146/annurev.pp.03.060152.001231

Zipori, I., R. Erel, U. Yermiyahu, A. Ben-Gal, and A. Dag. 2020. Sustainable management of olive orchard nutrition: A review. Agriculture 10(1): 11. https://doi.org/10.3390/agriculture10010011

Zipori, I., U. Yermiyahu, R. Erel, E. Presnov, I. Faingold, A. Ben-Gal, and A. Dag. 2015. The influence of irrigation level on olive tree nutritional status. Irrigation Science 33:277–287. https://doi.org/10.1007/s00271-015-0465-5

9

Canopy Management

Richard Rosecrance, Professor, California State University, Chico

William Krueger, UC Cooperative Extension (UCCE) Farm Advisor Emeritus

Louise Ferguson, Professor of Extension in the Department of Plant Sciences at UC Davis

Luke Milliron, UCCE Orchard Systems Advisor, Butte County

Highlights

- The objectives of pruning are to achieve optimal yields of high-quality olive oil; facilitate orchard operations, pest management, and harvesting; and achieve consistent yields by mitigating alternate bearing.

- An olive tree canopy should be managed to maximize canopy exposure to at least 50 percent full sun averaged over the day, as lower light exposure decreases oil quantity and quality.

- Mechanical pruning is typically conducted in high-density and super-high-density orchards because of the high labor costs of hand-pruning.

- Preferred timing for hedging is early spring to produce moderate vegetative growth for the following year. The preferred timing for topping is mid- to late summer to minimize regrowth that interferes with harvest and shades the lower canopy the following year.

- Heading cuts on branches greater than ¼ inch (6.4 mm) in diameter should be minimized because these cuts result in vigorous vegetative regrowth at the expense of reproductive growth.

- A hedgerow olive model predicts that the highest yields and fruit quality will be achieved when the canopy vertical depth is equal to the free alley width.

- The model predicts yields can be increased by sloping both sides of the canopy. The slope increases light interception down the canopy and allows for a closer row spacing, resulting in higher yields.

anopy management, critical for producing high-quality fruit, is defined as the manipulation of tree form by cultural practices, such as training and pruning management. Tree spacing density as well as tree shape and porosity (light infiltration through the canopy) determine canopy light interception, which affects vegetative growth, flowering, fruit yield, and olive oil quality. Furthermore, as olives are borne on the previous year's growth, canopy management includes removing older limbs so that younger ones replace them and fruit production is maintained.

Growing olive in hedgerows

Tree canopy size and structure are two primary factors in growing olives for oil production. Larger canopies typically produce more olives than smaller canopies; however, a tree structure that allows light distribution within the canopy is critical for fruit production. Olives are produced on 1-year-old shoots exposed to sunlight. In unpruned trees, most of the production is concentrated in a shell of new, light-exposed shoots around the canopy periphery (fig. 9.1). In contrast, trees in many modern oil olive orchards are pruned to control tree size, remove unwanted branches, and create a hedge shape. In super-high-density (SHD) and high-density (HD) orchards, the canopies of the trees form a dense fruiting wall (fig. 9.2).

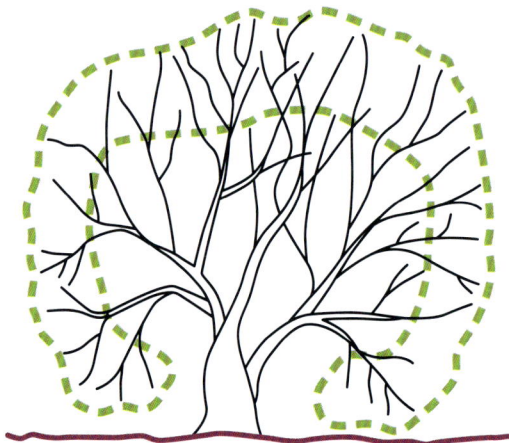

Olive oil production economics make mechanical harvesting almost mandatory. Hedgerow systems with trees more closely spaced within the row than between the rows facilitate mechanical harvesting because the trees can be managed as a fruiting wall rather than as individual trees. Growers typically harvest oil olive orchards using a canopy contact harvester or a trunk-shaking harvester. Over-the-row harvesters are the most common type of canopy contact harvester; they straddle the olive row, shaking the canopy with individual rods or bow rods to remove the fruit. Over-the-row harvesters impose height and width restrictions on the canopy, which influence light interception, yield, and oil quality.

Varieties suited for hedgerow systems

Olive varieties vary in their suitability for planting in SHD systems. Researchers have reported two main limitations for the adoption of new varieties in SHD systems: excessive vegetative growth (Rallo et al. 2018); and incompatible tree growth habit, characterized by the presence of a few thick branches that typically remain vegetative following hedging and can interfere with over-the-row harvesters (Rosati et al. 2013). Many trials have been initiated to evaluate local varieties for their suitability for SHD orchards (see chapter 4, Establishing a Super-High-Density Olive Orchard).

Figure 9.1

The dotted line indicates the productive area of an unpruned olive tree.

Figure 9.2

The canopies of SHD orchards (A) and HD orchards (B) are pruned to a hedge shape. The objective is to get maximum sunlight exposure on the east and west vertical sides of the rows oriented north to south. *Photos:* Richard Rosecrance

Briefly, varieties desirable for SHD systems have the following characteristics (Rosati et al. 2013, 2017):

* a growth habit producing many branches and inflorescences per node on small lateral branches
* early production after planting (precocity), followed by high and consistent yields of high-quality oil
* low vegetative vigor and compact growth to facilitate mechanical harvesting

Thus far, only three varieties—Arbequina, Arbosana, and Koroneiki—have been widely planted in SHD orchards. Other low-vigor varieties, such as Oliana and Sikitita, are being evaluated.

In contrast to SHD systems, a wide range of varieties have been planted in HD systems. The main HD varieties in Argentina are Arauco, Barnea, Coratina, Frantoio, Hojiblanca, Manzanilla de Sevilla, and Picual (Gómez-del-Campo 2010); in Australia, they are Barnea, Coratina, Correggiola, Frantoio, Koroneiki, Manzanillo, and Picual (Torres et al. 2017); and in California, Ascolano, Frantoio, Hojiblanca, Manzanillo, Mission, Sevillano, and Picual are among those commonly grown.

As hedgerow production systems develop, it is likely that varieties adapted to particular densities, planting patterns, and harvesting methods will emerge. Successful variety adaptation will require concerted efforts between researchers and growers to evaluate many aspects of hedgerow structure and function.

Methods to reduce scion growth

Ideally, olive growers could select varieties that fill the allotted space of their orchard and thereafter grow only enough vegetation each year to bear the next crop. Unfortunately, no such varieties exist, but there are a number of growth control techniques available that the orchardist can apply. There is no single, generally applicable way to control vegetative growth. Olive growers must blend the best techniques available based on their site, variety, and costs of intervention.

Regulated deficit irrigation

Regulated deficit irrigation (RDI) is a common practice in tree fruit crops. Basically, irrigation is reduced to below crop water requirements during the most drought-resistant phenological stages, with minimal or no reduction in fruit production. In olive, RDI is typically applied around pit hardening; in California, this occurs between 6 and 9 weeks after full bloom. Shoot growth typically stops during pit hardening, as carbohydrates are used to support fruit growth. Studies have been mixed on the effects of RDI on olive oil production, with some indicating decreased oil production and others indicating increased production (Rosecrance et al. 2015). More research is needed on timing, length, and severity of RDI under California conditions. For more information on RDI, see chapter 7, Irrigation Management.

Dwarfing rootstocks

The advent and widespread use of size-controlling rootstocks have been critical in many fruit crops to the adoption of compact high-density orchards. Olive researchers have been evaluating rootstocks to reduce scion vigor for many years. Indeed, it was over 60 years ago that Hartmann identified two rootstocks, *Olea verrucosa* and *O. ferruginea*, that significantly reduced scion growth (Hartmann 1958a). Unfortunately, there seems to have been no further research on these selections.

Recently, the interest in dwarfing rootstocks has increased to allow for a greater number of varieties to be planted in SHD systems. Most researchers have reported that dwarfing in olive may arise from reduced root hydraulic conductance (Gascó et al. 2007). Root hydraulic conductance is the ability of roots to conduct water from the root surface to the xylem in the stem. The ability of dwarfing rootstocks to reduce vegetative growth while maintaining optimal fruit growth, however, needs further study.

In Spain, Arbosana and Limoncillo rootstocks reduced the vigor of Arbequina I-18 and promoted higher productivity (Tous et al. 2011). Similarly, Nana, Redonda Mora de Toledo, and Verdal de Manresa rootstocks reduced Arbequina I-18 by about 25 percent compared with the own-grafted Arbequina I-18 (Romero et al. 2014). Potential dwarfing rootstocks identified from research trials include Habichuelero, Hojiblanca, Picual, and Redondilla de Logroño (Therios 2009); Leccino Dwarf (Rugini et al. 2016); and Nana, Redonda Mora de Toledo, and Verdal de Manresa (Romero et al. 2014).

Mutagenesis (the production of genetic mutations) may prove an effective method to develop dwarfing olive rootstocks. Exposing flower buds to gamma radiation can cause mutations in the seed that change the plant from a diploid, having two sets of chromosomes, to a tetraploid, having four sets of chromosomes (Rugini et al. 1996). Tetraploid Leccino and Frantoio plants were significantly shorter and their trunk diameter smaller than the standard diploid genotypes (Rugini et al. 2016). Furthermore, when used as a rootstock, the tetraploid Leccino significantly reduced the size of the scion. These results appear promising but need to be substantiated in the field.

Trunk constriction

Another method to reduce olive shoot growth, aside from dwarfing rootstocks, is to constrict the trunk (Tombesi et al. 2018). Plastic straps, approximately 1.5 inches (3.8 cm) in diameter, applied in December and removed a year later significantly reduced midday stem water potential as well as trunk and vegetative growth (Tombesi and Farinelli 2016). Indeed, the constriction of the trunk in the research trial caused almost complete cessation of vegetative growth, which severely reduced yield during the following year. More research needs to be conducted to evaluate constriction for shorter periods of time; it could be a promising technique for regulating excessive vigor without the need of dwarfing rootstocks.

Plant growth regulators

The use of plant growth regulators can also reduce olive tree size. Soil applications of a gibberellin 3 (GA_3) inhibitor, uniconazole, changed the growth habit of vigorous, upright Barnea, Leccino, and Picholine varieties to a droopy and willowy habit, suitable for SHD orchards (Avidan et al. 2011). Moreover, uniconazole significantly decreased tree growth in Koroneiki, although yields also appreciably decreased. This reduction in yields may have occurred because the chemical did not affect the number of new leaves on a shoot, resulting in a dense canopy and less light interception (Schneider et al. 2012). More work needs to be done to evaluate plant growth regulators to control the tree size of olive.

Limiting nitrogen

There are conflicting results regarding the effects on olive vegetative growth from limiting nitrogen (N). One study found that limiting N fertilization reduced vegetative growth in olive orchards planted in shallow, infertile soils (Hartmann 1958b). Olive vegetative growth, however, did not differ when applying 0, 36, 70, and 135 pounds per acre (0, 40, 78, and 151 kg per ha) of N in a mature HD Barnea orchard; only at an application of over 250 pounds per acre (280 kg per ha) did vegetative growth increase compared to the 0 N control (Haberman et al. 2019). Varying N fertilization levels did not affect shoot growth in an HD Picual orchard, and shoot growth increased only in the final year of a 3-year trial in an SHD Arbequina orchard (Centeno et al. 2017). It appears that N limitation may reduce vegetative growth in shallow, infertile soils but not in fertile soils. Thus, N limitation is currently not recommended as an approach to limit vegetative growth.

Light management

Over the past 20-plus years, olive orchard density has increased, spurring research on optimizing light environments in HD and SHD plantings. Yields in mature HD or SHD orchards are optimized when trees are spaced at the greatest density that still allows them to intercept adequate sunlight for annual shoot growth throughout the canopy periphery (Connor et al. 2014). Insufficient light exposure may lead to inferior flowering and may reduce olive oil quality. The current interest in light effects on flowering and oil quality of olive trees also stems from the introduction of light, frequent mechanical pruning in modern olive orchards, which makes a denser hedge.

Adequate light in the canopy

Olive tree shoots are independent units. There is only minimal translocation of carbohydrates and nutrients from one main branch to another, particularly during summer (Granado-Yela et al. 2011). The fruit rely primarily on nearby leaves for their carbohydrate needs. Therefore, maximizing light penetration into the canopy to support photosynthesis of nearby leaves is critical for optimal yield.

The relationships between light intensity, yield, oil quality, and vegetative growth are increasingly important as orchard tree density increases. In low-density, hand-pruned olive orchards, tree canopies generally have high light penetration, and shading effects are relatively unimportant on fruit growth and oil content (Acebedo et al. 2000; Connor et al. 2014). Mechanically pruned trees produce a very dense canopy because multiple shoots grow near the cuts, decreasing light penetration behind the cuts (fig. 9.3). In HD and SHD orchards, managing light levels of the vertical hedgerow wall is critical to maximizing bloom and yield.

Light impacts on yield and quality

By producing carbohydrates through photosynthesis, sunlight strongly influences flower and fruit production. Sunlight is required for flower bud initiation, induction, and differentiation (Rapoport et al. 2013). Olive fruit and oil yield increase linearly with increasing light levels on bearing shoots exposed to up to 50 to 60 percent full sun (calculated over the whole day), then level off or decrease at very high light levels (Caruso et al. 2017; Connor et al. 2009). Shaded fruit (that is, receiving less than 50% full sun) develop at a slower rate, are smaller at harvest, and contain less olive oil than outer canopy fruit. Therefore, an olive tree canopy should be managed to maximize the number of bearing shoots exposed to 50 percent or more full sun. SHD orchard management practices have resulted in fruit located mostly in the upper canopy. Pruning

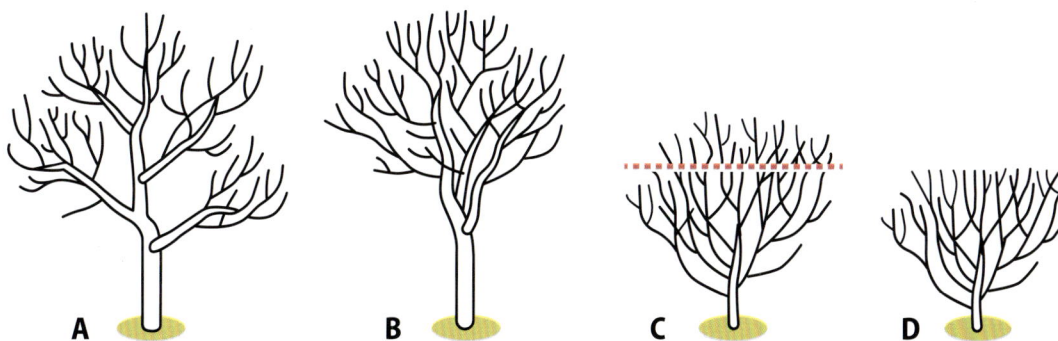

Figure 9.3

A hand-pruned tree (A) has the highest light penetration; less light penetrates the canopy of a mechanically pruned tree (B). The canopy of an untopped tree (C) is more open and thus receives more light than that of a topped tree (D).

techniques need to focus on increasing light levels, particularly in the lower canopy (fig. 9.4).

In an SHD Arbequina orchard study, the upper part of the canopy, over 6 feet (1.8 m), the middle part, from 3 to 6 feet (0.9 to 1.8 m), and the lower part, below 3 feet (0.9 m), averaged about 45, 35, and 28 percent full sun, respectively, and accounted for 54, 30, and 16 percent of total tree yield, respectively (Larbi et al. 2015). In another study, the upper canopy also typically produced the largest fruit with the highest percentage of oil (Cherbiy-Hoffmann et al. 2012).

An SHD orchard planted at 670 trees per acre (1,656 trees per ha) at a spacing of 13 by 5 feet (4 by 1.5 m) intercepts about 50 percent of the sunlight at midday. Some SHD growers have decreased the alley width to 12 or even 11 feet (3.7 or 3.4 m) to intercept more light per acre. In California, SHD growers typically plant at a spacing of 12 by 5 feet (3.7 by 1.5 m) because they have found that decreasing the alley width increases light interception per acre and subsequently fruit yield. In a 12-year Spanish study, an orchard planted at 912 trees per acre (2,253 trees per ha) at a spacing of 11.7 by 4.1 feet (3.6 by 1.3 m) intercepted more light and the fruit accumulated more oil per acre than an orchard planted at the traditional spacing (Díez et al. 2016). At this high density, however, most production was concentrated in the upper part of the canopy.

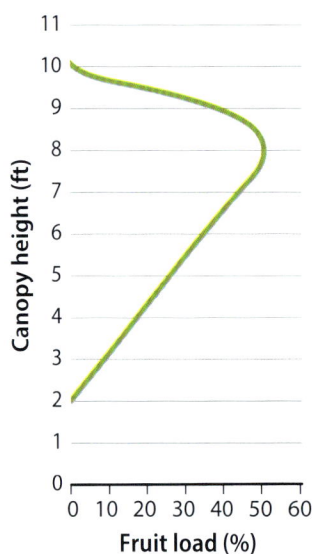

Figure 9.4

Fruit distribution relative to canopy height is a key concern in SHD orchards. *Photo:* Richard Rosecrance.

Shading can decrease inflorescence number, flowers per inflorescence, fruit set, and fruit size (Rapoport et al. 2012). Olive trees produce perfect (hermaphrodite) and imperfect (staminate) flowers; only perfect flowers have an ovary and can produce fruit. Shaded canopies typically produce more imperfect flowers, which do not develop into fruit. Research has shown that the period prior to bloom is critical for floral differentiation and fruit yield. Shading and water stress prior to bloom decrease flower number and ovule development and increase the number of imperfect flowers.

Sunlight not only influences flowering and fruit set, but it also enhances olive oil quality. Branches borne on the outer canopy typically produce fruit with more stable oil and higher polyphenol and fatty acid levels than shaded branches. Fruit grown in high light environments also contain higher palmitic and linoleic acid levels than shaded fruit (Caruso et al. 2017).

Light impacts on vegetative growth

In olive, flower buds develop in spring on shoots that grew the previous year. Therefore, the previous year's shoot growth strongly influences the current year's fruit yield. Because longer shoots have more potential flowering sites, the previous season's vegetative shoot growth is among the strongest determinants of yield. Fruiting branches partition their carbohydrates primarily into fruit growth at the expense of vegetative growth. This partitioning is more pronounced under shaded conditions, resulting in little or no vegetative growth, few flowering sites in the following year, and reduced oil quality and yield (Cherbiy-Hoffmann et al. 2013).

Light levels affect the physiology of shaded leaves. An artificial shade experiment (60% shade) found that shaded Arbosana leaves had a greater surface area and were thinner than leaves exposed to the sun (Ajmi et al. 2018). In addition, shade changed the leaf structure in Arbosana (Ajmi et al. 2018) and Arbequina trees (Larbi et al. 2015), reducing the thickness of the palisade and spongy parenchyma tissues. Stomatal density, net photosynthetic rate, stomatal conductance, and transpiration rate were also reduced by shade (Ajmi et al. 2018). These morphological changes in the leaf are permanent; exposing shaded leaves to full sun can result in sunburning of the leaves.

Partial shade, however, can increase photosynthetic rates. Photosynthesis in olive saturates at around 1000 µmol (photon) m^{-2} s^{-1} (Proietti and Famiani 2002), which is about 50 percent full sun in summer in California. Sofo et al. (2009) found that Coratina trees exposed to 34 percent artificial shade (that is, 66% full sun) had higher rates of photosynthesis and lower stomatal conductance than trees exposed to full sun. In other words, shade provided more efficient light use because less of the light was dissipated as heat.

Olive model yield estimates from orchard design

An olive model to better understand and estimate the relationship between light interception and yield, and to improve orchard design, was developed in Spain with the Arbequina variety (fig. 9.5) (Connor et al. 2014). This model uses canopy depth (*d*, tree height minus the height of the leaf-free area beneath the canopy) and free alley width (*a*, row spacing minus the canopy width) to predict yield. According to the model, the highest yield and fruit quality are achieved when the canopy depth is equal to the free alley width, that is, $d \div a = 1$.

The model, which was developed after years of evaluating olive yield at different in-row and between-row intervals for Arbequina, functions between latitudes 30° and 40°. It applies only to orchards oriented north–south; the canopy is somewhat symmetrical due to morning sun on the east side and afternoon sun on the west side (an east–west planting develops an asymmetrical canopy due to greater sun exposure on the south side).

A more recent study of SHD orchards indicated that the highest yields resulted when the canopy depth was closer to twice the free alley width rather than equal to it; however, in these very high density plantings, the quality of fruit, especially in the lower canopy, suffered (Díez et al. 2016). The best ratio of canopy depth to free alley width for a particular orchard may not be $d \div a = 1$, but the model provides a useful framework to evaluate the effects of orchard design on fruit yield and quality.

The model also predicts that yield can be increased by applying a slope to the canopy, usually between 5 and 15 degrees. Many growers apply a straight cut when hedging; however, an angled cut slightly increases the length of the canopy area exposed to sun and improves light exposure in the lower canopy (fig. 9.6). Connor et al. (2012) found that applying a slope of about 5 to 10 degrees to the hedging cut increases the area of the fruiting wall by 4 to 8 percent. This additional area exposed to sun allows for a closer row spacing and, therefore, increased yields. A sloped cut, however, may result in vigorous regrowth at the top of the canopy (producing a mushroom top), which decreases light penetration. Research is ongoing to evaluate sloping canopies and further refine, improve, and confirm the predictability of the model.

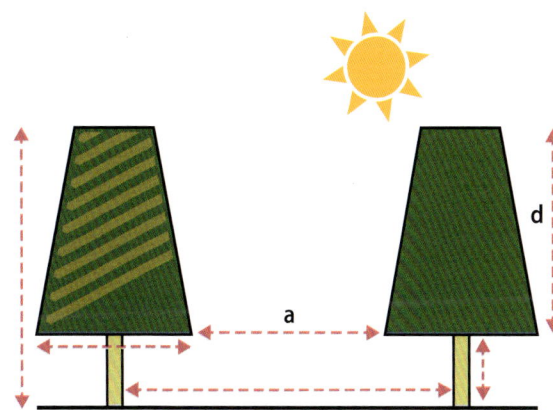

Figure 9.5

According to a model developed in Spain, the highest yield and fruit quality are achieved when canopy depth, *d*, is equal to the free alley width, *a*. *Source:* Adapted from Connor et al. 2014.

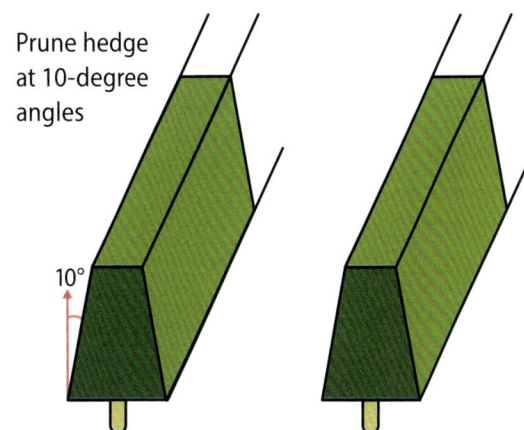

Prune hedge at 10-degree angles

10°

Figure 9.6

Sloping a hedge 10 degrees increases the canopy area exposed to sun and improves light exposure in the lower canopy.

Types of pruning cuts

How a tree responds to pruning depends to a great extent on the type of cut: thinning or heading (fig. 9.7). A thinning cut removes a branch at its point of origin, whereas a heading cut removes only the distal portion of the branch. With thinning cuts, the growth response is distributed among the remaining branches. Thinning cuts improve light distribution within the canopy, control tree size, and minimize excessive vegetative growth (Gucci and Cantini 2000). Heading cuts, which are more invigorating than thinning cuts, result in multiple strong, vigorous, upright shoots near the cut (fig. 9.8).

Two main factors cause the vigorous growth after heading cuts: the terminal bud, removed by the pruning, was producing a natural growth-inhibiting substance; and the bottom half of a branch is usually greater in diameter and often contains 2 to 3 times more carbohydrates per unit of dry weight than the upper half, which means that the remaining lateral buds on a headed branch have a large supply of carbohydrates that supports regrowth of the tree. The response is proportional to the severity of the cut. Mechanical pruning primarily results in heading cuts, which usually remove less wood and stored carbohydrates than thinning cuts.

Goals of pruning

Once the hedgerow is established through training, pruning is used to ensure light penetration within the canopy for optimum yield and quality and to maintain tree size for mechanical harvesting. Pruning in early spring during the "on" year can reduce alternate bearing: it will reduce the crop in the "on" year, improve fruit size, and promote vegetative growth and flower buds for the crop in the "off" year. Pruning can also reduce pests and diseases, for example, by removing olive knot inoculum in infected wood and by creating a more open tree canopy, which increases black scale insect mortality from heat exposure.

Pruning in SHD and HD orchards can be done by hand or mechanically. Mechanical hedging and topping are inherently nonselective and almost exclusively make heading cuts, resulting in a strong localized growth response proportionate to the severity of the cut. Hand-pruning consists

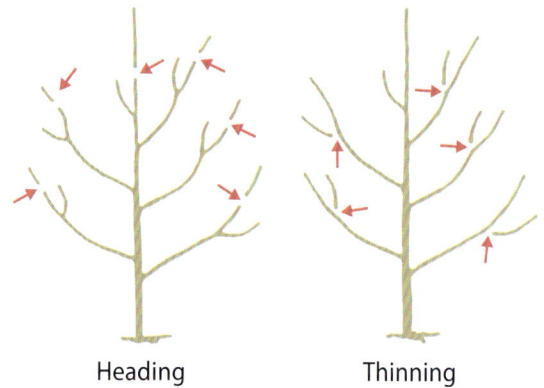

Heading   Thinning

Figure 9.7

A heading cut removes the end of a shoot or branch, whereas a thinning cut removes the entire shoot or branch.

Heading cut

First year   1 year later

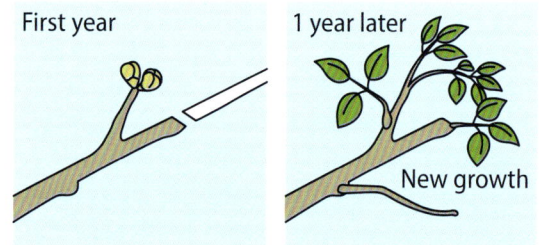

New growth

Thinning cut

First year   1 year later

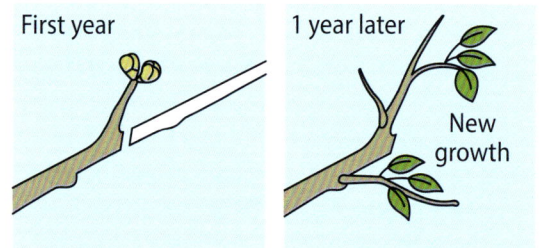

New growth

Figure 9.8

Heading and thinning cuts have different effects on subsequent growth.

primarily of thinning cuts, which do not stimulate a strong localized growth response and produce better light distribution.

Hand-pruning

Hand-pruning, which is inherently selective, offers some advantages over mechanical pruning. It can be focused on problem areas in the tree, as well as on generally increasing light distribution within the canopy. To open the canopy, pruners typically use thinning cuts to remove large branches (fig. 9.9). Most SHD growers

Figure 9.9

Hand-pruning uses thinning cuts to open up and increase light penetration into the canopy, and improve yield. *Photo:* Louise Ferguson.

mechanically prune to reduce high labor costs, but some rely on hand-pruning crews, particularly in orchards planted in low-vigor varieties, such as Arbosana, Oliana, and Sikitita. Crews move quickly through the orchard removing branches that protrude into the row middle, grow too high for the harvester, or block the harvester closure mechanism (skirting). Some growers are experimenting with self-propelled moving platforms for pruning crews, which can improve worker efficiency, decrease worker fatigue, and reduce costs.

Large lateral branches that grow into the tree alley can significantly reduce fruit removal by the harvester. Over-the-row harvesters force these branches against the neighboring tree, shielding the canopy from the bow rods, which results in poor fruit removal. Hedging these lateral branches does not fix the problem because they grow back (fig. 9.10). Hand-pruning is required to remove these branches with thinning cuts. Crews can use loppers or pole chainsaws to make the desired cuts.

Researchers compared mechanical pruning to hand-pruning in an SHD Arbequina orchard in Italy (Tombesi and Farinelli 2017). For the hand-pruning treatment, they used loppers to remove limbs that were determined not to be elastic enough to bend down into the harvester. They found that both mechanical pruning and selective hand-pruning removed similar amounts of material, but the mechanical treatment removed more of the 1-year-old shoots and thus reduced yield. Hand-pruning resulted in a larger

tree volume and better light infiltration lower in the canopy. However, no data on relative costs were presented, and it is unclear whether the additional cost of hand-pruning would be offset by increased yield.

Although hand-pruning techniques vary depending on many factors (for example, plant age, training system, crop load, soil fertility), there are some rules of thumb:

- Make large cuts before small ones; it allows a fresh estimate of light distribution within the canopy, minimizing needless cuts.
- Thin shoots in the upper part of the canopy.
- Eliminate suckers and water sprouts.
- Correct differences in vigor among branches by more severely pruning back less-vigorous branches.
- Identify the maximum tree height and width that will accommodate the tree harvester. Use thinning cuts to reduce tree size.
- When in doubt, remember that less pruning is better, especially when the trees are young. Minimal pruning encourages fruiting at an early age. After the trees fill their allotted space, prune only to increase light distribution within the tree and facilitate mechanical harvest.
- Use thinning cuts to remove branches that extend into the alley so they don't interfere with mechanical harvesters.
- Remove damaged or dead shoots and branches, for example, branches infected with olive knot disease.

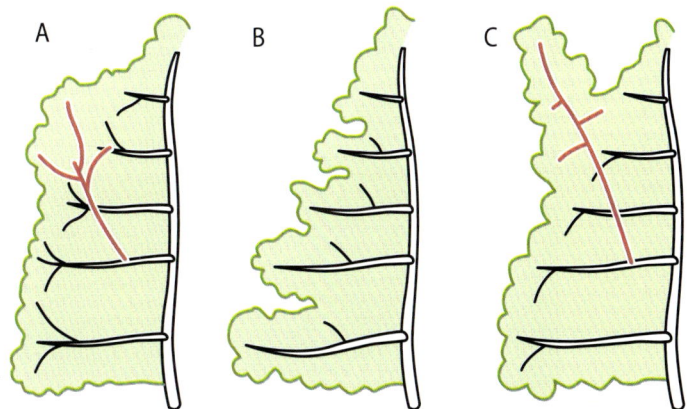

Figure 9.10

Branches growing into the alley need to be removed because they reduce harvesting efficiency (A). These branches are preferably removed with thinning cuts, that is, hand-pruned (B). If mechanically hedged, regrowth will occur at the cut surface and the problem will continue (C).

• To decrease the cost of hand-pruning, among the most expensive cultural practices, keep in mind that pruning should be rapid and simple and not require ladders. Using pole chainsaws can decrease the number of cuts required by taking out a thick branch in a single cut. Using pruning platforms can decrease worker fatigue and pruning costs.

Mechanical pruning

The goals of mechanical pruning are to stimulate fruiting wood and to maintain or reduce tree size to allow for efficient mechanical harvesting and orchard operations (fig. 9.11). However, mechanical pruning makes nonselective heading cuts, which promote vegetative growth that can decrease oil production and quality by limiting light penetration into the canopy. After mechanical pruning, dense shoots form on the outer edge of hedgerows, preventing light penetration into the inner canopy.

Timing of pruning

The time of year that a tree is pruned strongly affects its response. Winter pruning is more invigorating than summer pruning because there are fewer active growing points on the tree (Lodolini et al. 2018). After winter pruning, the relative amount of carbohydrates and nutrients available to each growing point is at a maximum, which results in vigorous regrowth. In California's colder winter climates, such as the Sacramento Valley,

Figure 9.11

A mechanical pruner tops the canopy of an HD orchard. The blades can be turned perpendicular to the ground to hedge the trees as well. *Photo: Richard Rosecrance.*

pruning treatments that thin and open the canopy should be avoided during the winter because they increase the risk of severe freeze injury in both SHD and HD orchards.

In summer, there is a greater number of active growing points and increased carbohydrate demand from the developing fruit, so, after pruning, there are fewer resources available to each growing point. Removing leaves and developing branches in summer also means that the large amounts of carbohydrates and nutrients they store in summer are not available for fruit and shoot growth. Summer pruning reduces the vigor of the tree, and it produces little vegetative regrowth from the cut branches.

Pruning in spring produces an intermediate growth response. Not all of the stored carbohydrates and nutrients have been translocated to new growth, so they are not lost during pruning, and there is time for regrowth to occur during the growing season before dormancy.

The limited research available on the seasonal timing of pruning in olive suggests that hedging should be done in early spring and topping in mid- to late summer. Hedging early in the year stimulates new growth and fruiting points. Topping in mid- to late summer results in minimal regrowth compared with topping earlier in the season (Lodolini et al. 2018). This regrowth usually is not large enough to impede mechanical harvesting.

If severe pruning is necessary, pruning in early spring in an "on" year is preferred to pruning in an "off" year (Albarracín et al. 2018). This timing promotes shoot growth that will result in improved fruit production in the "off" year. In contrast, severe pruning in early spring of an "off" year can result in excessive vegetative growth that will need pruning in the following year.

Severity of pruning

Severe pruning may be necessary if vigorous canopy development has resulted in excessive tree shading and reduced yield and the tree no longer fits into its allocated space. If heading cuts made during mechanical hedging and topping are too severe, however, the trees will respond with vigorous vegetative regrowth, and that growth will require severe hedging to increase light distribution, leading to a vicious cycle favoring vegetative growth over reproductive growth. Trees subjected to severe cuts may remain vegetative for 3 years before they flower and fruit again. Severe pruning cuts also promote the growth of unproductive water sprouts on older wood, losing potential resources for flowering.

Research has shown that heading cuts on branches greater than ¼ inch (6.4 mm) in diameter reduce flowering and cropping on the resulting shoot growth (Rosecrance and Krueger, unpublished data). Moreover, severe pruning of young trees decreases the shoot-to-root ratio, reducing resources for root growth. Consequently, pruning should be limited and gradual during the first 4 or 5 years after planting.

Frequency of hedging

How often olive trees require mechanical hedging depends largely on their vigor. Vigorously growing trees need more frequent hedging than slower-growing trees. Hedging should be frequent enough to avoid large heading cuts, which result in excessive tree vigor and reduced bloom.

Some olive growers have adopted a 3-year hedging cycle (a somewhat similar practice is common in walnuts): alternate rows are hedged on both sides in the first 2 years, and hedging is skipped in the third year. Other growers choose alternate-side hedging, a 2-year cycle: one side of the alley is hedged in the first year, the other side of the alley is hedged the next year. In alternate-side hedging, only one side of the tree canopy is cut each year, which typically results in a heavier yield on the nonhedged side. However, the fruit on the hedged side are typically larger than the fruit on the nonhedged side; additionally, the size of the fruit on the unhedged side may be improved because of light penetrating through the hedged side to the unhedged side. Alternate-side hedging tends to reduce alternate bearing because only one side of the tree heavily bears at a time (Trentacoste et al. 2018).

Even a 2-year hedging cycle may be too infrequent for vigorous HD olive orchards. Some HD growers lightly tip every year, cutting only small branches, to minimize the vegetative growth response, and then, as necessary, periodically hand-prune excessive interior canopy wood to increase light distribution within the canopy.

Costs of hand-pruning, mechanical pruning

Hand-pruning is significantly more expensive than mechanical pruning. A University of California table olive cost study (UC ANR 2016a) looked at hand-pruning in an HD planting of 180 trees per acre (450 trees per ha); it assumed comprehensive "ladder and lopper" pruning one year followed by less-detailed pruning with pole saws from the ground the next year. The average time required for the combined 2 years was about 25 hours per acre (62 h per ha), and the cost was $425 per acre ($1,050 per ha).

Experienced growers in California and Australia have reported that a less detailed hand-pruning in HD orchards, done primarily from the ground, can be accomplished in 4 to 10 hours per acre (10 to 25 h per ha). In SHD orchards planted in low-vigor varieties, a few growers can hand-prune with loppers at about 15 hours per acre (37 h per ha), but their pruning crews work in orchards that have never been mechanically pruned, and they remove only problem branches (C. Chavez, Cobram Estate Olives, personal communication).

A 2016 SHD oil olive cost study estimated mechanical pruning at $90 per acre ($222 per ha) (UC ANR 2016b). A commercial mechanical pruning firm (E. Nielsen, Erick Nielsen Enterprises, personal communication) in 2021 estimated custom mechanical hedging an SHD olive orchard takes 25 to 30 minutes and costs $100 to $120 per acre ($247 to $297 per ha), topping takes 12 to 15 minutes and costs $50 to $70 per acre ($124 to $173 per ha), and skirting costs another $40 per acre ($99 per ha).

Maximizing harvesting efficiency

For a financially viable oil orchard, the olive tree canopies must be managed to maximize fruit removal by a mechanical harvester. Harvesting efficiency is the percentage of the fruit on the tree that is harvested and delivered to the processing plant. The minimum average harvesting efficiency should be at least 85 percent (Miglietta et al. 2019). Leaving fruit on the tree obviously reduces harvesting efficiency, but it also significantly decreases the potential crop in the following season (fig. 9.12). Many factors, including variety, fruit maturity, crop load, and fruit size influence harvesting efficiency.

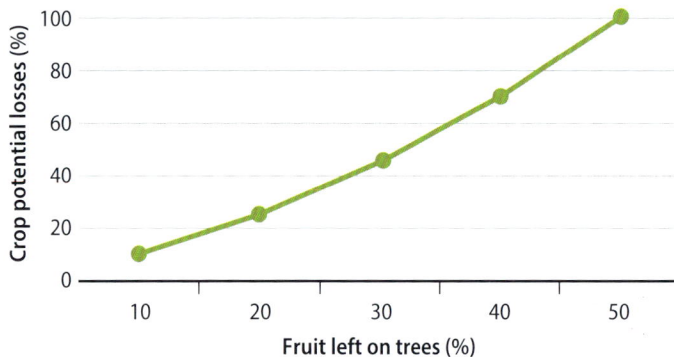

Figure 9.12

Losses in next season's crop potential occur when fruit is left on the trees. *Source:* Australian Government and RIRDC 2008.

Trunk-shaking harvesters

Trunk-shaking harvesters must transfer energy from the harvester through the trunk and branches to the fruit with enough force to detach it from the stem (fig. 9.13). The shorter the distance between the main trunk and the fruit-bearing branches, the stronger the shaking force. Short, upright branches more efficiently transmit energy to the fruit abscission zone than branches at angles greater than 45 degrees from vertical. Willowy shoots that cascade down from canopy branches do not efficiently transmit energy to the fruit, especially if the shoots are clumped together (Ferguson et al. 2010). The ideal shape

for trunk-shaking harvesters is a hedgerow tree 6 feet (1.8 m) wide and no more than 13 feet (4 m) tall with a lower canopy that is about 3 to 4 feet (0.9 to 1.2 m) from the ground. Trunk-shaking harvesters are suitable for HD but not SHD orchards.

Over-the-row harvesters

Over-the-row, or straddle, harvesters use rotating combs, spindles, or bow rods to shake the canopy and remove fruit. They are used in HD and SHD orchards (see chapter 15, The Olive Harvest). The trees are pruned into a hedgerow so that the canopy fits into the harvester. In SHD systems, the trees are topped between 8 and 10 feet (2.5 and 3 m) and hedged to maintain a canopy width of 5 to 8 feet (1.5 to 2.5 m), depending on the row width (fig. 9.14). The largest over-the-row harvester (Colossus) used in HD orchards has a harvesting tunnel adjustable to 16.1 feet (4.9 m) tall by 13.1 feet (4 m) wide, which will accommodate most HD orchards in California.

Tree skirting

Tree skirting is the removal of branches low on the trunk that would impede the harvester from forming a tight seal around the tree. It is often done mechanically. Depending on the harvester, skirting removes limbs from the bottom 24 to 48 inches (2 to 4 ft) of the trunk. Without skirting, fruit losses increase, and harvesting efficiency declines, as fruit falls through openings in the catch frame of the harvester onto the orchard floor. Additionally, fruit and limbs near the ground are often damaged

Figure 9.13

In an HD orchard harvested by a trunk-shaking harvester, the canopy must be shaped and pruned so the shaking force transmits well to the outer branches and the fruit drops. *Photo:* Richard Rosecrance.

Figure 9.14

For an over-the-row harvester, the canopy must be shaped and pruned to fit entirely into the harvester. *Photo:* Paul Vossen.

by equipment and herbicide sprays. Skirting also facilitates microirrigation system inspection and reduces disease incidence because it improves air circulation. In addition to mechanical skirting, some growers apply herbicides to control suckers emanating from the tree trunk.

Regenerative pruning

Regenerative pruning, or dehorning, is done when there are obvious signs of aging—when the wood is old, the top of the tree thins out, resulting in branches with sparse leaves and flower buds. Regenerative pruning consists of cutting the tree back to its main branches or even to its trunk to promote new growth (fig. 9.15). Olive trees will resprout vigorously from epicormic buds, which lie dormant beneath the bark. In the months following

Figure 9.15

Regenerative pruning, at right, can make old trees more productive. *Photo:* Richard Rosecrance.

regenerative pruning, it is important to thin out the new branches to train the tree and to ensure enough light reaches the interior of the canopy. Depending on cultural practices, a tree typically fruits 3 to 5 years after regenerative pruning.

Mulching tree prunings

Pruning large trees results in significant quantities of woody and leafy debris, a rich source of carbon and other elements that are lost when prunings are burned. The availability of very powerful brush shredders and the banning of burning prunings because of pollution have increased the practice of shredding or chipping olive prunings (fig. 9.16). Olive wood chip and leaf mulch increases soil organic matter, which improves soil fertility, soil structure, water infiltration, and nutrient availability, and decreases soil surface crusting. Olive mulch breaks down quickly in California.

Pest and disease considerations

In general, avoid pruning during rainy periods. Olive knot (see chapter 12, Diseases of Olive), a bacterial disease spread by rain, enters the tree through leaf scars, frost cracks, and pruning wounds. Research has shown that pruning wounds made in spring remain susceptible to infection for up to 2 weeks (Krueger et al. 1997). Pruning wounds made in summer under warm, dry conditions are unlikely to become infected. Hedging in early spring, however, is optimal for

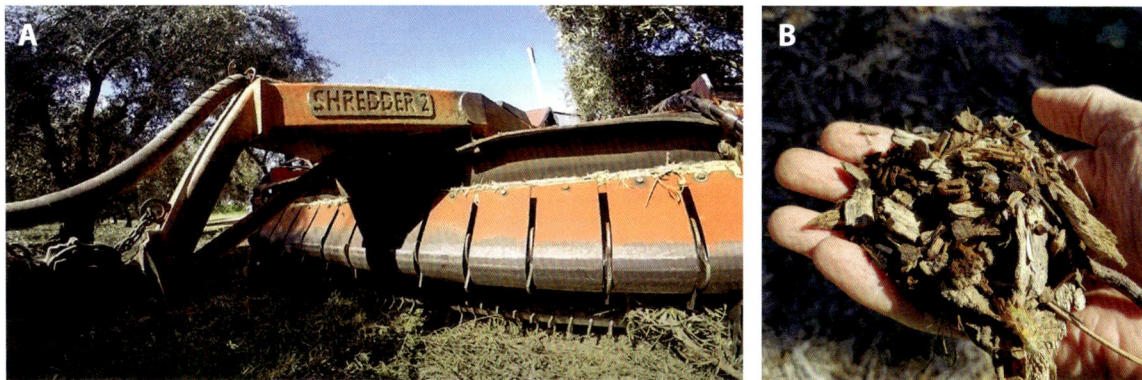

Figure 9.16

A brush shredder (A) turns olive branches and leaves into organic material (B), which improves the soil. *Photos:* Richard Rosecrance.

maximum yield and return bloom. Thus, hedging in spring during dry periods is recommended. If rain is expected, apply prophylactic copper sprays immediately after pruning. For more information, see the UC IPM Olive Pest Management Guidelines, ipm.ucanr.edu/agriculture/olive.

Black scale (see chapter 10, Arthropod Pests of Olive) feeds on the leaf mesophyll, producing a sticky honeydew, which is colonized by sooty mold fungus, darkening the leaf surface, restricting photosynthesis, and potentially significantly reducing production the following year. The insect prefers unpruned portions of the tree and is rarely seen in exposed, airy parts of the tree. In California's Central Valley, where summers are hot and trees are pruned to have open airy canopies, black scale is susceptible to heat and is seldom an economic problem. In dense canopies, however, black scale is more problematic and often requires chemical control measures.

References

Acebedo, M. M., M. L. Cañete, and J. Cuevas. 2000. Processes affecting fruit distribution and its quality in the canopy of olive trees. Advances in Horticultural Science 14:169–175.

Ajmi, A., S. Vázquez, F. Morales, A. Chaari, H. El-Jendoubi, A. Abadía, and A. Larbi. 2018. Prolonged artificial shade affects morphological, anatomical, biochemical, and ecophysiological behavior of young olive trees (cv. Arbosana). Scientia Horticulturae 241:275–284. https://doi.org/10.1016/j.scienta.2018.06.089

Albarracín, V., A. J. Hall, P. S. Searles, and M. C. Rousseaux. 2018. Impact of simulated mechanical hedge pruning and wood age on new shoot demography and return flowering in olive trees. Trees 32:1767–1777. https://doi.org/10.1007/s00468-018-1749-1

Australian Government and Rural Industries Research and Development Corporation (RIRDC). 2008. Guide to efficient olive harvesting.

Avidan, B., R. Birger, F. Abed-El-Hadi, O. Salmon, O. Hekster, Y. Friedman, and S. Lavee. 2011. Adopting vigorous olive cultivars to high density hedgerow cultivation by soil applications of uniconazol, a gibberellin synthesis inhibitor. Spanish Journal of Agricultural Research 9(3):821. https://doi.org/10.5424/sjar/20110903-336-10

Caruso, G., R. Gucci, M. I. Sifola, R. Selvaggini, S. Urbani, S. Esposto, A. Taticchi, and M. Servili. 2017. Irrigation and fruit canopy position modify oil quality of olive trees (cv. Frantoio). Journal of the Science of Food and Agriculture 97:3530–3539. https://doi.org/10.1002/jsfa.8207

Centeno, A., J. M. García Martos, and M. Gómez-del-Campo. 2017. Effects of nitrogen fertilization and nitrification inhibitor product on vegetative growth, production and oil quality in 'Arbequina' hedgerow and 'Picual' vase-trained orchards. Consejo Superior de Investigaciones Científicas. https://doi.org/10.3989/gya.0441171

Cherbiy-Hoffmann, S. U., A. J. Hall, and M. C. Rousseaux. 2013. Fruit, yield, and vegetative growth responses to photosynthetically active radiation during oil synthesis in olive trees. Scientia Horticulturae 150:110–116. https://doi.org/10.1016/j.scienta.2012.10.027

Cherbiy-Hoffman, S. U., P. S. Searles, J. A. Hall, and M. A. Rousseaux. 2012. Influence of light environment on yield determinants and components in large olive hedgerows following mechanical pruning in the subtropics of the Southern Hemisphere. Scientia Horticulturae 137:36–42. https://doi.org/10.1016/j.scienta.2012.01.019

Connor, D. J., A. Centeno, and M. Gómez-del-Campo. 2009. Yield determination in olive hedgerow orchards. II. Analysis of radiation and fruiting profiles. Crop Pasture Science 60:443–452. https://doi.org/10.1071/CP08253

Connor, D. J., M. Gomez-del-Campo, and J. Comas. 2012. Yield characteristics of N–S oriented olive hedgerow orchards cv. Arbequina. Scientia Horticulturae 133:31–36. https://doi.org/10.1016/j.scienta.2011.10.008

Connor, D. J., M. Gómez-del-Campo, M. C. Rousseaux, and P. S. Searles. 2014. Structure, management and productivity of hedgerow olive orchards: A review. Scientia Horticulturae 169:71–93. https://doi.org/10.1016/j.scienta.2014.02.010

Díez, C. M., J. Moral, D. Cabello, P. Morello, L. Rallo, and D. Barranco. 2016. Cultivar and tree density as key factors in the long-term performance of super high-density olive orchards. Frontiers in Plant Science 7:1226. https://doi.org/10.3389/fpls.2016.01226

Ferguson, L., U. A. Rosa, S. Castro-Garcia, S. M. Lee, J. X. Guinard, J. Burns, W. H. Krueger, N. V. O'Connell, and K. Glozer. 2010. Mechanical harvesting of California table and oil olives. Advances in Horticultural Science 24(1):53–63.

Gascó, A., A. Nardini, F. Raimondo, E. Gortan, A. Motisi, M. A. Lo Gullo, and S. Salleo. 2007. Hydraulic kinetics of the graft union in different *Olea europaea* L. scion/rootstock combinations. Environmental and Experimental Botany 60(2):245–250. https://doi.org/10.1016/j.envexpbot.2006.10.011

Gómez-del-Campo, M. 2010. Physiological and growth responses to irrigation of a newly established hedgerow olive orchard. HortScience 45(5):809–814. https://doi.org/10.21273/HORTSCI.45.5.809

Granado-Yela, C., C. García-Verdugo, K. Carrillo, R. Rubio de Casas, L. A. Kleczkowski, and L. Balaguer. 2011. Temporal matching among diurnal photosynthetic patterns within the crown of the evergreen sclerophyll *Olea europaea* L. Plant, Cell & Environment 34:800–810. https://doi.org/10.1111/j.1365-3040.2011.02283.x

Gucci, R., and C. Cantini. 2000. Pruning and training systems for modern olive growing. Clayton South VIC, Australia: Csiro Publishing. https://doi.org/10.1071/9780643101302

Haberman, A., A. Dag, N. Shtern, I. Zipori, R. Erel, A. Ben-Gal, and U. Yermiyahu. 2019. Significance of proper nitrogen fertilization for olive productivity in intensive cultivation. Scientia Horticulturae 246:710–717. https://doi.org/10.1016/j.scienta.2018.11.055

Hartmann, H. T. 1958a. Rootstock effects in the olive. Proceedings of the American Society for Horticultural Science 72:242–51.

———1958b. Some responses of the olive to nitrogen fertilizers. Proceedings of the American Society for Horticultural Science 72:257–266.

Krueger, W. H., B. L. Teviotdale, and M. N. Schroth. 1997. Improvements in the control of olive knot disease. Acta Horticulturae 474:567–572. https://doi.org/10.17660/ActaHortic.1999.474.117

Larbi, A., S. Vázquez, H. El-Jendoubi, M. Msallem, J. Abadía, A. Abadía, and F. Morales. 2015. Canopy light heterogeneity drives leaf anatomical, eco-physiological, and photosynthetic changes in olive trees grown in a high-density plantation. Photosynthesis Research 123:141–155. https://doi.org/10.1007/s11120-014-0052-2

Lodolini, E. M., S. Polverigiani, D. Grossetti, and D. Neri. 2018. Pruning management in a high-density olive orchard. Acta Horticulturae 1199:385–390. https://doi.org/10.17660/ActaHortic.2018.1199.61

Miglietta, P. P., R. Micale, R. Sciortino, T. Caruso, A. Giallanza, and G. La Scalia. 2019. The sustainability of olive orchard planting management for different harvesting techniques: An integrated methodology. Journal of Cleaner Production 238:117989. https://doi.org/10.1016/j.jclepro.2019.117989

Proietti, P., and F. Famiani. 2002. Diurnal and seasonal changes in photosynthetic characteristics in different olive (*Olea europaea* L.) cultivars. Photosynthetica 40(2): 171–176. https://doi.org/10.1023/A:1021329220613

Rallo, L., C. M. Díez, A. Morales-Sillero, H. Miho, F. Priego-Capote, and P. Rallo. 2018. Quality of olives: A focus on agricultural preharvest factors. Scientia Horticulturae 233:491–509. https://doi.org/10.1016/j.scienta.2017.12.034

Rapoport, H. F., S. B. M. Hammami, P. Martins, O. Pérez-Priego, and F. Orgaz. 2012. Influence of water deficits at different times during olive tree inflorescence and flower development. Environmental and Experimental Botany 77:227–233. https://doi.org/10.1016/j.envexpbot.2011.11.021

Rapoport, H. F., D. Perez-Lopez, S. B. M. Hammami, J. Agueera, and A. Moriana. 2013. Fruit pit hardening: Physical measurement during olive fruit growth. Annals of Applied Biology 163(2):200–208. https://doi.org/10.1111/aab.12046

Romero, A., J. F. Hermoso, and J. Tous. 2014. Olive rootstocks to control 'Arbequina IRTA-I18' clone vigour: Results from a second comparative trial. Acta Horticulturae 105:577–584. https://doi.org/10.17660/ActaHortic.2014.1057.74

Rosati, A., A. Paoletti, S. Caporali, and E. Perri. 2013. The role of tree architecture in super high density olive orchards. Scientia Horticulturae 161:24–29. https://doi.org/10.1016/j.scienta.2013.06.044

Rosati, A., A. Paoletti, G. Pannelli, and F. Famiani. 2017. Growth is inversely correlated with yield efficiency across cultivars in young olive (*Olea europaea* L.)trees. HortScience 52(11):1525–1529. https://doi.org/10.21273/HORTSCI12321-17

Rosecrance, R. C., W. H. Krueger, L. Milliron, J. Bloese, C. Garcia, and B. Mori. 2015. Moderate regulated deficit irrigation can increase olive oil yields and decrease tree growth in super high density 'Arbequina' olive orchards. Scientia Horticulturae 190:75–82. https://doi.org/10.1016/j.scienta.2015.03.045

Rugini, E., M. Ceccarelli, R. Muelo, and V. Cristofori. 2016. Mutagenesis and biotechnology techniques as tools for selecting new stable diploid and tetraploid olive genotypes and their dwarfing agronomical characterization. HortScience 51(7):779–804. https://doi.org/10.21273/HORTSCI.51.7.799

Rugini, E., G. Pannelli, M. Ceccarelli, and M. Muganu. 1996. Isolation of triploid and tetraploid olive (*Olea europaea* L.) plants from mixoploid cv. 'Frantoio' and 'Leccino' mutants by *in vivo* and *in vitro* selection. Plant Breeding 115(1):23–27. https://doi.org/10.1111/j.1439-0523.1996.tb00865.x

Schneider, D., M. Goldway, R. Birger, and R. A. Stern. 2012. Does alteration of 'Koroneiki' olive tree architecture by uniconazole affect productivity? Scientia Horticulturae 139:79–85. https://doi.org/10.1016/j.scienta.2012.03.006

Sofo, A., B. Dichio, G. Montanaro, and C. Xiloyannis. 2009. Photosynthetic performance and light response of two olive cultivars under different water and light regimes. Photosynthetica 47:602–608. https://doi.org/10.1007/s11099-009-0086-4

Therios, I. 2009. Rootstocks. Olives. Crop Production Science in Horticulture 18. Cambridge: CABI. 43–48.

Tombesi, S., and D. Farinelli. 2016. Trunk constriction effects on vegetative vigour and yield efficiency in olive tree (*Olea europaea* L.). Journal of Agricultural Science and Technology 18:1667–1680.

———. 2017. Canopy management in super high-density olive orchards: Relationship between canopy light penetration, canopy size and productivity. Acta Horticulturae 1177: 87–91. https://doi.org/10.17660/ActaHortic.2017.1177.9

Tombesi, S., D. Farinelli, A. Palliotti, S. Poni, and T. M. DeJong. 2018. Xylem manipulation techniques affecting tree vigour in peach and olive trees. Acta Horticulturae 1228:91–96. https://doi.org/10.17660/ActaHortic.2018.1228.12

Torres, M., P. Pierantozzi, P. Seales, M. C. Rousseaux, G. García-Inza, A. Miserere, R. Bodoira, C. Contreras, and D. Maestri. 2017. Olive cultivation in the Southern Hemisphere: Flowering, water requirements and oil quality responses to new crop environments. Frontiers in Plant Science 8:1830. https://doi.org/10.3389/fpls.2017.01830

Tous, J., A. Romero, J. F. Hermoso, and A. Ninot. 2011. Influence of different olive rootstocks on growth and yield of the 'Arbequina IRTA-I.18' clone. Acta Horticulturae 924:315–320. https://doi.org/10.17660/ActaHortic.2011.924.39

Trentacoste, E. R., F. J. Calderón, C. M. Puertas, A. P. Banco, O. Contreras-Zanussi, W. Galarza, and D. J. Connor. 2018. Vegetative structure and distribution of oil yield components and fruit characteristics within olive hedgerows (cv. Arbosana) mechanically pruned annually on alternating sides in San Juan, Argentina. Scientia Horticulturae 240:425–429. https://doi.org/10.1016/j.scienta.2018.06.045

[UC ANR] UC Davis Agriculture and Natural Resources Agricultural Issues Center. 2016a. Sample costs to produce table olives.

———. 2016b. Sample costs to produce oil olives.

Suggested reading

Albarracín, V., A. J. Hall, P. S. Searles, and M. C. Rousseaux. 2017. Responses of vegetative growth and fruit yield to winter and summer mechanical pruning of olive trees. Scientia Horticulturae 225:185–194.

Connor, D. J., and M. Gómez-del-Campo. 2013. Simulation of oil productivity and quality of N–S oriented olive hedgerow orchards in response to structure and interception of radiation. Scientia Horticulturae 150:92–99. https://doi.org/10.1016/j.scienta.2012.09.032

Proietti, P., L. Nasini, L. Reale, T. Caruso, and F. Ferranti. 2014. Productive and vegetative behavior of olive cultivars in super high-density olive grove. Scientia Agricola 72:20–27. https://doi.org/10.1590/0103-9016-2014-0037

Rallo, L., D. Barranco, S. Castro-García, D. J. Connor, M. Gómez-del-Campo, and P. Rallo. 2013. High-density olive plantations. Horticultural Reviews 41:303–384. https://doi.org/10.1002/9781118707418.ch07

10

Arthropod Pests of Olive

Houston Wilson,
UC Cooperative Extension (UCCE) Specialist, Department of Entomology, UC Riverside

Kent M. Daane,
UCCE Specialist, Department of Environmental Science, Policy and Management, UC Berkeley

Frank G. Zalom,
Professor, Department of Entomology and Nematology, UC Davis

Highlights

- The invasive olive fruit fly is the insect pest of greatest concern for most California olive growers. Management currently requires a combination of cultural practices, insecticides, and mass trapping. Ongoing projects to introduce parasitoids could expand the potential for biological control. Registered chemical control options are somewhat limited.

- Many scale insects are known to attack olive trees, although not all pose a serious threat to production. Growers should most closely monitor black scale, which is sensitive to hot temperatures and dry conditions, and can therefore be managed through cultural practices to increase canopy air flow.

- Secondary pests include the olive mite, western flower thrips, black vine weevil, and borers, all of which are sporadic pests that can largely be controlled through cultural practices.

- The olive psyllid is a newly arrived invasive pest that currently remains confined to Southern California. Damage can occur at high densities, and efforts are underway to establish natural enemies of this new pest.

- While not direct pests, ants can contribute to the proliferation of scales and psyllids by feeding on their honeydew and protecting them from natural enemies. Ant controls rely on a combination of cultural practices and the use of environmentally friendly bait stations.

nsect and mite pests can have a significant impact on olive production by reducing yield and quality. Before the start of this century, damage from pests in California olive orchards could be kept below economic levels through a combination of biological and cultural controls, and only a few insecticides were registered for use in olive. Any insecticide treatments targeted specific pests with an emphasis on achieving control without disturbing the natural regulation of other insect and mite pests.

Historically, the key pest of olive was black scale, but the establishment of the olive fruit fly in California in the late 1990s has made annual pesticide treatment essential in many commercial orchards. An important component of the integrated pest management (IPM) approach for olive is monitoring pest and beneficial species to determine if and when pesticides are necessary and choosing controls that prevent the occurrence of secondary pest outbreaks. Therefore, controls for olive fruit fly must be integrated with other pest management decisions.

Olive fruit fly

The olive fruit fly, *Bactrocera oleae* (Gmelin) (Diptera: Tephritidae), is the major insect pest of olive in California. A key olive pest in the Mediterranean region of southern Europe, northern Africa, and the Middle East for thousands of years, it was first found in North America in 1998, infesting olive landscape trees in Los Angeles County. The olive fruit fly is now established in all California production areas, with higher populations more typical in coastal areas than in the Central Valley. It commonly infests ornamental trees grown for landscaping and roadside borders. When left unmanaged, these ornamental olive trees serve as an important source of invasion for nearby commercial orchards. Genetic studies indicate that the California olive fruit fly population is most closely related to flies found in the eastern Mediterranean countries of Cyprus, Israel, and Turkey.

Description

The olive fruit fly is a member of the insect family Tephritidae, commonly known as fruit flies. The adult fly is approximately ³⁄₁₆ inch (4.8 mm) long with clear wings, dark venation, and a small dark spot at each wing tip (fig. 10.1). The head, thorax, and abdomen are brown with darker markings, and several white or yellow patches are present on the top and sides of the thorax. Adult flies can live from 2 to 6 months, depending on food availability and temperature. Male flies are polygamous, and females are normally monogamous. The sexes are easy to distinguish as adults; the male abdomen is blunt while the female has a stout, black ovipositor at the end of her abdomen.

Olive fruit fly females have been reported to lay from ten to forty eggs per day, and from 200 to 500 eggs during their lives, making the

Figure 10.1

The olive fruit fly lays from ten to forty eggs a day and lives from 2 to 6 months. It is the primary pest of olive.

reproductive potential extremely high if host fruit is available for oviposition. Olive fruit fly eggs are small and difficult to see because they are inserted under the fruit surface. However, the oviposition scar, also known as a sting, is more distinctive, appearing as a depression with necrotic brown tissue at its center on the surface of the fruit (fig. 10.2).

The larva is a maggot with a pointed head; its color varies from pale white to light yellow (figs. 10.3 and 10.4). There are three larval instars, with the mature larva about ¼ inch (6.4 mm) long and ¹⁄₁₆ inch (1.6 mm) wide. Unlike most other tephritid species, mature olive fruit fly larvae pupate in fruit during summer (fig. 10.5), but larvae that mature in fall and winter leave the fruit to pupate in the soil under the tree. The pupa, initially a pale yellow color but changing to dark reddish-brown as it matures, is about ³⁄₁₆ inch (4.8 mm) long and ³⁄₃₂ inch (2.4 mm) wide.

Figure 10.2

An oviposition scar marks where olive fruit fly has laid eggs under the fruit surface. *Photo:* Cindy Kron, UCCE IPM Advisor.

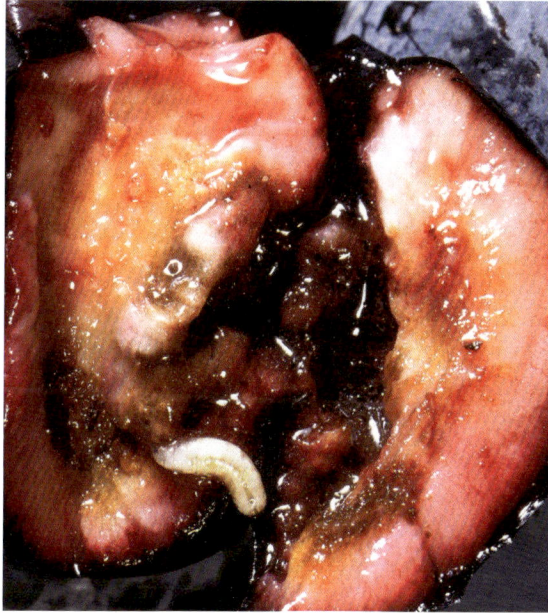

Figure 10.4

Olive fruit fly larvae are light colored, almost white to pale yellow.

Figure 10.3

Olive fruit fly larvae have a pointed head.

Figure 10.5

Mature larvae of olive fruit fly pupate in the fruit in summer.

Field biology

The only reproductive host of the olive fruit fly in California is olive, and the pest's seasonal cycle is generally synchronized with the olive tree and fruit development. There are three to as many as five generations of olive fruit fly in California each year, depending on temperature, host availability, and region. Four generations are common in the Central Valley. Olive fruit flies overwinter in the adult or pupal stages. In areas with warmer winters, adults can lay eggs on fruit remaining on trees throughout the winter; however, this overwintered adult population typically declines to low levels by February or March. In cooler regions, pupae are the most important overwintering stage. New adults from these overwintering pupae emerge in spring, from March through May, and feed on honeydew and other sources of sugar and protein.

Mated females can begin to lay eggs when they are 6 days old, and they can continue to lay eggs for as long as 90 days. Peak oviposition occurs when females are 13 to 37 days old. Fruit size is critical to the success of larval development; it must exceed about ⅜ inch (9.5 mm) before the fly's life cycle can be completed. The next generation of adults, which represents offspring from overwintered adults and pupae, appears between June and August, depending on regional temperatures.

Additional generations of olive fruit flies are produced from late August through December, depending on temperature, fruit size, and fruit availability. Olives that remain on trees following harvest can produce large numbers of flies from late fall to early spring if the fruit are allowed to mature and drop to the ground. Significant differences in the seasonal phenology and biology of olive fruit fly are seen across California due to the numerous microclimates found within and between coastal and interior regions. For example, in coastal regions with mild winter temperatures, such as San Diego and Santa Barbara Counties, olive fruit fly development is continuous throughout the year if old olive fruit remain on trees through the winter and into early summer. In the Central Valley, adults appear to have two peaks of activity: the first in spring, when adults of the overwintering generation occur; and a second, much larger peak in fall, when most olives have reached sufficient size for larval development and temperatures are consistently below the level that suppresses olive fruit fly development in summer.

Summer temperatures exceeding 95°F (35°C) during late June, July, and August have the largest effect on the abundance and distribution of the olive fruit fly in California. Populations are generally greatest in the cooler, more humid coastal areas and lowest in hot, dry areas of the Central Valley, especially the central and southern San Joaquin Valley. The high temperatures kill eggs and small larvae, while adult mortality and longevity increase and females lay fewer eggs. However, adult flies are very mobile and can seek protection from the heat in more humid areas in or near olive orchards or urban landscapes, particularly heavily irrigated ones. The adult fly has been reported to move from 650 feet (198 m) in the presence of olive hosts to as much as 2.5 miles (4 km) to find hosts.

The adult olive fruit fly activity threshold is approximately 60°F (16°C). The species can complete a generation in as few as 26 days under moderate temperatures of 70° to 79°F (21° to 26°C). The eggs hatch in 1 to 5 days, and larvae develop in about 17 days under optimal temperatures of about 79°F (26°C). Development is slower at lower and much higher temperatures. Development of the pupal stage can be as short as 8 to 10 days during summer but may last for 6 months in winter, when larvae leave the fruit to pupate in the soil.

Damage

The larva of the olive fruit fly, which feeds exclusively on olive fruit, is the life stage that causes the most significant damage. Heavy infestations can cause 100 percent damage to varieties with large fruit. Although damaged fruit are not acceptable for table olive production, commonly cited European sources indicate that olive oil can be made if the damage level is below about 10 percent. Research in California has shown that much greater levels of damaged fruit can be used to produce extra virgin olive oil, as long as the fruit shows no signs of rot. However, the impact of olive fruit fly damage, while minimal when the oil is fresh, can become evident in the same oil as it ages; thus, larval feeding should be minimized as much as possible. Larval feeding increases the acidity of the oil, allowing microorganisms to invade the fruit and cause rot that lowers oil quality and value. Even in the absence of larvae, the adult female's sting on the fruit surface can also be economically important for table olives as this damage compromises appearance.

Larger-fruited olive varieties tend to be more attractive to female flies, and they are able to support the development of greater numbers of larvae per fruit. Additionally, larger fruit incur greater damage because they reach a size suitable for oviposition earlier in the summer than smaller-fruited varieties. Ascolana, Coratina, Hojiblanca, Manzanillo, Mission, Picual, and Sevillano are examples of medium- to large-fruited varieties planted in California.

Monitoring

Adult olive fruit flies can be monitored with a pheromone lure or an ammonia-based lure placed in an appropriate trap. For a dry trap option, yellow sticky panel traps can be baited with a sex pheromone (spiroketal) or with an ammonium carbonate, ammonium bicarbonate, or diammonium phosphate bait. The sex pheromone attracts only male flies. The ammonium carbonate and related compounds produce ammonia, which is associated with protein decomposition. Females are attracted to the volatile ammonia because they need protein for egg production. Both sexes are attracted to the yellow color of the trap and become stuck on the sticky surface.

Olive fruit flies can also be monitored effectively with glass or plastic McPhail traps. These traps contain torula yeast pellets in water; during fermentation, they produce a suite of volatiles including ammonia. Although McPhail traps capture more flies than yellow sticky traps do, these wet traps can be more difficult to use. McPhail traps made of plastic are easier to use than ones made of glass. Either can be reused for several years.

In warmer locations, traps should be placed in orchards by March 1 and hung midcanopy in the shade (north side of the tree), in an open area to avoid leaves blocking the trap. The value of trapping is primarily for detection, especially in areas where olive fruit flies do not require treatment every year. While there is no formal trap-based threshold on which to base initiation of control

actions, traps can help to evaluate treatment efficacy by comparing trap captures before and after treatment.

Surveying fruit for stings or the presence of larvae is another monitoring approach, and also provides some indication of the severity of an infestation. It is easier to detect an infestation if samples are collected in areas of an orchard where adult flies have been observed or dimples on fruit are suspected to be oviposition stings. Collecting greater numbers of fruit samples increases the chances of detecting a low-level infestation. Fruit samples can be placed in a single layer in a storage container and covered with cloth to allow ventilation and slow the development of mold. A hardware cloth–type screen made of wire or plastic can be suspended just above the bottom of the container to allow the larvae to exit the fruit and drop to the bottom, where they can be counted.

Management

Prior to the widespread establishment of the olive fruit fly in California, insecticide applications were rare and few products were registered. The introduction of this insect pest created an urgent problem for olive growers as none of the insecticides registered for use on the crop in the United States was effective for its control. Further, the organophosphate insecticides that were commonly used for olive fruit fly control in Europe at that time could not be registered for use in olive in the United States due to regulatory concerns. To complicate matters, olive is considered a minor use market for insecticide manufacturers, so there is little incentive for them to register their products for use on olive. As a result, the number of effective insecticides for olive fruit fly control remains limited, and growers and their consultants typically adopt an IPM approach, which uses a combination of control tactics.

Cultural control

Timing the harvest to remove most of the fruit before large adult populations develop in the orchard during fall helps prevent damage and losses. Trapping in late summer and fall, as described in the "Monitoring" section, above, can be especially helpful to determine when adult populations are starting to rapidly increase in density. Olives harvested earlier in fall, when they are still hard and green, generally have a lower incidence of infestation than when harvested later in fall. It is especially critical that fly-damaged olives be processed as soon as possible after harvest, as quality may decline between harvest and processing, and the damaged olives are vulnerable to some fungal and bacterial pathogens.

Sanitation following harvest can reduce overall olive fruit fly densities. Fruit left on trees or on the ground after harvest will sustain the resident population in an orchard from season to season. Olives remaining on trees serve as an oviposition site for adults, and infested olives remaining on trees and on the ground allow for continued development of the flies. Homeowners and commercial growers should collect fallen olives and remove as much remaining fruit as possible from trees. Sanitation of landscape and unmaintained trees that serve as a habitat for olive fruit flies can help reduce the overall densities present in an area. Unused or fallen fruit should be disposed of in landfills, composted, or buried at least 4 inches (10 cm) deep.

A source of water, especially during hot periods, is necessary to ensure survival of adult olive fruit flies. Therefore, standing water in an orchard should be eliminated when feasible. Leaking irrigation systems or similar sources of standing water can provide adult flies with sufficient moisture to thrive.

Chemical controls

Controls currently registered for olive fruit fly in the United States include a spinosad bait, a pyrethroid cover spray that is applied to the entire tree canopy and provides protective coverage for susceptible fruit, and a kaolin clay repellent spray. In 2004, the bait formulation of spinosad became the first product specifically registered on an emergency basis for control of olive fruit fly in California. The bait contains an attractant and insecticide liquid mixture to lure adult flies, which must feed to prolong their life and produce eggs. GF-120 NF Naturalyte Fruit Fly Bait, which contains spinosad as its active ingredient, is currently registered and approved for use in organic production as well; it has been the primary insecticide used for control to date.

The first application is best applied when fruit size is ⅜ inch (10 mm), the point at which

females begin laying eggs in fruit and larvae successfully develop. Coarse spray application methods that produce large droplets about 3/16 inch (5 mm) in diameter are preferred so that the bait does not dry out too quickly. Application at dawn or in the early evening will also help reduce evaporation and extend residual efficacy. The bait is best applied at a rate of 3 to 6 droplets per square foot (32 to 65 per m²) of foliage to a minimum of a 2-foot (0.6-m) diameter area within the upper part of the canopy on the north or east side of each tree. Once initiated, bait sprays must continue, according to label directions, to protect the crop until harvest. Unfortunately, olive fruit fly resistance to the spinosad bait has been documented in a number of growing areas, and some growers have reported complete loss of efficacy and greatly increased fruit damage.

The only effective cover spray currently registered for olive fruit fly control is the pyrethroid insecticide fenpropathrin. Because pyrethroids are widely regarded as highly toxic to beneficial insects, particularly insect parasitoids that keep scale insects under good biological control, it is recommended that growers avoid its routine use and apply only once, after fruit infestation begins in late summer.

Kaolin clay is a repellent rather than a toxicant. It is first sprayed when olives reach ⅜ inch (10 mm), the size at which the fruit become attractive to females and larvae can successfully develop in the fruit. While kaolin can be very effective, it is rather difficult to use because large amounts of the product must be applied to completely cover the fruit and leaves. It is typically applied every 5 to 6 weeks, as needed, to maintain complete coverage as new leaves appear and fruit enlarge. After the spray dries, it turns into a white powdery coat that must be rinsed from the fruit before processing.

Attract and kill devices, mass trapping

Attract and kill panels baited with olive fruit fly attractants and coated with a long-residual insecticide have shown promise for control in some studies. Flies attracted to the panels die after they contact the insecticide-treated surface. While several attract and kill products are available in

Europe, relatively few have been available in the United States, and none are currently registered.

For mass trapping, large numbers of traps are needed, typically one trap per tree, making this approach most effective when olive fruit fly numbers are low. In Spain, some growers utilize the Olipe trap to suppress populations in organic orchards and in sensitive areas near homes and parks with ornamental olives. The trap consists simply of a ¼- to ½-gallon (1- to 2-L) plastic bottle with approximately ³⁄₁₆-inch (4.8-mm) holes melted or drilled into the bottle shoulders. It is filled about two-thirds full with a 3-to-5-percent solution of an ammonia-producing substance and water and hung in a tree; the flies are attracted to the trap, crawl inside, and die. The McPhail-type liquid trap and yellow sticky panel traps used as monitoring tools can also be employed for mass trapping; ammonia-producing attractants and pheromone lures can be added to increase attraction (see the "Monitoring" section, above). Hang one Olipe or McPhail trap per tree in a shady location. Change the bait in each trap monthly, and clean out the dead flies at this time.

Mass trapping is more effective in isolated plantings or when other olive trees in an area are managed for olive fruit fly or also use this approach. In one Sonoma County study, mass trapping reduced damage levels to an average of around 30 percent compared with almost 90 percent damage in untreated controls.

Biological control

Currently, there are no consistently effective biological control agents for the olive fruit fly in California, although a species of *Pteromalus*, a generalist ectoparasitoid, has been found to parasitize up to 30 percent of collected larvae in some coastal locations. University of California and state scientists have imported and released two species of parasitic wasps from sub-Saharan Africa. One of these parasitoids, *Psyttalia lounsburyi* (Silvestri) (fig. 10.6), has established in some of California's coastal regions, where up to 33 percent larval parasitism on ornamental olive has been observed. *P. lounsburyi* is an endoparasitoid that coevolved with the olive fruit fly in Africa. Parasitoids released in California were collected in Kenya and South Africa from 2002 to 2007, studied in

Figure 10.6

A parasitoid, *Psyttalia lounsburyi* (Silvestri), of olive fruit fly has established in some coastal areas. *Photo:* Kent Daane.

quarantine, and then cleared for release in the United States.

Economic control of the olive fruit fly in olive orchards using only biological controls may prove difficult because of the commercial requirements of very low infestation levels and the need for growers to use insecticides that are incompatible with parasitoid survival. One disadvantage of *P. lounsburyi* is that its temperature tolerances do not match exactly those of the olive fruit fly, thus establishment in California's warmer interior valleys has been poor. Another aspect impacting all the coevolved parasitoids is the large fruit size of the cultivated olive varieties compared with the smaller fruit produced by wild olive in Africa; the olive fruit fly larvae feeding deep inside the larger olive is out of reach of *P. lounsburyi*'s shorter ovipositor.

Black scale

Black scale, *Saissetia oleae* (Olivier) (Hemiptera: Coccidae), was the major insect pest of olive in California from 1970 to 2000, prior to the arrival of the olive fruit fly and after the successful biological control of olive scale (see "Olive scale" section, below). Although black scale is still of concern, the olive fruit fly has displaced it as the primary insect pest of olive. However, it is important that controls, particularly pesticides, for olive fruit fly are used within an IPM framework so that they do not disrupt the many natural enemies of black scale and cause secondary pest outbreaks.

Like the olive fruit fly, black scale is native to Africa and is now found in most Mediterranean and semitropical regions of the world. Hosts include almond, apple, apricot, citrus, coyote brush, fig, fuchsia, grape, grapefruit, oleander, peppertree, plum, prune, and rose. With such a wide host range, black scale is found throughout California and readily infests olive and citrus orchards, where it causes serious economic losses. Also in California is the close relative *S. miranda*, often called Mexican black scale; although the two species look similar, *S. miranda* has not yet been identified as a pest of olive in California.

Description

Adult females are pale brown when young, changing to black with a pronounced hemispherical shape as they mature. Adult females deposit their eggs under the scale cover (fig. 10.7). The eggs are tiny, 1/64 inch (0.4 mm), and light colored when first laid,

becoming pinkish 2 or 3 days later, and red-orange a few days before hatching.

First-instar nymphs, called crawlers, about 1/32 inch (0.8 mm) long, are pale yellow to light brown or orange with dark eyes (fig. 10.8). After emerging, crawlers search for a suitable feeding site for up to 7 days, depending on temperature, before they insert their strawlike mouthparts and begin to feed. The nymphs double in size before their first molt, which occurs 3 to 8 weeks after hatching, depending on temperature and host plant condition. Second-instar nymphs are about 3/64 inch (1.2 mm) long when an H pattern on the top of the scale begins to form.

The next stage, the sexually immature adult (third instar), is quite different in appearance. It is dark ash-gray to brown and about 3/32 inch (2.4 mm) long, with the legs hidden beneath the body. The H is quite distinct (fig. 10.9). Although this stage is often referred to as the rubber stage, it

Figure 10.8

Black scale first-instar nymphs are pale yellow to light brown or orange.

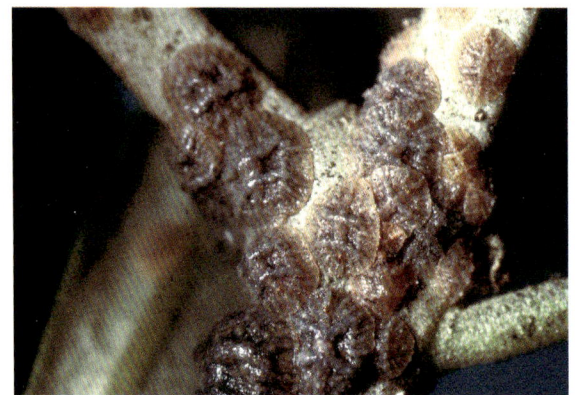

Figure 10.7

Black scale has many hosts in California and readily infests olive orchards.

Figure 10.9

Black scale third-instar nymphs have an H on their back; this is known as the rubber stage.

could be more correctly called the preovipositional adult stage because the reproductive organs have formed and egg production is about to begin.

Although black scale is categorized as a soft scale, in the adult or ovipositional stage, its outer shell cover (integument) turns black and becomes quite hard. Adult scale size varies greatly, from ⅛ to ³⁄₁₆ inch (3 to 4.8 mm) long and up to ⁵⁄₃₂ inch (4 mm) high. The number of eggs per female varies with adult size and host condition and can range from 300 to 3,000, with the upper range indicating the explosive nature of scale outbreaks.

Male black scales are rarely found in California. Young males are similar in appearance to the females until the second-instar stage, when they become more elongated. After 4 to 6 weeks, in warm weather, the male scale molts to a prepupal and then to a pupal stage, which is characterized by a red head with black eyes; the antennae, legs, and wing pads become minute. The adult male is light yellow and winged.

Field biology

In California's interior valleys, black scale typically overwinters as a second- or third-instar nymph. As temperatures increase in spring, a period of rapid growth and development is accompanied by a large amount of honeydew excretion. By April, most black scales have progressed to the preovipositional and adult stages. Egg hatch and crawler emergence begin shortly thereafter, peak in May or June, and are often not completed until July. Crawlers settle and begin to feed on leaves and twigs. In late summer and fall, most of the scales that are still feeding on leaves migrate to more favorable feeding and overwintering sites on twigs, where they develop to the second or third instar.

Black scale typically has one generation per year; however, because development is driven by temperature and host quality, there can be two generations when conditions permit. In California's interior valleys, scale development is slower during winter and summer because temperatures approach the scale's intolerable levels, resulting in one generation per year. Prolonged periods of high temperatures cause significant mortality in the early developmental stages. Therefore, cultural practices that lower canopy temperatures in summer (for example, pruning that leads to a dense canopy with little air circulation, or flood irrigation, which acts like a swamp cooler) reduce scale mortality and can result in a second or partial second generation. On the coast, where temperatures are milder, there are often two generations per year.

Damage

Black scale feeds by inserting its strawlike mouthparts into a leaf or twig to extract plant juices. As the scale feeds, it excretes carbohydrate-rich honeydew, which accumulates on leaves, forming a sticky, sugary fluid. The honeydew provides a substrate for sooty mold, which can form a dense black covering that shades leaves and reduces photosynthesis and respiration. The combination of insect feeding and buildup of honeydew can reduce fruit bud formation, cause leaf drop and

Figure 10.10

Black scale excretes honeydew and feeds on leaves and twigs, causing reduced fruit bud formation, leaf drop, and twig dieback.

twig dieback, and reduce the following year's crop (fig. 10.10).

Monitoring and control

The presence of honeydew droplets on olive leaves in March and April, corresponding to a rapid increase in scale size, is often the earliest signal of increased scale density in the orchard. To monitor, in April check about forty trees per block for honeydew. With new infestations, the scale is often found on just a few trees, so it is important to check trees throughout the block. Be sure to check the inner canopy, where the scale population typically starts to build.

If honeydew is found, sample for adult scales in early May, focusing on two or three areas in each block, particularly those that have had scale problems in the past. At least ten trees per 10-acre (4-ha) section should be searched. On each tree, count the number of mature scales (third instar to adult) on the terminal 20 inches (51 cm) of ten branches; again, be sure to include the inner sections of the tree. Add the numbers of black scales in each ten-branch sample and divide by ten to get the average infestation level. Infestations can be categorized into four levels: light (less than one per branch), moderate (one to three per branch), heavy (four to ten per branch), and severe (more than ten per branch).

Light infestation. If no scale is found, no treatment is needed. Light infestations typically do not require treatment in open-canopy orchards (see "Cultural control" section, below). Closed-canopy orchards should be pruned to increase air circulation and raise temperatures. If trees cannot be pruned or if spring and summer temperatures are lower than average, consider applying a dormant oil as a safeguard.

Moderate infestation. A small infestation can build to a moderate infestation following a cool spring and summer or within a closed-canopy orchard. A moderate scale infestation typically does not cause damage; however, the offspring of adult scales found in spring can substantially damage the crop. In trees with open canopies, the scale population should decrease or remain stable, depending on summer temperatures; however, consider applying a dormant oil to reduce scale densities. Prune orchards with closed canopies;

if trees cannot be pruned, consider an insecticide application.

Heavy infestation. Heavy infestations cause economic damage if left untreated. If the orchard has not been pruned for years and the tree canopy has closed, use an in-season insecticide treatment to prevent further damage at harvest. After harvest, prune trees, if possible, to open the canopy and make the environment less hospitable for scale survival.

Severe infestation. Severe infestations typically occur in closed-canopy orchards in which treatment of moderate or heavy scale infestation was delayed. Economic loss can be extensive because the adult scale population in spring represents a potentially great increase during summer. An in-season insecticide treatment is needed to protect the crop and trees. After harvest, prune to open the canopy and remove severely infested or damaged branches.

Cultural control

Crawlers and second-instar nymphs cannot survive prolonged periods of hot, dry weather because they cannot feed fast enough to replace the fluids they lose. The small scales desiccate and, in an open canopy, as much as 70 to 90 percent of these small scales can die during spring and summer. Mortality is higher in the warmer interior valleys than in coastal regions. Canopy structures that affect microclimate influence black scale mortality and development. A

Figure 10.11

Black scale dessicates in an open canopy but can survive hot summers in trees left unpruned for many years. *Photo:* Kent Daane.

canopy left unpruned for many years becomes dense, or closed, moderating spring and summer heat (fig. 10.11); in this protected environment, black scales can better survive hot summers and develop to outbreak levels in mild summers. Regular pruning opens the canopy (fig. 10.12) and exposes the scales to higher temperatures and drier conditions. This is the best cultural control available. To update this practice, a study is needed of mechanical pruning and how it impacts black scale mortality and development.

Biological control

Many natural enemies have been imported and released in California to control black scale. About fifteen parasitoid species have become established. The most common parasitic wasps are *Metaphycus helvolus* Compere, *M. anneckei* Guerrieri and Noyes, *M. hageni* Daane and Caltagirone, *Coccophagus ochraceus* Howard, *C. lycimnia* (Walker), *C. scutellaris* (Dalman), and *Scutellista caerulea* (Fonscolombe). Also present are insect predators, such as green lacewings (*Chrysoperla* spp.) and lady beetles, including *Hippodamia convergens* Guérin-Méneville and *Hyperaspis* species. These predators feed primarily on young black scales.

Unfortunately, parasitoids and predators have had only partial success in controlling black scale below economic injury levels. One problem is that the parasitoids cannot reproduce without suitable black scale host stages. Because black scale often

Figure 10.12

Keeping an orchard regularly pruned is the best cultural control for black scale. *Photo:* Kent Daane.

has only one generation per year, there can be periods when the proper host stage, or size, is not available, causing parasitoid numbers to drop to very low levels.

Chemical control

Dormant oil treatments are effective against light to moderate infestations, especially when used in conjunction with pruning to open the orchard canopy. However, for heavy or severe infestations, more toxic pesticides, such as a carbamate, should be considered. Juvenile hormone mimics, which disrupt development, are also being used. Pesticides are most effective against crawlers and second-instar nymphs found in summer. Before application, check twenty to thirty adult scales to make sure egg hatch is near complete and the crawlers are out from under the protective adult scale cover, so they will be exposed to the pesticide—simply flip over the scale or squish it to see if eggs are still present underneath. Postharvest treatments are also possible until the rubber (third-instar) stage is reached the following spring.

Olive scale

A pest of many crops, olive scale, *Parlatoria oleae* (Colvee) (Hemiptera: Diaspididae), has been collected from over 200 plant species. This hard, or armored, scale is widely distributed throughout the world's olive-growing regions, including Argentina, India, the Mediterranean, the Middle East, Russia, and Turkey. In the United States, it was first noted on privet in Maryland in 1924, although it may have been in Phoenix, Arizona, as early as the 1890s. In California, it was found near Fresno in 1934 and spread rapidly through olive-growing areas in the Central Valley, reaching south of the Tehachapi Mountains by 1961. It became a major pest of olives and required annual pesticide treatments to protect fruit from economic damage. Today, olive scale ranges from San Diego to the northern Sacramento Valley. However, due to successful biological control (see the section "Monitoring and control," below), this pest is rarely a problem.

Description

The adult female scale's cover is almost circular, about 1/32 inch (0.8 mm) across, slightly convex,

Figure 10.13

A young female olive scale has a reddish to deep purple body.

and light to dark gray with a raised brownish spot. Under the scale cover, the young female's body ranges from reddish to deep purple (fig. 10.13). The male's cover is white, flat, and elongated with a brownish or blackish spot closer to one end rather than centrally positioned.

Field biology

There are two generations of olive scale each year. This scale overwinters as an immature, mated, third-instar female. Overwintered females mature and begin laying eggs in late April and early May. Crawlers begin emerging in May and wander for a short time before settling on twigs, leaves, and, in some cases, newly set fruit. Emergence of first-generation crawlers is usually completed by June or early July. Both males and females are produced. Females from this generation mature and begin egg production in August and early September. The eggs of this fall generation hatch, and crawlers settle on leaves and twigs, but to a much greater extent on fruit. Males complete their development in fall and fertilize immature females before winter.

Damage

High olive scale densities can cause defoliation and twig death and frequently reduce tree productivity. The major damage results from scales settling on fruit. Fruit infested by the first generation become badly misshapen. Infestation by the second generation causes pronounced dark purple spotting of the green fruit, rendering it worthless for pickling, although the impact of this and other hard scale pests on olive oil has not been determined (fig. 10.14). It is thought that olive scale does not change the ripening process, oil production, or oil acidity; however, infested fruit may cause some oxidative deterioration of oil and lower the oleic to linoleic ratio, indicating a lower oxidative stability.

Monitoring and control

The best time to detect an increasing olive scale population is at harvest because of the characteristic discoloration of the infested fruit. Chemical control is normally not required. Two species of introduced parasites provide excellent control of olive scale. *Aphytis maculicornis* (Masi) (shown throughout its immature development stages in fig. 10.15) was introduced in 1952 and provided good control in some coastal orchards, but it was less successful in California's interior valleys, where it was inhibited by the hot, dry summers. *Coccophagoides utilis* (Doutt) was introduced in 1957. In combination, the two parasites provide widespread biological control of olive scale in California.

Chemical treatment to control olive scale is rarely needed unless biological control is disturbed by treatments for other pests. If olive scale was detected in the previous season or if chemicals are used in the orchard or on nearby crops, watch closely to detect crawlers moving onto the fruit in summer. If treatment is needed, apply pesticides, such as oils or insect growth regulators,

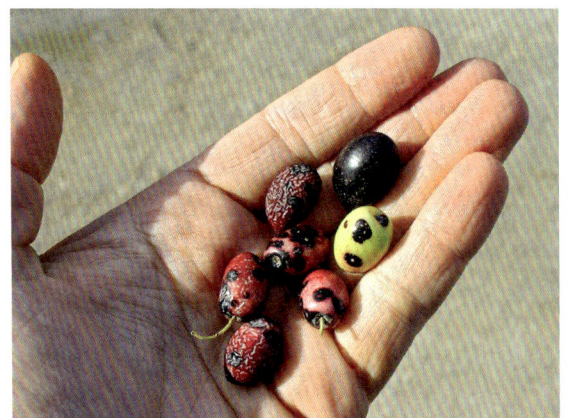

Figure 10.14

Olive scale causes major damage to olive fruit, but the impact on the oil is not known. *Photo:* Jack Kelly Clark.

Figure 10.15

Olive scale is effectively controlled by two parasites—*Aphytis maculicornis* and *Coccophagoides utilis*. Where insecticides such as carbamates organophosphates and pyrethroids are used, biological control will be disrupted. Be sure to monitor scale populations if disruptive insecticides are used.

against the first generation in May or June. A postharvest treatment is also effective.

Oleander scale

Oleander scale, *Aspidiotus nerii* (Bouche) (Hemiptera: Diaspididae), also known as ivy scale, occurs throughout the warmer parts of the United States and is one of the most common scale insects in California. It attacks a wide range of plants, including acacia, aloe, avocado, azalea, Boston ivy, boxwood, cactus, camellia, cherry, grape, grapefruit, lemon, magnolia, mistletoe, Monterey pine, oleander, rose, sago palm, and yucca. It rarely requires treatment on other hosts, but, on olive, infestations occasionally become sufficient to damage the fruit.

Description

Oleander scale resembles greedy scale (see the "Greedy scale" section, below), except its scale cover is less convex and the raised spot is almost central. Oleander scale is about ¹⁄₁₆ inch (1.6 mm) across, flat, and gray with a yellow or light brown spot (fig. 10.16). The male's cover is smaller than the female's and more elongated, with the raised spot at one end. The mature scale, found under the cover, and the crawlers are yellow.

Figure 10.16

Oleander scale has a flat gray cover. Underneath, the crawlers are yellow. *Photo:* Jack Kelly Clark.

Field biology

Little is known about the life cycle of oleander scale. The majority of overwintering individuals are adult females. Egg production and crawler emergence begin in April, and females continue producing progeny for almost 2 months. A second generation is produced in July and August, but because progeny production is spread over such a long period, there is often an overlap of the first and second generations.

Damage

Oleander scale is generally found on leaves and, to a lesser extent, twigs in the lower, inner

Figure 10.17

Leaves and twigs can tolerate heavy populations of Oleander scale before any reduction in crop yield occurs. Oleander scale only occasionally reaches an infestation level that damages olive fruit. *Photo:* Jack Kelly Clark.

canopy (fig. 10.17). Leaves and twigs can tolerate heavy populations before any reduction in crop yield occurs. When population density is high, oleander scale infests olive fruit. The infestation is characterized by green spots on purple fruit, as maturity of the tissue around the scale is delayed. Early infestations seriously deform fruit and later cause spotting. No conclusive work has been done on the impact of hard scale pests on olive oil, although one European study indicated that the olive scale did not significantly harm oil production.

Monitoring and control

The best way to decide whether treatment may be needed is to examine the previous season's crop yield and quality assessments to determine the amount of culled fruit caused by oleander scale. This is a good guideline for all the hard scales. Oleander scale can also be monitored with double-sided sticky tape, placed around branches where adult scales are present, to detect crawler emergence. However, it is difficult to determine which of the hard scales are present from the small crawlers caught on sticky tape. As with the other hard scales, there is no economic injury caused by oleander scale. Parasitoids and predators most often keep oleander scale in check. When monitoring, look for holes caused by parasite emergence to estimate the level of biological control. If pesticides are required, treat first-generation crawlers in late May or June, when they are first seen moving onto fruit.

Latania scale

Latania scale, *Hemiberlesia lataniae* (Signoret) (Hemiptera: Diaspididae), has been found on a wide range of hosts in tropical and subtropical areas. It is very common in California, recorded on such diverse hosts as acacia, avocado, bramble, cedar, euonymus, fuchsia, gladiolus, Kentia palm, kiwifruit, philodendron, rose, willow, and yucca.

Description

The adult scale cover is convex and gray or white, with a darker spot toward one side (fig. 10.18). The body of the female under the scale cover is yellow to orange, in contrast to the purple color of the olive scale. Apparently, male scales do not occur in California, although they have been observed

Figure 10.18

Latania scale has a convex cover, with an off-center spot.

in other parts of the United States. The bodies of the crawlers through second-instar nymphs are orange.

Field biology

Latania scale overwinters as a second-instar nymph and matures in early spring. Females initially lay eggs in batches of fifteen to twenty, and then in batches of three to five a day until dying. Eggs are laid beneath the scale cover, where they hatch.

Young crawlers migrate a short distance from the mature female, settle down (often clumped together because of the short distance they move), insert their mouthparts, and begin feeding. The legs of the crawlers become functionless after they settle. The first generation may complete its development in as little as 2 months. Crawlers are reported to be active in May, July, and September, indicating that two or three generations per year occur in California.

Damage

Latania scale feeds on leaves, bark, and fruit. If the population density is high, twig dieback may occur, reducing olive yield (fig. 10.19). At lower population densities, most damage occurs when latania scale infests olive fruit, rendering it worthless for table-olive processing, but it is unlikely that this cosmetic damage has an impact on olive oil.

Figure 10.19

In a major infestation of latania scale, twig dieback may occur. *Photo:* Jack Kelly Clark.

Figure 10.20

Greedy scale has a thin, light gray cover and is easily confused with latania scale. *Photo:* Jack Kelly Clark.

Monitoring and control

Pesticides are rarely needed because parasitoids are commonly present and can significantly reduce populations of latania scale. However, as with most of the hard scales, if the previous season's crop yield and quality assessments indicated an economic infestation, closer monitoring is needed and pesticides should be considered. Time the pesticide application to the emergence of crawlers in the first generation, typically in May or June. Double-sided sticky tape is useful to determine when crawlers are active.

Greedy scale

Greedy scale, *Hemiberlesia rapax* (Comstock) (Hemiptera: Diaspididae), is the most common and widely distributed species of hard scale in California. It infests innumerable hosts throughout the United States and may attack almost any woody plant. However, it is not as damaging to olive as many of the other scale species described in this chapter.

Description

The scale cover is usually light gray, circular, and very thin and convex in side view (fig. 10.20), with a yellow to dark brown spot that is slightly off-center. It measures about $\frac{1}{16}$ inch (1.6 mm) across when fully grown. The body of the mature female under the scale cover is yellow and circular or pear-shaped. This scale is easily confused with latania scale and can be distinguished only by microscopic examination. Adults also resemble ivy or oleander scale but are less convex, and the spot is more centrally positioned.

Field biology

Little is known about the biology of this insect. Most overwinter as adult females, although other development stages have been found overwintering. As this scale may be found in various stages of development, presumably it has several overlapping generations per year.

Damage

Greedy scale may become abundant on the bark of branches, especially on older suckers; like other hard scales, it inflicts little damage on wood. However, if it develops on the fruit, it can cause deformation and discoloring marks.

Monitoring and control

Parasites and predators play a prominent role in regulating greedy scale populations. If the previous season's crop yield and quality assessments indicated that greedy scale is causing fruit loss, monitor crawlers in the following generation with double-sided sticky tape and apply treatments when crawlers are present.

California red scale

Although California red scale, *Aonidiella aurantii* (Maskell) (Hemiptera: Diaspididae), is a significant pest of citrus, infestations on olive rarely require treatment. It is widely distributed throughout Central and Southern California on citrus, and it attacks a diverse range of hosts, including fruitless mulberry, grape, nightshade, rose, and walnut.

Description

The adult female has a thin, round scale cover about $\frac{5}{64}$ inch (2 mm) across with a raised spot located centrally or just off-center. The reddish body is visible through the scale cover (fig. 10.21). The male scale covering is gray and elongated.

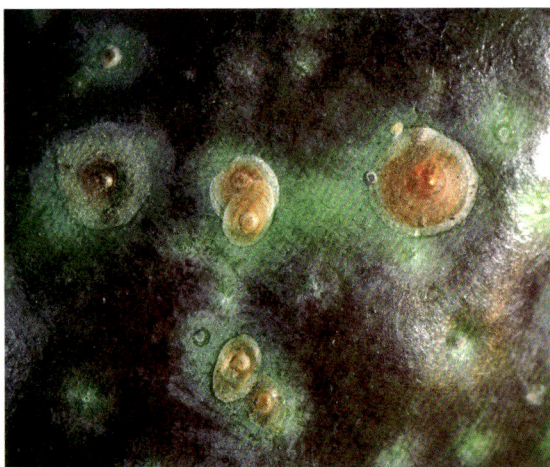

Figure 10.21

California red scale infestations on olive rarely require treatment.

Field biology

The female gives birth to young that remain under the scale cover for 1 or 2 days before emerging as crawlers. Crawlers seek a favorable site to settle, insert their mouthparts, and begin feeding. The female molts twice before becoming mature. Male and female development are similar until after the first molt, when the male cover becomes elongated rather than round. The male passes through a prepupal, pupal, and winged adult stage. There are two or three generations per year, and any stage in the life cycle may be found at any time of the year.

Damage

All parts of the olive tree can be infested, but only on rare occasions are treatments needed. California red scale does not discolor fruit, which distinguishes its damage from that of olive and oleander scales. As is the case with the other hard scales, fruit infested by this species is worthless for table olives, but it is unlikely that minor cosmetic damage to the fruit has an impact on olive oil quality. Therefore, the population density, or number of scales, is the most important consideration in terms of the amount of damage caused.

Monitoring and control

Depending on the climate, growing season, and treatment for other pests, biological control by parasitoids can be effective against California red scale in most coastal areas of the state, although not in the San Joaquin Valley, where warmer temperatures change both scale and parasitoid development and survival parameters. Red scale can be monitored by examining fruit, twigs, and leaves, or pheromone traps. If treatments are required, time pesticide applications of oils, insect growth regulators, or other materials against the first-generation crawlers in June or the second generation in late July and August.

Olive mite

The olive mite, *Oxycenus maxwelli* (Keifer) (Acari: Eriophyidae), is a native of the Mediterranean region and is now found throughout olive-growing regions globally. Among olive varieties grown in California, Ascolano is the most susceptible, followed by Sevillano, Manzanillo, and Mission.

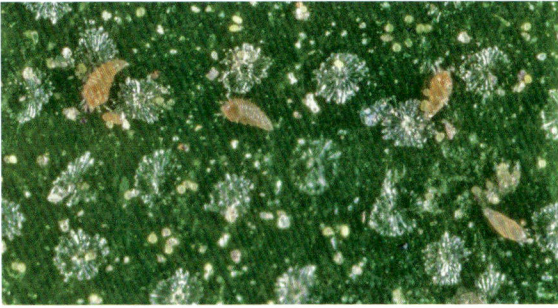

Figure 10.22

Olive mite has a yellowish to orange body. Females are broadest at the front and taper to the rear.

Description

The olive mite is a tiny, four-legged eriophyid mite with a yellowish to orange body (fig. 10.22). Females are broadest at the front of the body and taper to the rear. They are very small, about ½₀₀ inch (0.13 mm) long, and extremely difficult to see without a microscope or a hand lens of at least 20X power.

Field biology

The olive mite overwinters as an adult in bark crevices. Egg laying begins in late winter or early spring and continues until summer, when the mite estivates in response to hot and dry conditions. Sudden periods of relative humidity below 20 percent and high temperatures cause high mortality. The mite resumes activity in fall until cold weather arrives.

Damage

Normally found on the surface of immature terminal olive leaves, the olive mite usually causes no damage. However, when present in large numbers on young leaves, the mite silvers the leaves and causes longitudinal curling (fig. 10.23), but this does not damage the tree or its productivity. In spring, the mites collect on the developing inflorescences, and high density populations can cause pistil abortion and subsequent crop losses. Dead and discolored floral buds, bud drop, blossom blast, and inflorescence abscission are also symptoms of bud damage associated with the olive mite.

Monitoring and control

Treatment is not recommended unless fruit set and crop have been below normal for several years and large numbers of mites are found.

Figure 10.23

Olive mite damage occurs only with heavy infestations on young leaves. This pest is not a major problem. Olive mites feed on succulent stem and bud tissues and on the upper surface of leaves. Gross symptoms of mite damage include sickle-shaped leaves, dead vegetative buds in spring, discoloration of flower buds, bud drop, blossom blasting, inflorescence abscission, and reduced shoot growth. *Photo:* Paul Vossen.

Populations can be monitored by examining shoot tips and developing bloom for the presence of olive mites. If high density populations occur on developing inflorescences, insecticide treatment is recommended. Both wettable and dusting sulfur can be used for the control of olive mite, as they can with many eriophyids, and these products are acceptable for use on organically certified crops. Under high temperatures, dusting sulfur is safer to use than wettable sulfur.

Western flower thrips

The western flower thrips, *Frankliniella occidentalis* (Pergande) (Thysanoptera: Thripidae), is native to and widely distributed throughout western North America. It has a very broad host range, including more than 100 species of plants in California. Western flower thrips is attracted to olive during bloom. Research in table olives found that Ascolano is the most susceptible variety, although other varieties can be injured. No insecticides are registered for this pest in olive, so cultural practices are critical to avoid outbreaks.

Description

The western flower thrips adult is a tiny insect, about ½₃₂ inch (0.8 mm) long, with two pairs of fringed wings (fig. 10.24). While adult males

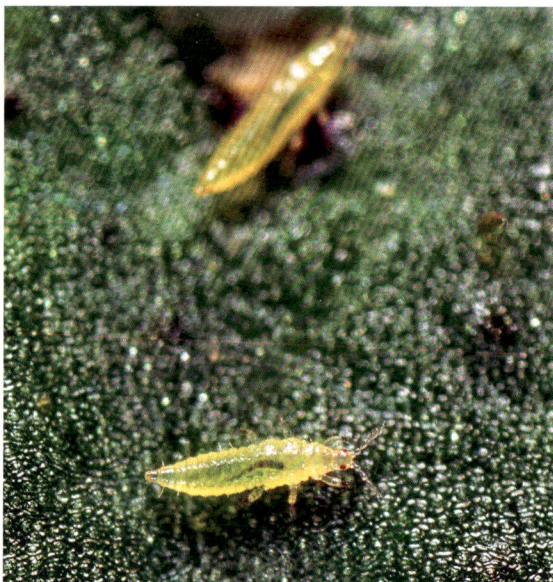

Figure 10.24

Western flower thrips adults overwinter in weeds and grasses. In early spring, they move into the olive trees, infesting leaves, flowers, and then fruit. *Photo:* Jack Kelly Clark.

are pale, adult females vary in color, from white to yellow with slight brown spots on top of the abdomen, to yellowish with an orange thorax and brown abdomen, to dark brown all over. Different color forms predominate according to the time of year, with darker forms more frequent after cold periods. First-instar nymphs are opaque or light yellow, turning golden yellow after the first molt. The pupae are soft bodied, with visible wing pads.

Field biology

Adults overwinter in weeds, grasses, and other sites within or outside the orchard. In early spring, females use a sawlike ovipositor to insert opaque, kidney-shaped eggs under the epidermis of leaves, flower parts, and fruit. Eggs hatch in 5 to 15 days, depending on temperature. Nymphs feed on developing leaves, fruit, and shoots. After completing their development, the nymphs drop to the ground, where they pass through two non-feeding pupal stages before emerging as adults.

Populations usually peak in May or June, as wild areas dry up and adults migrate to cultivated areas, including olive orchards. Typically, there are five or six generations of western flower thrips per year, although they can develop anytime temperatures exceed the minimum threshold of 46° to 50°F (8° to

10°C). The ideal temperature range for rapid development is between 77° and 86°F (25° and 30°C).

Damage

Because the western flower thrips migrates to olive orchards in spring from adjoining vegetation, trees near drying grain fields or dry weeds in the orchard are most susceptible to damage. Although the insect feeds on leaves and tender shoots, it causes the most damage when it feeds on fruit. While damaged fruit is scarred and dimpled and must be culled before processing for table olives, this type of cosmetic damage (fig. 10.25) is likely less important in oil production, although the impact on oil quality has not been specifically evaluated.

Monitoring and control

There are no insecticides registered for control of thrips on olive, but cultural practices can help prevent infestations. Avoid disking vegetation while trees are in bloom. Open areas adjacent to the orchard should also be disked as early as possible to prevent a buildup of thrips populations, which can later migrate to olive trees.

Branch and twig borer

The branch and twig borer, *Melalgus* (=*Polycaon*) *confertus* (LeConte) (Coleoptera: Bostrichidae), attacks a number of crops in California. Damage to olive, seldom economically harmful, is generally limited to areas adjoining the preferred hosts. The adults are known to bore

Figure 10.25

Western flower thrips damage to fruit is superficial. The impact on the oil is unknown.

Figure 10.26

Adult branch and twig borers bore round holes in the axils of small twigs and branches.

into a wide range of healthy hosts. Reproduction occurs in deadwood, primarily of grape, madrone, and oak.

Description

The adult branch and twig borer is a slender brown beetle about ½ to ¾ inch (13 to 19 mm) long; its body is oblong, with the head and thorax narrower than the abdomen (fig. 10.26). The larvae are C-shaped, white, and covered with fine hairs.

Field biology

The adult lays eggs in dead and dying wood of many native and cultivated trees, shrubs, and vines (including grape) outside the orchard. The larvae bore into the heartwood of the host and feed until development is complete. Pupation occurs inside infested wood, and adults emerge early in spring. Adults fly to nearby trees, where they bore into branches and twigs. There is one generation per year.

Damage

Adults bore round holes into the axil of small twigs and branches or into the base of buds. Small branches die, and injured branches frequently break off at the bored holes. Borer damage is sporadic and not usually of economic importance, although it can be more problematic in young orchards, where trees are still establishing.

Monitoring and control

To minimize the branch and twig borer, burn all infested wood inside and around the orchard to destroy developing larvae and get rid of potential host material. There are no chemical controls for this pest.

American plum borer

The American plum borer, *Euzophera semifuneralis* (Walker) (Lepidoptera: Pyralidae), is distributed throughout North America and attacks a wide range of host plants, including cultivated fruit and nut trees and ornamental plants. Adults prefer to oviposit at wound sites or wherever there is callous tissue. There are multiple generations per year, and adult moths are most active in spring.

Description

The adult moth has gray forewings with brown and black markings and a wingspan of about ¾ inch (19 mm) (fig. 10.27). The eggs are oval and

Figure 10.27

The American plum borer lays eggs in callous tissue that forms at pruning wounds, crown galls, and scaffold crotches. *Photo:* Jack Kelly Clark.

Figure 10.28

The American plum borer overwinters as an immature or mature larva. *Photo:* Jack Kelly Clark.

white when first laid, turning a dull red shortly afterward. They are laid singly or in small clusters, usually in cracks or bark crevices where callous tissue has developed, for example, at pruning wounds, crown galls, and scaffold crotches. Young larvae are white with a large, dark brown head. Mature larvae are about 1 inch (25 mm) long and vary in color from dusky white, to pink, to dull green (fig. 10.28). Pupae are olive green when they first form and become dark brown before emergence.

Field biology

The American plum borer overwinters as either immature or mature larvae, depending on the olive region. Mature larvae pupate in early spring, and adult moths emerge in April and May. There are multiple generations in California. In late fall, the larvae construct loose silken cocoons under bark scales, at the entrance to or near the feeding sites in which they overwinter.

Damage

The larvae attack the soft, spongy callus tissue that occurs in galls (which are frequently caused by the olive knot bacterium), at graft unions, and in tree wounds. They can continue to feed into normal tissue, girdling limbs, which may be weakened or killed. Most of the damage is to new grafts.

Monitoring and control

The activity of the American plum borer can be detected and differentiated from that of other borers by the presence of brownish frass and webbing at feeding sites. Pheromone lures are available for this species and can be used with sticky traps to detect activity of adult moths in the orchard.

If larvae are present, remove and destroy infested wood, if possible. If the wood cannot be removed, chemical control (carbaryl) may be necessary. Spray trees from 1 foot (0.3 m) above the scaffold crotch to 1 foot (0.3 m) below, two or three times during the growing season, starting in mid- to late April. Count and remove frass piles prior to spraying, and continue to monitor the area for evidence of larval activity after spraying. Painting the trunk with water diluted with white latex paint after spraying will prevent sunburn and preserve the insecticide over a longer period.

Black vine weevil

The black vine weevil, *Otiorhynchus sulcatus* (Fabricius) (Coleoptera: Curculionidae), has more than 80 hosts. It has long been a serious pest of grape in Europe and of ornamentals, especially container-grown plants, in the western United States. Occasionally, it attacks olive, primarily in the San Joaquin Valley.

Description

The adult black vine weevil is oblong, about ⅜ inch (10 mm) long, and dark brown to black, with a somewhat bumpy surface (fig. 10.29). The body is covered with dense, short, light-colored hairs with small patches of yellow hairs on the elytra, or hard wings, that cover the abdomen. A long,

Figure 10.29

Black vine weevil is a serious pest of other plants but only occasionally attacks olive.

Figure 10.30

Black vine weevil larvae have a long, broad snout. *Photo:* Jack Kelly Clark.

broad snout projects from the front of the head. The legless larvae are about ⅜ inch (10 mm) long when mature, with a dirty white body and a brownish head (fig. 10.30).

Field biology

The black vine weevil overwinters as larvae in the soil and subsequently pupates in the soil in spring. Adult females emerge around March in the San Joaquin Valley and, about 2 to 4 weeks later, begin laying eggs in cracks and crevices in the soil. The black vine weevil is parthenogenetic (that is, there are only females, no males). After the eggs hatch, young larvae work their way into the soil, where they feed on roots until the following spring. There is one generation per year.

Damage

Damage caused by the black vine weevil is attributed primarily to larval feeding on roots. However, little is known about the impact of this feeding on olive roots. Adult black vine weevils are nocturnal, hiding beneath loose bark, in debris, or in other protected places during the day and then moving up the tree to feed at night. Feeding by adults is what is most apparent, and concerning, to growers in California. The adults feed primarily along leaf margins, causing a characteristic notched or scalloped appearance (see fig. 10.29); heavy infestations can cause considerable leaf loss. In general, established trees are more resistant to damage than young trees.

Monitoring and control

Adults are difficult to detect as they hide under loose bark and debris at the base of trees during the day, although it may be possible to monitor them by checking trees at night with a flashlight. If signs of feeding damage appear, further infestation can be prevented by minimizing weevil access to the canopy as follows: prune lower limbs to ensure they do not touch the soil, thus stopping weevils from using branches to climb into the tree; wrap the trunk tightly with a 6- to 8-inch (15- to 20-cm) band of plastic wrap or duct tape; apply a 3- to 4-inch (8- to 10-cm) band of sticky barrier on top of the wrap (do not apply sticky barrier directly to the tree as it can soften the bark); reapply the sticky material when it becomes dirty or loses effectiveness; and remove the bands before winter. Young or newly grafted trees may also be sprayed with insecticide to help control leaf damage.

Olive psyllid

The olive psyllid, *Euphyllura olivina* (Costa) (Hemiptera: Psyllidae), first appeared in Southern California in 2007. Thus far, populations remain limited to this region, although the pest has been reported in coastal regions as far north as Monterey. In Europe, the olive psyllid has been reported from more interior regions, suggesting that range expansion in California may be possible. The insect's host plants are olive, Russian olive (*Elaeagnus angustifolia* L.), and mock privet (*Phillyrea latifolia* L.).

Description

Olive psyllid adults are 3/32 inch (2.4 mm) long and strong jumpers; they are light tan and have forewings marked with a few small dark spots (fig. 10.31). The nymphs are flat and green to tan, and they secrete a white waxy coating that covers the entire colony (fig. 10.32). There are five nymphal stages, which are 1/100 to 1/16 inch (0.3 mm to 1.6 mm) long. The eggs are elliptical, 1/100 inch (0.3

Figure 10.31

Olive psyllid grows to an adult in about 3 months. It is a strong jumper. *Photo:* Marshall W. Johnson.

Figure 10.32

Olive psyllid secretes a white waxy coating. *Photo:* Marshall W. Johnson.

Figure 10.33

The waxy secretions from olive psylid cause flower and small fruit drop. *Photo:* Marshall W. Johnson.

mm) long, pale yellow, and attach to the substrate by a pedicel.

Field biology

Depending on temperature, the olive psyllid can grow from egg to adult in about 3 months, with 68° to 77°F (20° to 25°C) being the optimal temperature range for growth. There are three generations per year. In temperatures above 81°F (27°C), the pest is less active. Mortality increases at temperatures above 90°F (32°C).

Adults overwinter in sheltered areas of the olive trunk. In spring, females lay eggs on the new shoots and buds, with a single female able to lay 1,000 eggs or more. The second generation develops on buds and flowers in May and June. The third generation is often unnoticed, appearing in September and October. In Southern California, the population declines after June, most likely due to hot temperatures. Numbers remain low until the following spring, when buds appear on the trees.

Damage

Trees that are heavily infested can have yield losses of 30 to 60 percent. The olive psyllid damages trees through direct feeding on buds, flowers, tender shoots, and small fruit, as well as through the production of honeydew, which increases

sooty mold development. During olive flowering and fruiting, waxy secretions from the pest cause flower and small fruit drop and yield reductions (fig. 10.33). Large populations may inhibit the growth of young trees.

Monitoring and control

While the impact of the psyllid on olive trees is not fully understood, it appears that trees can tolerate low numbers of the pest with minimal economic damage. Monitor for the psyllid by inspecting flower clusters between March and May. Losses may occur if more than ten psyllids per flower cluster are encountered. While the second generation (May and June) causes the most damage, reducing the population size of the first generation in early spring will help keep the second generation small. Control measures should be taken before psyllids start secreting their heavy waxy coating, which protects them from pesticides.

Due to the olive psyllid's recent introduction in California, exotic natural enemies have not been released, and there are few potential natural enemies in the state that provide effective control. Researchers are seeking exotic natural enemies from locations in Europe, where the psyllid is found, and have identified the Mediterranean parasitoid *Psyllaephagus euphyllurae* (Masi) as a potential candidate for release.

Since the olive psyllid is sensitive to heat, pruning out center limbs to enhance air circulation and expose the psyllid to heat may reduce the population in areas with hot temperatures. If chemical control is necessary, use of nonresidual contact insecticides such as azadirachtin, neem oil, insecticidal soaps, and narrow-range oils can be effective if good coverage can be achieved, although that can be difficult in SHD orchards. Alternatively, carbamate can be applied, but it has a lot of residual activity and is highly toxic to natural enemies, which may lead to secondary pest outbreaks.

Ants

Ant species (Hymenoptera: Formicidae) commonly encountered in olive orchards include Argentine ant, *Linepithema humile* (Mayr); pavement ant, *Tetramorium caespitum* (Linnaeus); southern fire ant, *Solenopsis xyloni* (McCook); and native gray ants, *Formica aerata* (Francoeur) and *F. perpilosa* (Wheeler). All these species can be extremely disruptive to IPM, especially of scale insects, as they feed on their honeydew and protect them from natural enemies. Ants may become a problem in olive orchards that are not frequently tilled. Manage ants when they are interfering with biological control of scales or other pests. A combination of cultural practices and baits can be used in an integrated program.

Description

The Argentine ant (fig. 10.34) is common in most California olive-growing areas. It is about ⅛ inch (3.2 mm) long, uniformly deep brown to light black, and does not bite or sting. The Argentine ant has one petiole node (hump) (fig. 10.35) between the thorax and the gastor (swollen part of abdomen right behind the petiole).

The pavement ant (fig. 10.36) is ³⁄₃₂ inch (2.4 mm) long and has a dull, blackish brown body covered with coarse hairs. The head and thorax have many parallel furrows. Pavement ants have two petiole nodes between the thorax and the gastor (fig. 10.37).

Figure 10.34

The Argentine ant is common in olive orchards and disrupts IPM programs. *Photo:* Jack Kelly Clark.

Figure 10.35

Argentine ants have a petiole node, or hump, between the thorax and gastor. *Photo:* Jack Kelly Clark.

Figure 10.36

The pavement ant is covered with coarse hairs. *Photo:* Jack Kelly Clark.

Figure 10.37

Pavement ants have two petiole nodes between the thorax and gastor. *Photo:* Jack Kelly Clark.

Figure 10.38

The southern fire ant is covered with golden hairs. *Photo:* Jack Kelly Clark.

Figure 10.39

The native gray ant is larger than other ants found in olive orchards. It disrupts IPM programs. *Photo:* Jack Kelly Clark.

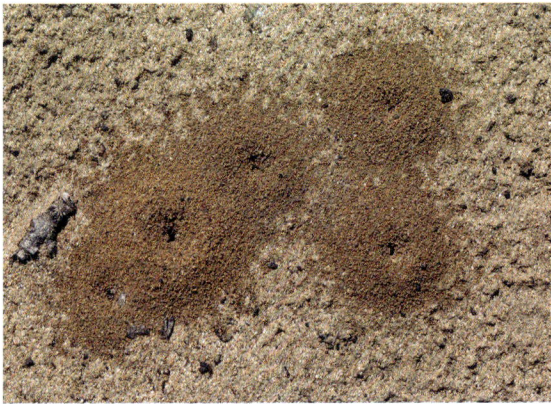

Figure 10.40

Pavement ants prefer to nest in sandy or loam soils. *Photo:* Jack Kelly Clark.

Figure 10.41

Southern fire ants build nests of loose mounds or craters in wetted areas. *Photo:* Jack Kelly Clark.

The southern fire ant (fig. 10.38), also called the California or native fire ant, is light reddish brown with a black abdomen. The entire body is covered with golden hairs and has two petiole nodes between the thorax and the gastor. Individuals range in size from ³⁄₃₂ to ³⁄₁₆ inch (2.4 to 4.8 mm) long.

Native gray ants (fig. 10.39), also called field ants, are larger than the other ants, measuring up to ⁵⁄₁₆ inch (8 mm). Like the Argentine ant, they have one petiole node.

Field biology

Argentine ants travel in characteristic trails on the ground and up the trunk and scaffolds. They forage during daylight hours. Their nests are very shallow, usually within 2 inches (5 cm) of the soil surface.

Pavement ants move in slow deliberate motion. They prefer to nest in sandy or loam soils (fig. 10.40).

Southern fire ants do not usually travel in conspicuous trails and will swarm over the ground when disturbed. They sting when provoked. Southern fire ants build nests of loose mounds or craters (fig. 10.41) near bases of trees around wetted areas and do not aggregate in colonies as large as those of the Argentine ant. They forage in the morning and early evening and remain underground during hot periods.

Native gray ants nest in topsoil or under rocks and debris, move in an irregular jerky manner, and generally do not travel in trails or sting. *F. aerata* is more common in the San Joaquin Valley, whereas *F. perpilosa* occurs primarily in the Coachella Valley.

Damage

Ants, especially Argentine and native gray ants, can be extremely disruptive to IPM programs. They feed on honeydew excreted by scales and also protect the honeydew-producing insects from predators and parasites, thus disrupting biological control.

Monitoring and control

Monitor the orchard in spring when scales begin producing honeydew. Check the abdomen of ants descending the trunks to see if they are swollen and translucent; this helps identify them as honeydew-collecting species. Periodically inspect for ants on branches and leaves.

Planting a cover crop of common vetch (*Vicia sativa* L.) can help to keep native gray ants off the trees. Common vetch has an abundance of nectaries that attract the ants away from honeydew-producing insects. In grape vineyard studies, common vetch was planted in an 80:20 mixture with 20 percent Merced rye so that it could establish in late fall and winter to attract the ants in spring and early summer. The addition of rye to the mixture helps to provide structure and support for the vetch in the cover crop.

Tilling the soil to control weeds disturbs the nesting sites of ants and helps to reduce their populations. Because the Argentine ant is well adapted to disturbance, it is probably best controlled with baits. While baits are the preferred chemical method for control of all ant species, their use can be labor intensive and may not always be considered economically feasible. Effective bait insecticides have slow-acting toxicants that worker ants collect and feed to other ants, including nest-building immatures and queens. For the most effective and economical control, treat when ants are active in early spring following winter rains and again in late August. To determine which bait to use, identify the primary ant species; the southern fire ant is predominantly a protein feeder, whereas most gray and black ants are sugar feeders.

Solid baits utilize corncob grits mixed with soybean oil as the food attractant plus an insecticide. These are effective for the primarily protein-feeding southern fire ant. The toxicants tend to degrade in light, so apply baits early in the morning or late in the day, when ants are active and will take the bait into the nest. Generally, corncob grit baits are broadcast over the acreage that needs to be treated. However, spot application of baits at the location of the ant nest is preferred over widely spreading the bait because it concentrates the food where the ants are.

Liquid baits use a toxicant mixed in sugar water, which disguises the toxicant as well as attracts the ants. These baits are most useful for the liquid sugar-feeding Argentine and native gray ants. Evaporation of the bait can cause the concentration of the toxicant to increase to a level in the bait that becomes repellant to ants. All liquid baits must be used in an EPA-approved bait station.

Suggested reading

Burrack, H. J., and F. G. Zalom. 2008. Olive fruit fly (Diptera: Tephritidae) ovipositional preference and larval performance in several commercially important olive varieties in California. Journal of Economic Entomology 101(3):750–758. https://doi.org/10.1093/jee/101.3.750

Burrack, H. J., R. Bingham, R. Price, J. H. Connell, P. A. Phillips, L. Wunderlich, P. M. Vossen, N. V. O'Connell, L. Ferguson, and F. G. Zalom. 2011. Understanding the seasonal and reproductive biology of olive fruit fly is critical to its management. California Agriculture 65(1):14–20.

Daane, K. M., and L. E. Caltagirone. 1989. Biological control of black scale in olives. Hilgardia 43(1):9–11.

Daane, K. M., and M. W. Johnson. 2010. Olive fruit fly: Managing an ancient pest in modern times. Annual Review of Entomology 55:151–169. https://doi.org/10.1146/annurev.ento.54.110807.090553

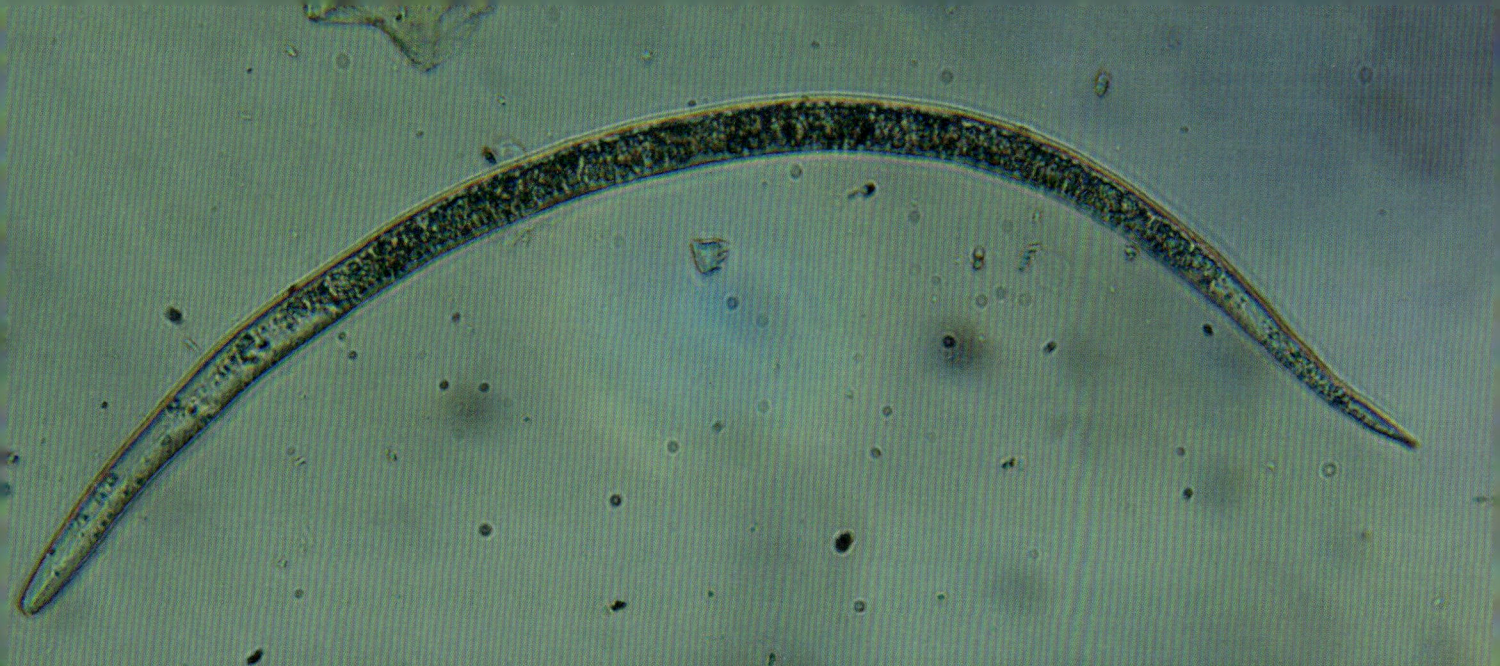

11 Nematodes of Olive

Amanda Hodson,
Assistant Professional Researcher, Department of Entomology and Nematology, UC Davis

Highlights

- Plant-parasitic nematodes reduce productivity by damaging olive roots.

- Economic damage due to nematodes is likely underestimated.

- Nematodes of concern include some species of root lesion nematodes (*Pratylenchus* spp.), citrus nematodes (*Tylenchulus semipenetrans*), and root knot nematodes (*Meloidogyne* spp.).

- Nematode infection may increase tree susceptibility to Verticillium wilt.

- Nematodes can be diagnosed only through soil sampling.

- Preplant is the best time to consider management of nematodes.

- Fumigants registered for use in olive include 1,3-dichloropropene and 1,3-dichloropropene plus chloropicrin, but environmental and human health concerns as well as regulatory restrictions limit their use.

- Nonchemical alternatives that may reduce nematode populations include solarization, biosolarization, and biofumigation.

The ability of nematodes to injure olive trees has not been studied extensively in the field, particularly in California. More than 20 years ago, root knot and citrus nematodes were estimated to cause up to 10 percent of all olive production losses in California (Koenning et al. 1999). With the adoption of higher olive orchard planting densities and increased irrigation (which favors nematode populations), losses due to nematodes have likely increased substantially. However, since nematodes require soil sampling for diagnosis, and soil samples are not taken unless nematodes are suspected to be a problem, the extent of the economic damage caused by nematodes in olive production is unknown and probably underestimated.

Nematodes of concern

Nematodes are microscopic true roundworms found in all soils; most feed on bacteria, fungi, or other nematodes and can be considered neutral or beneficial for the orchard. Of primary economic concern for olive growers are plant-parasitic nematodes, which use a hollow mouth stylet to feed on live olive root cells and disrupt root functioning, limiting water and nutrient absorption (Karssen and Moens 2006). After feeding on roots, adults reproduce and lay eggs, which hatch into juvenile nematodes that search for new roots to infect. Nematodes of particular concern include certain species of root lesion nematodes (*Pratylenchus* spp.), citrus nematodes (*Tylenchulus semipenetrans*), and root knot nematodes (*Meloidogyne* spp.). Other nematodes that might occur and are known to parasitize olive roots include spiral nematodes (*Helicotylenchus* spp.) and dagger nematodes (*Xiphinema* spp.).

Root lesion nematodes

Several species of root lesion nematodes can occur in olive orchards. It is important to know which species is present when attempting to diagnose a field problem as nematode related. The two main species that damage olives trees are *Pratylenchus vulnus* and *P. penetrans* (Castillo et al. 2010). These nematodes are classified as migratory endoparasites, which means they can feed, move, and reproduce within the root system. Nematode feeding destroys root cells, causing dark lesions along the root that are often invaded by rot-causing pathogens. Studies from Europe found that *P. vulnus* and *P. penetrans* reduced olive seedling growth in laboratory experiments (Lamberti and Baines 1969; Nico et al. 2002; Nico et al. 2003a). Other species of root lesion nematodes, such as *P. neglectus* and *P. thornei*, may also occur in orchards (Siddiqui et al. 1973), but they likely feed on weed roots, not olive roots.

Pratylenchus vulnus

P. vulnus occurs on commercial crops such as walnut, almond, stone fruit, grape, and olives (Siddiqui et al. 1973). In olive, aboveground signs of damage include poor growth and die-back of small branches, or stunting in younger trees. Longitudinal cracking of the root cortex occurs in larger roots of infested trees; the area underneath these cracks is often darkened and necrotic (McKenry 2004). Nematodes also attack and frequently kill small feeder roots. This nematode is more prevalent in the warmest regions of California, but it may also be found in California's northern olive-producing regions (McKenry 2004).

Pratylenchus penetrans

The very wide host range of *P. penetrans* includes the roots of woody perennials, clovers, and grasses (Siddiqui et al. 1973) (fig. 11.1). This species is most common in the cooler climates of the northern United States and the northern half of California. Still, it also occurs south of the Sacramento–San Joaquin River Delta region, especially at higher elevations on cherry and apple trees (McKenry 2004). Generally, Northern California olive growers deal primarily with *P. penetrans*, whereas growers in the southern San Joaquin olive district deal primarily with *P. vulnus* (McKenry 2004).

Citrus nematode

The citrus nematode, *Tylenchulus semipenetrans*, has regional strains, or biotypes, known to infest grape, persimmon, and olive roots (Inserra et al. 1980) (fig. 11.2). It is classified as a semi-endoparasite, which means it partially buries its body in the root cortex. Lamberti and Baines (1970) showed that, in California, the citrus nematode strain isolated from olive reproduced more quickly on olive trees than a *T. semipenetrans* strain isolated from citrus. This suggests that there are strains of *T. semipenetrans* particularly adapted to reproduce on olive roots. Infested root systems are characterized by the disintegration of small feeder roots (Baines 1951); aboveground symptoms may be similar to those seen in citrus trees: reduced growth, lack of vigor, and a thin foliar canopy (Verdejo-Lucas and McKenry 2004). Infested roots may appear dirty and enlarged because soil adheres to the gelatinous matrix that protects the nematode eggs. This nematode occurs in olive orchards of Southern and Northern California and can be common even in clay soil. It is known to develop very high population levels (McKenry 2004).

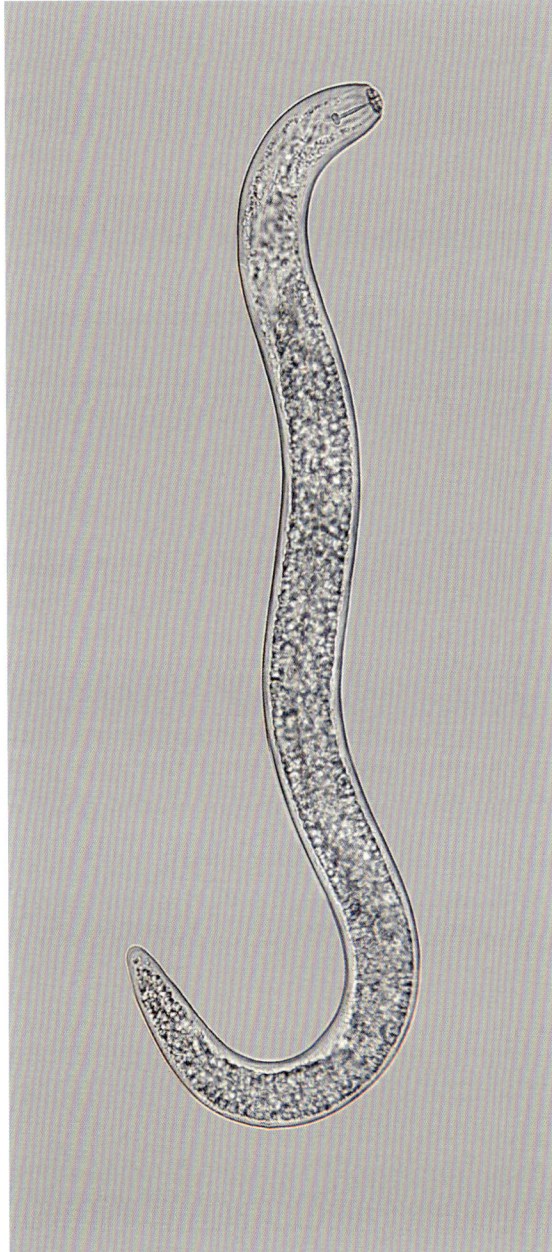

Figure 11.1

Root lesion nematode *(Pratylenchus penetrans)* is an endoparasite that migrates within olive roots. *Photo:* J. D. Eisenback.

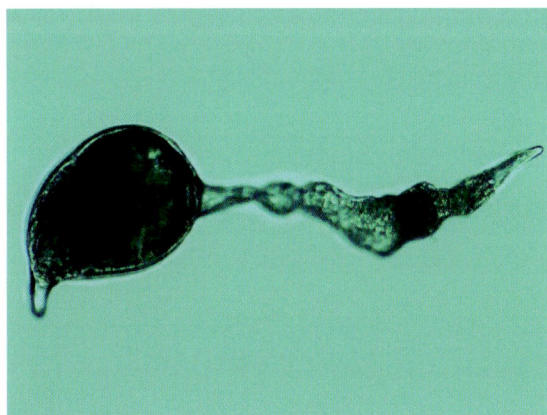

Figure 11.2

This swollen female citrus nematode (*Tylenchulus semipenetrans*) has been removed from a root, but normally the slender head would be embedded within root cells to feed while the bulging posterior would protrude from the root to release eggs. *Photo:* J. D. Eisenback.

Root knot nematodes

At least three *Meloidogyne* species, *M. incognita*, *M. javanica,* and *M. arenaria,* can damage olive rootstocks (Sasanelli et al. 1997; Sasanelli et al. 2002) as well as self-rooted Arbequina and Picual varieties (Nico et al. 2003a) (fig. 11.3). All self-rooted olive varieties grown in California are susceptible to root knot nematodes (McKenry 2004). Classified as endoparasites, root knot nematodes invade growing root tips and establish permanent feeding sites. Their feeding stimulates the root tissue to enlarge, forming giant cells called galls. Females partially protrude from the galls, releasing eggs contained in gelatinous masses (fig. 11.4). Root knot nematodes are very common and especially problematic in sandy soils. Infested trees may show reduced vigor and fruit size, or seem healthy. Roots appear bumpy and swollen, without any fine feeder roots.

Nematodes and Verticillium wilt

Nematode feeding may increase a tree's susceptibility to Verticillium wilt *(Verticillium dahliae)*. When both Javanese root knot nematode *(Meloidogyne javanica)* and *V. dahliae* were inoculated simultaneously into the soil, olive seedlings showed reduced growth and more symptoms, such as more chlorotic, wilted, and necrotic leaves, than when either pathogen was inoculated alone (Saeedizadeh et al. 2003). However, others have found no relationship between *V. dahliae* and root lesion nematodes (*Pratylenchus vulnus*), *V. dahliae* and root knot

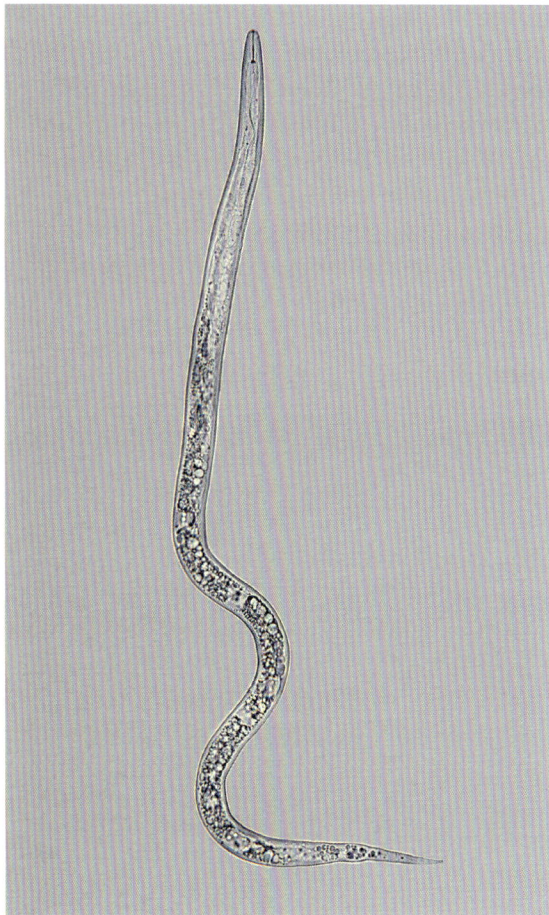

Figure 11.3

A juvenile of *Meloidogyne incognita*, a species of root knot nematode, searches through the soil for new roots to infest. *Photo:* J. D. Eisenback.

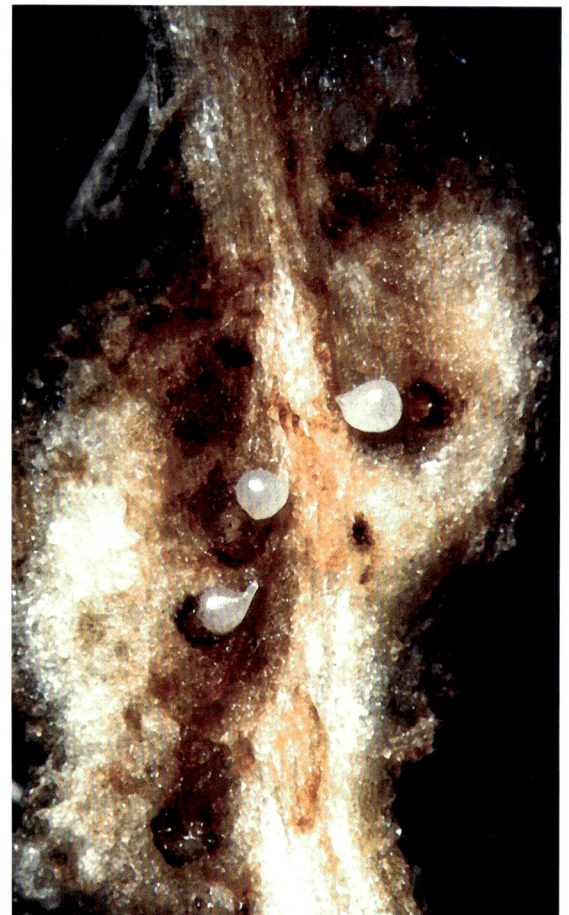

Figure 11.4

Females of root knot nematode, *Meloidogyne* species, protrude from a root gall, so that they can release eggs into the surrounding soil. *Photo:* J. D. Eisenback.

nematodes (*M. incognita*), or *V. dahliae* and both nematodes together (Lamberti et al. 2001). Castillo et al. (2010) suggested that while it is unknown exactly how plant-parasitic nematodes and pathogenic fungi interact, effects probably vary among olive varieties.

Soil samples

To determine the species present and its abundance, soil samples must be taken for nematode analysis. The following sampling recommendations are based on protocols developed by McKenry (2004) and Doll (2009). To collect samples, visually divide the orchard site into sampling blocks representing differences in soil texture, drainage patterns, and cropping history. For samples to be representative, blocks should not be more than 5 acres (2 ha) each. Within these blocks, collect at least ten subsamples randomly from areas where nematode activity is likely the highest, either in areas showing tree symptoms, such as wilting, or in areas known to be sandier. Mix the subsamples gently but thoroughly, and make a composite sample of about 1 quart (1 L) for each block.

While plant-parasitic nematodes can infest roots at least 5 feet deep, sampling at that depth may not be feasible. If the previous planting is still present, sample to at least 1.5 feet (0.5 m). But if the field has been fallow, leaving a dried-out top layer of soil, consider deeper sampling. Take samples near the wetting zone. This area will include feeder roots, where nematodes are likely to be present (ideally, include the roots in the sample). In an established orchard, it can be helpful to take separate subsamples from around trees that show symptoms and around adjacent healthy-looking trees for comparison.

For diagnostic analysis, label each bag with a sample ID name and note the location of the orchard, the sampling block within the orchard, current and previous crops, and, if applicable, next crop. While more than one sample from each sampling block will improve nematode distribution estimates in the field, processing samples is costly. Keep samples cool but not frozen, perhaps in a cooler with ice packs, preferably at 40° to 55°F (4° to 13°C). Be sure to keep samples out of the sun; even brief exposure of a soil sample in a plastic bag to direct sunshine can heat the soil to nematode-killing temperatures and result in an underestimate of the number of nematodes present. The samples should be transported to a diagnostic laboratory as soon as possible. There, the nematodes will be extracted, identified, and counted to form the basis for management decisions. University of California Cooperative Extension farm advisors can provide more details about sampling, locate a laboratory for extracting and identifying nematodes, and help in interpreting sample results.

Damage thresholds have not been established for the different nematode pest species in olive. If soil samples indicate that root knot, root lesion, or citrus nematodes are present, and the land to be planted has a crop history of woody perennials or broadleaf plants, management options for nematodes should be strongly considered. Ideally, diagnostics should distinguish between multiple root lesion nematode species to determine if the nematodes present are likely to feed on olive.

Management of nematodes

No treatment will completely eradicate nematode pests, and populations will rebuild gradually. Still, management may buy time for young, more vulnerable trees to develop healthy root systems. Nematicides are commonly recommended in perennial crops. However, preplant fumigation with nematicides in olive is uncommon, perhaps because the extent of damage by nematodes is poorly understood (Castillo et al. 2010) and the cost return of managing for nematodes in olive orchards is unclear. The two recommended fumigants are 1,3-dichloropropene and 1,3-dichloropropene plus chloropicrin, but environmental and human health concerns as well as regulatory restrictions limit their use. Efficacy depends on the moisture level, temperature, and texture of the soil. For example, because fumigants travel through soil air spaces, applying them to wet soil results in poor penetration and does not provide long-term control (Lembright 1990). Postplant nematicides that can be applied through the irrigation system, such as fluopyram and spirotetramat, have been used successfully in other tree crops but are not currently registered for use in olive.

Nonchemical alternatives may reduce nematode populations when applied prior to planting.

These approaches are less regulated than fumigants and are more environmentally friendly, but they may also provide less reliable control. In Spain, covering moist soil with a clear plastic tarp for 4 to 6 weeks during the hottest time of the year to induce heating (solarization) reduced *Meloidogyne incognita* populations in soil piles for olive nursery production (Nico et al. 2003b) and the incidence of Verticillium wilt in olive orchards (López-Escudero and Blanco-López 2001). Biosolarization, which combines solarization with the application of organic amendments, has shown promise in controlling plant-parasitic nematodes in other cropping systems, such as strawberry (Talavera et al. 2019) and almond (Shea et al. 2022).

Biofumigation involves plants releasing pesticidal chemicals as they break down, which can kill nematodes; biofumigation may be combined with tarping and/or tillage to increase the effect. Preplant biofumigation with Brassica species (Halbrendt 1996) and Sorghum species (Nyczepir and Rodriguez-Kabana 2007) decreased populations of plant-parasitic nematodes in a study with other orchard crops. Greenhouse studies showed that soil inoculation with arbuscular mycorrhizal fungi reduced root knot nematode damage and reproduction in olive seedlings, suggesting that such an approach could potentially be scaled up to the field level (Castillo et al. 2006).

References

Baines, R. C., and O. F. Clarke. 1951. Citrus-root nematode. The California Citrograph 37:60–86.

Castillo, P., A. I. Nico, C. Azcón-Aguilar, C. Del Río Rincón, C. Calvet, and R. M. Jiménez-Díaz. 2006. Protection of olive planting stocks against parasitism of root-knot nematodes by arbuscular mycorrhizal fungi. Plant Pathology 55:705–713. https://doi.org/10.1111/j.1365-3059.2006.01400.x

Castillo, P., A. I. Nico, J. A. Navas-Cortés, B. B. Landa, R. M. Jiménez-Díaz, and N. Vovlas. 2010. Plant-parasitic nematodes attacking olive trees and their management. Plant Disease 94(2):148–162. https://doi.org/10.1094/PDIS-94-2-0148

Doll, D. 2009. Sampling for plant parasitic nematodes. The Almond Doctor blog.

Halbrendt, J. M. 1996. Allelopathy in the management of plant-parasitic nematodes. Journal of Nematology 28:8–14.

Inserra, R. N., N. Vovlas, and J. H. O'Bannon. 1980. A classification of *Tylenchulus semipenetrans* biotypes. Journal of Nematology 12:283–287.

Karssen, G., and M. Moens. 2006. Root-knot nematodes. In R. N. Perry and M. Moens, eds., Plant Nematology. Wallingford, UK: CABI. 59–90.

Koenning, S. R., C. Overstreet, J. W. Noling, P. A. Donald, J. O. Becker, and B. A. Fortnum. 1999. Survey of crop losses in response to phytoparasitic nematodes in the United States for 1994. Journal of Nematology 31:587–618.

Lamberti, F., and R. Baines. 1969. Effect of *Pratylenchus vulnus* on the growth of "Ascolano" and "Manzanillo" olive trees in a glasshouse. Plant Disease Reporter 53:557–558.

———. 1970. Infectivity of three biotypes of the citrus nematode (*Tylenchulus semipenetrans*). Plant Disease Reporter 54:717–718.

Lamberti, F., F. Ciccarese, N. Sasanelli, A. Ambrico, T. D'Addabbo, and D. Schiavone. 2001. Relationship between plant-parasitic nematodes and *Verticillium dahliae* on olive. Nematologia Mediterranea 29:3–9.

Lembright, H. 1990. Soil fumigation: Principles and application technology. Journal of Nematology 22:632–644.

López-Escudero, F. J., and M. A. Blanco-López. 2001. Effect of a single or double soil solarization to control Verticillium wilt in established olive orchards in Spain. Plant Disease 85:489–496. https://doi.org/10.1094/PDIS.2001.85.5.489

McKenry, M. 2004. Nematodes of olive. In G. S. Sibbett and L. Ferguson, eds., Olive production manual. 2nd ed. Oakland: UC Agriculture and Natural Resources Publication 3353.

Nico, A. I., R. M. Jiménez-Díaz, and P. Castillo. 2003a. Host suitability of the olive cultivars Arbequina and Picual for plant-parasitic nematodes. Journal of Nematology 35:29–34.

———. 2003b. Solarization of soil in piles for the control of *Meloidogyne incognita* in olive nurseries in southern Spain. Plant Pathology 52:770–778. https://doi.org/10.1111/j.1365-3059.2003.00927.x

Nico, A. I., H. F. Rapoport, R. M. Jiménez-Díaz, and P. Castillo. 2002. Incidence and population density of plant-parasitic nematodes associated with olive planting stocks at nurseries in southern Spain. Plant Disease 86:1075–1079. https://doi.org/10.1094/PDIS.2002.86.10.1075

Nyczepir, A. P., and R. Rodriguez-Kabana. 2007. Preplant biofumigation with sorghum or methyl bromide compared for managing *Criconemoides xenoplax* in a young peach orchard. Plant Disease 91(12):1607–1611. https://doi.org/10.1094/PDIS-91-12-1607

Saeedizadeh, A., A. Kheiri, M. Okhovat, and A. Hoseininejad. 2003. Study on interaction between root-knot nematode *Meloidogyne javanica* and wilt fungus *Verticillium dahliae* on olive seedlings in greenhouse. Communications in Agricultural and Applied Biological Sciences 68 (4 Pt A):139–143.

Sasanelli, N., T. D'Addabbo, and R. M. Lemos. 2002. Influence of *Meloidogyne javanica* on growth of olive cuttings in pots. Nematrópica 32:59–63.

Sasanelli, N., G. Fontanazza, F. Lamberti, T. D'Addabbo, M. Patumi, and G. Vergari. 1997. Reaction of olive cultivars to *Meloidogyne* species. Nematologia Mediterranea 25:183–190.

Shea, E. A., J. D. Fernandez-Bayo, A. K. Hodson, A. E. Parr, E. Lopez, Y. Achmon, J. Toniato, J. Milereit, R. Crowley, J. J. Stapleton, J. S. VanderGheyst, and C. W. Simmons. 2022. Biosolarization restructures soil bacterial communities and decreases parasitic nematode populations. Applied Soil Ecology 172. https://doi.org/10.1016/j.apsoil.2021.104343

Siddiqui, I. A., S. A. Sher, and A. M. French. 1973. Distribution of plant parasitic nematodes in California. State of California Department of Food and Agriculture, Division of Plant Industry.

Talavera, M., L. Miranda, J. A. Gómez-Mora, M. D. Vela, and S. Verdejo-Lucas. 2019. Nematode management in the strawberry fields of southern Spain. Agronomy 9:252. https://doi.org/10.3390/agronomy9050252

Verdejo-Lucas, S., and M. V. McKenry. 2004. Management of the citrus nematode, *Tylenchulus semipenetrans*. Journal of Nematology 36(4):424–432.

Suggested reading

Lamberti, F., and R. Baines. 1969. Pathogenicity of four species of *Meloidogyne* on three varieties of olive trees. Journal of Nematology 1:111–115.

UC IPM. 2014. Pest management guidelines: Olive. UC Agriculture and Natural Resources Publication 3452. Accessed July 2020. www2.ipm.ucanr.edu/agriculture/olive/Nematodes/

Vovlas, N., P. Castillo, H. F. Rapoport, and R. M. Jiménez-Díaz. 2002. Parasitic nematodes associated with olive in countries bordering the Mediterranean Sea. Acta Horticulturae 586:857–860.

12

Diseases of Olive

Florent P. Trouillas,
Associate Professor
of Cooperative
Extension,
Department of Plant
Pathology, UC Davis

James E. Adaskaveg,
Professor,
Department of
Microbiology and
Plant Pathology, UC
Riverside

Highlights

- Olive knot is the most serious disease of olive trees in California and can be a limiting factor for economic olive oil production. Severe outbreaks of the disease may result in substantial yield reductions, as well as off-flavors in harvested fruit that may render orchards unprofitable. The widespread adoption of mechanized cultural practices has taken the disease to epidemic levels in some locations, and systematic and strict control measures are required in olive orchards to avoid substantial losses.

- Olive leaf spot (peacock spot) is a chronic disease and Pseudocercospora (Cercospora or Mycocentrospora) leaf spot is of increasing concern in high- and super-high-density olive orchards in California; these diseases can lead to defoliation and weakening of trees under favorable environmental conditions.

- The highly mechanized production practices in olive orchards have led to the emergence of Neofabraea leaf and shoot lesions, which are associated with wounds caused by harvesting equipment.

- Much-needed new bactericides and fungicides to manage leaf and shoot diseases are pending registration for use in mechanized olive production. Timing treatment applications after harvest, during the dormant season, and after leaf drop in spring prior to flowering will ensure high-quality, residue-free California olive oil.

- Verticillium wilt is a persistent disease of olive in California. It can have a substantial economic impact when orchards are planted in areas where highly susceptible crops, such as cotton, tomato, pepper, potato, eggplant, or squash, previously were grown for several years. Management of the disease is difficult and relies mainly on planting in soils with no history of susceptible crops, although inoculum of the pathogen can be reduced with chemical fumigation, solarization, flooding fallow fields, growing grass crops for several seasons, or any combination of these treatments.

- Phytophthora root and crown rot can be a problem when planting orchards in heavy loam or clay soils with poor water infiltration and draining capacities. Proper water management, particularly avoiding overirrigation, is the basis of successful control of the disease.

- The introduction of devastating diseases such as olive quick decline syndrome must be prevented. Monitoring of orchards should be done routinely, and affected trees should be destroyed immediately.

uper-high-density (SHD) and high-density (HD) olive orchards have become the standard for olive cultivation in California. These planting designs can lead to increases in leaf wetness duration and relative humidity within the canopy, providing microclimates that are more favorable to foliar diseases. The broad adoption in SHD and HD orchards of mechanized cultural practices has reduced manual labor and allowed for expansion of crop acreage as well as growth of the entire industry. However, mechanical harvesting and pruning inevitably create wounds on twigs, branches, and trunks that may become infection sites for bacterial and fungal pathogens, thus increasing the risks for disease occurrence.

Harvesting of olives occurs in autumn, which coincides with the onset of the rainy season in California. Rain is generally the main contributing factor for disease spread, including the release and dispersal of bacterial and fungal inocula. Furthermore, wetness allows for infection and colonization of the olive host. In California, temperatures are fairly conducive for most diseases year-round, except during the hot, dry summer months.

Olive varieties vary widely in their susceptibility to different diseases. Most are susceptible to at least one or several diseases. Resistant varieties may be considered when planting olive orchards, as part of an overall integrated pest management strategy; see chapter 4, Establishing a Super-High-Density Orchard, and chapter 5, Establishing a High-Density Orchard, for information on varietal susceptibility to various diseases. However, host resistance does not guarantee complete immunity to diseases, and even resistant varieties can succumb to diseases under severe environmental conditions and stress.

California has a long history of perennial and annual crop production systems, which may perpetuate inoculum reservoirs. Some fungal pathogens that occur in other crops can potentially infect olive, causing a risk for new disease emergence. In contrast, some diseases are endemic to olive but do not pose a problem under the current climate and cultural practices. Changing environmental conditions, however, including rising temperatures, drought, and exceptionally wet years or seasons (for example, abundant spring precipitation), are affecting California agriculture. These factors may contribute to a deviation in disease occurrence, expression, and distribution.

Disease outbreaks and epidemics in perennial crops generally arise in response to drastic changes in production practices, increases in monoculture farming, loss of effective chemicals by regulatory cancellation or development of pathogen resistance, climate change, continuing adaptation of pathogens to new environments, and, most important, global movement of plant material—or any combination of these factors. Disease prevention by avoiding pathogen introductions and monitoring orchards for early disease symptoms and infectious agents, as well as minimizing their spread, is necessary for the long-term health and security of the industry.

Olive knot

Olive knot is a bacterial disease caused by *Pseudomonas savastanoi* pv. *savastanoi*. The bacterium is an epiphyte living on the bark of twigs but can cause infections when injuries occur to the branch that expose internal tissue. Once inside the plant, the pathogen produces and releases indoleacetic acid and cytokinins (Comai and Kosuge 1980, 1982). These phytohormones stimulate hyperplasia and hypertrophy of the surrounding olive tissue, leading to the development of knots, or galls, that can be long-lived and harbor high concentrations of bacteria (Surico et al.

1985). The knots interfere with water and nutrient flow causing the distal (from the gall upward) portions of twigs, and sometimes larger branches and scaffolds, to die back.

All varieties of olive are susceptible; however, some are more susceptible than others. If unmanaged, the disease can be devastating to olive crop production. The disease generally does not cause tree death, but severe outbreaks may result in substantial yield reductions as well as off-flavors in harvested fruit that may render orchards unprofitable. Olive knot has become more severe in California due to the planting of susceptible varieties, increased planting density, mechanical harvesting and pruning practices, and harvests that extend into rainy periods.

Symptoms

Knot size is based on the tissue infected. Olive knot galls can be very small, as tiny as 0.125 inch (0.5 cm) on small twigs, and 2 to 3 inches (5 to 7 cm) or more in diameter on scaffold branches or the main trunk. They may develop on natural wounds, such as leaf and bud scars (fig. 12.1); on environmental injuries from freeze or hail; or on mechanical injuries from pruning, harvesting, or other orchard equipment (fig. 12.2).

Knots that develop as a result of freeze injury are longitudinally spindle-shaped galls that spread because the bacterium moves into tissue damaged by ice expansion and subsequent water soaking. This symptom gives the appearance of a spreading or systemic infection, but infections are generally nonsystemic, localized to parenchyma and nonvascular tissues. Less commonly, knots develop on leaves and fruit or at the nodes of branch emergence points. Transpiration and nutrient flow are disrupted by expanding gall tissue, which may cause defoliation, death of twigs and branches, and crop reduction. Symptoms do not develop on roots.

Comments on the disease

Olive knot occurs in olive-growing regions worldwide. It has been reported in California since the late 1800s. Rainfall favors dissemination of the pathogen, and any precipitation resulting in wetness allows infection of wounds. In California, rainfall typically occurs in late fall, winter, and spring, whereas frost injury may occur in the winter and hail injury with spring and summer

Figure 12.1

These leaf scar infections were caused by olive knot. *Photo:* James E. Adaskaveg.

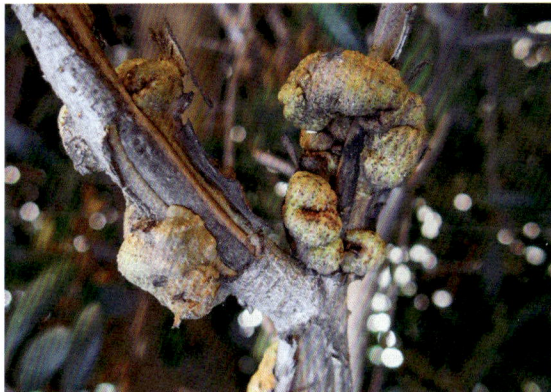

Figure 12.2

Olive knot galls developed after harvest injuries to a branch. *Photo:* James E. Adaskaveg.

thunderstorms. High precipitation and colder temperatures generally happen in the northern growing districts of the state; thus the disease occurs more frequently and at a higher incidence in these production areas. Temperature is not a limiting factor in disease development, and insects have not been associated with dissemination of the pathogen.

The bacterium survives as an epiphyte and inside the knots, where it can reproduce year-round. During wet events, such as rain or dew, the bacterium can multiply on exudates on the plant surface or extrude from inside the gall in high concentrations; it can be dispersed short distances by water splash, or moved with automated and manual pruning and harvesting equipment that becomes contaminated. Cardinal temperatures for growth of the pathogen are as follows: optimum, 72° to 75°F (22° to 24°C); maximum, 90°F (32°C); and minimum, 40° to 50°F (4° to 10°C).

P. savastanoi pv. *savastanoi* can only infect wounds and does not enter through natural openings, such as stomata, hydathodes, or lenticels. Leaf scars immediately after leaf drop and small wounds less than 1 inch (2.5 cm) in size are susceptible for 9 to 15 days. Larger wounds generally take longer to heal.

Five types of infection associated with injuries have been identified:

- **Fall harvest infection:** any precipitation or leaf wetness and/or mechanical dissemination that allows the pathogen to infect wounds
- **Winter freeze injury infection:** cold damage resulting in cracks in the bark and water-soaked tissue ideal for bacterial colonization
- **Spring leaf scar infection:** natural injuries occurring during leaf drop and rain events
- **Pruning injury infection:** conditions similar to fall harvest infection
- **Hail injury infection:** rainfall following hail injury resulting in dissemination of the pathogen and subsequent infection

Typically, with cold-winter events, it is not freeze injury but olive knot infection that causes the most damage to olive trees. The knots develop when the tree is actively growing in spring and summer. Thus, when infection occurs in late fall or winter, symptoms are delayed, whereas infection during olive growth periods may produce symptoms in 10 to 14 days (Nguyen et al. 2018a).

Management

Olive knot is a difficult disease to manage once it is established in the orchard (table 12.1).

Table 12.1

Summary of management strategies for olive diseases of importance or concern in California

Disease	Pathogen	Cultural practices			
		Site selection	Pruning time	Irrigation	Sanita
Olive knot	*Pseudomonas savastanoi* pv. savastanoi	---	+++	---	+++
Olive leaf spot	*Venturia oleaginea*	---	---	---	---
Pseudocercospora leaf spot	*Pseudocercospora cladosporioides*	---	---	---	---
Neofabraea leaf and shoot lesions and fruit rot	*Neofabraea kienholzii* and *Phlyctema vagabunda*	---	---	---	---
Phytophthora root and crown rot	*Phytophthora* spp.	+	---	++	
Armillaria root rot	*Armillaria mellea*	+++	---	---	+++
Verticillium wilt	*Verticillium dahliae*	+++	---	---	+++
Anthracnose[†]	*Colletotrichum acutatum* and *C. gloeosporioides*	---	---	---	---
Botryosphaeriaceae cankers	*Neofusicoccum mediterraneum, Diplodia mutila,* and *Botryosphaeria dothidea*	---	+++	---	---
Cytospora canker	*Cytospora oleicola* and *C. olivarum*	---	+++	---	---
Pleurostoma decline	*Pleurostoma richardsiae*	---	---	---	++
Olive leaf scorch	*Xylella fastidiosa* ssp. multiplex	---	---	---	---
Quick decline syndrome[‡]	*Xylella fastidiosa* ssp. pauca	---	---	---	---

KEY

--- = not effective   +/- = somewhat effective but inconsistent
+ = moderately effective   ++ = effective

+++ = highly effective

*Timing of protective treatments is based on environmental conditions and injury events: fall applications after harvest; wint

Cultural practices, sanitation, and prevention are the most effective strategies. Hand-pruning or mechanical-hedging operations during the dry season (spring to summer) allow for rapid wound healing, and dissemination of the bacterium from the galls to the wounds by rainfall is less likely. Pruning should be timed when no rain is forecast for at least 2 weeks.

Careful removal of knots under dry conditions reduces inoculum levels in the orchard. The petroleum distillates 2,4-xylenol and meta-cresol (the active ingredients in Gallex) kill galls and the bacterium, thus functioning as an eradicant. However, the product must be applied to individual galls because of potential phytotoxicity to healthy tissue. Therefore, application is labor intensive and costly when there are numerous galls on a tree. Treatment

is generally limited to large galls on scaffold branches or trunks.

Sanitizing harvesting and pruning equipment is also very important in preventing spread of the pathogen and disease. Sodium hypochlorite is commonly used for hand-pruning equipment, but generally not for expensive pruning or harvesting equipment because it is corrosive to metal. California and federal registrations have been obtained for a quaternary ammonium formulation (Deccosan 321, Master Label-Maquat) that can be used on equipment without the risk of corrosion. Quaternary ammonium compounds are registered for tools in contact with food, such as olive mills and handling equipment for extracting, moving, or storing oil, provided that a water rinse is used immediately after application. In the field, sanitizing automated equipment

Preplant			Postplant		Timing of bactericides/fungicides*			
arization	Soil fumigation	Flooding/ fallow	Monitoring	Insect management	Fall (after harvest)	Winter	Spring	Eradicant
---	---	---	+	---	+++	+++	+++	+
---	---	---	+	---	+++	+++	+/-	---
---	---	---	+	---	+++	++	++	---
---	---	---	+	---	+++	++	+/-	---
+/-	+++	++	+	---	---	---	---	---
+/-	+++	+/-	+	---	---	---	---	---
+/-	+++	+	+/-	---	---	---	---	---
---	---	---	+	---	(+++)	(+++)	(+++)	---
---	---	---	+	---	---	---	---	---
---	---	---	+	---	---	---	---	---
---	---	---	+	+	---	---	---	---
---	---	---	+	+	---	---	---	---
---	---	---	+	+	---	---	---	---

applications after significant winter rainfall or freeze events; and spring applications before significant rainfall and/or timed with leaf drop, especially for olive knot.

†Presently not of major concern, and management strategies are not suggested at this time.

‡Presently not in California but of major concern.

after each row or after passing through an area with visible knots on the trees is an ideal usage strategy (Nguyen et al. 2017).

The most effective strategy to minimize the risk of olive knot establishing in the orchard is applications of copper-based bactericides alone or in mixtures with kasugamycin or oxytetracycline (both antibiotic registrations are pending), which protect leaf scars and other injuries (Nguyen et al. 2018b). Copper residues from fixed copper (for example, copper hydroxide or copper oxide) mixed with a sticker adjuvant or with lime, or from a copper Bordeaux mixture (copper sulfate plus lime) applied in 100 to 150 gallons per acre (936 to 1,404 L per ha), will last several weeks before they are diluted and washed off the plant by rain or heavy dew. Antibiotics have shorter residual times, ranging from 3 to 4 days for oxytetracycline and 7 to 10 days for kasugamycin. Thus, mixtures of antibiotics and copper are the best way to obtain high efficacy, reduce the risk of resistance to either treatment, and have a long-lasting residual bactericide (copper) on the plant. With either antibiotics or copper, preventive applications should be made within 24 hours after injuries occur because the pathogen can multiply and establish itself inside wounds during this period. Applications after 24 hours are less effective, and later applications are often ineffective.

A minimum of two applications annually is usually required. The first application is made in fall immediately after harvest and before winter rains; a second application is made in spring at the onset of annual leaf drop (typically around late March to early May) prior to rain. A third application may be necessary in winter when low temperatures cause freeze injury or in spring or summer when hail and rainstorms cause injury and dissemination of the bacterial pathogen.

Harvesting and pruning should be avoided when the leaves and branches are wet or there is any chance of precipitation. In moist conditions, the bacterium oozes from the knots and can be carried by the harvesting or pruning equipment to abrasions or other mechanical injuries. Many growers who use automated hedging equipment schedule pruning to coincide with spring leaf drop, and they prune only every other or every third row each year. This means that growers can protect leaf scars and mechanical injuries with one bactericide application; it also allows for large acreages to be pruned rapidly in a forecast dry period of just 2 weeks.

Olive leaf spot

Olive leaf spot (OLS) is a fungal foliar disease of olive caused by *Venturia oleaginea* Rossman & Crous [syn. *Spilocea oleaginea* (Castagne) S. Hughes and *Fusicladium oleagineum* (Castagne) Ritschel & U. Braun]. Also known as pcacock spot and bird's-eye spot, OLS is one of the most important fungal diseases of olive. Its distribution is widespread, occurring in almost all olive-growing regions of the world. Known in Mediterranean countries and in California for over 100 years (Wilson and Miller 1949), OLS has been reported in all growing regions of California. Varieties vary in their susceptibility. Severely infected trees show defoliation, fruit blemishes, and poor twig growth. Outbreaks are sporadic, and the disease may take several years to build up and cause yield losses of up to 20 percent of the crop.

Symptoms

Although symptoms occur most commonly on leaves in the lower parts of the tree canopy, they can develop on leaves elsewhere in the canopy and on fruit and fruit stems. Leaf spots typically form on the upper leaf surface. At first, they consist of circular blotches 0.06 to 0.6 inch (2 to 15 mm) in diameter that become dark green to blackish with age (fig. 12.3). Some lesions may

Figure 12.3

Dark green to blackish leaf spots are symptomatic of olive leaf spot. *Photo:* Florent P. Trouillas.

be surrounded by concentric yellowish or pale brown halos. These symptoms resemble the eye-spots on the tail feathers of a peacock, hence the alternate names for the disease. Older leaf lesions may have a gray, crusty appearance (fig. 12.4). Multiple lesions can occur on a leaf, and infected leaves abscise prematurely, resulting in weakened twigs that often die back. This dieback eventually results in reduced orchard productivity.

Comments on the disease

High humidity, cool temperatures, and low light conditions during winter favor infection by the fungal pathogen causing OLS, whereas low humidity, warm temperatures, and high light conditions during summer restrict spore production and germination. Infection is associated with rainfall and high humidity, which occur in late fall, winter, and early spring in California. Rain drip typically carries spores in moving water to the lower canopy, resulting in higher disease levels in the lower parts of the tree. Lateral movement by rain splash is limited. In the early stages of the introduction of OLS into an orchard, adjacent trees may have different disease incidence and severity.

New infections are first seen in late winter and early spring. A significant amount of defoliation may occur by summer, although leaves remaining on the tree are mostly healthy. Still, some infected leaves stay on the tree and act as sources of inoculum for later infections. The fungus survives summer as old leaf lesions, which have a grayish, crusty appearance due to separation of the cuticle from the epidermal cells. The margins of these lesions enlarge in fall, and new spores are produced. In Spain, young leaves can be infected in spring but remain symptomless; typical symptoms only develop in fall and become inoculum sources for late fall and winter infections (Viruega and Trapero 1999; Viruega et al. 2011).

Conidia (asexual spores) germinate only in the presence of wetness, but temperatures above 86°F (30°C) restrict germination. Mycelial (that is, somatic) growth and infection proceed over a wide range of temperatures, but 70°F (21°C) is the optimum. In California, most of the infections take place in winter. Like other *Venturia* species, the fungus colonizes the leaf tissues, forming a dense mass of stromatic tissue that produces conidiophores (conidium-bearing hyphae or filaments). One- or two-celled conidia are brown to olivaceous brown. OLS progresses slowly once introduced into an orchard, and it may take several years to reach incidence and severity levels that cause economic losses.

Management

Management of OLS has been primarily done with a copper application in fall before late fall and winter rains (see table 12.1). A second application by mid-January may be necessary when high rainfall occurs in December. Later applications in spring are less effective or ineffective. Other fungicides, such as ziram, the premixture of difenoconazole and cyprodinil, polyoxin-D, dodine, and quinone outside inhibitor (QoI) fungicides, are pending registration.

Pseudocercospora leaf spot

Pseudocercospora (also called Cercospora or Mycocentrospora) leaf spot is a fungal foliar disease of olive trees caused by *Pseudocercospora cladosporioides* (Sacc.) U. Braun [syn. *Mycocentrospora cladosporioides* (Sacc.) P. Costa, *Cercospora cladosporioides* Sacc.]. It develops under similar conditions as olive leaf spot. The pathogen causes sooty mold–like symptoms on the underside of chlorotic leaves that result in

Figure 12.4

Older lesions of olive leaf spot are grayish. *Photo:* James E. Adaskaveg.

abscission. This disease has been documented in coastal areas of California under humid conditions, and more recently in the Central Valley, where most new olive plantings are SHD or HD. Historically, Pseudocercospora leaf spot occurs infrequently but may be a problem if left unmanaged in areas favorable for disease development.

Symptoms

Pseudocercospora leaf spot causes leaf chlorosis, with some olive varieties showing more yellowing than others. The undersides of some leaves become discolored and covered with a blackish dust consisting of the conidial reproductive stage of the fungus (fig. 12.5). Brown leaf spots and marginal necrosis may also develop (fig. 12.6). Infected leaves may fall, causing some defoliation. Fruit may also develop small brown lesions and mature unevenly.

Comments on the disease

Pseudocercospora leaf spot may take several years to build up to severity levels that cause serious economic damage, and epidemics are sporadic. Some infected leaves fall from the tree, and the fungus survives in leaves that remain on the tree. Leaf lesions turn brown and have a crusty appearance. The margins of these lesions enlarge in fall, and new conidia are produced and disseminated with rainfall, which mostly occurs during late fall and winter in California.

Infections occur primarily in winter, and most diseased leaves have fallen off the trees by summer. Defoliation of shoots may be incomplete, with some shoot tips having mostly healthy leaves. High temperatures inhibit conidial germination and mycelial growth, and thus the disease is inactive during California's warm, dry summers.

Management

Management of Pseudocercospora leaf spot is similar to that of olive leaf spot. In cool, wet regions, preventive foliar applications of copper compounds (fixed copper mixed with lime or a sticker adjuvant, or copper Bordeaux treatments) should be made immediately after harvest and before winter rains begin (see table 12.1). Additional copper applications may be necessary in spring before forecast rain.

Figure 12.5

Black dust (conidia) appears on the underside of leaves affected by Pseudocercospora leaf spot. *Photo:* F. P. Trouillas.

Figure 12.6

Brown leaf spots and marginal necrosis are symptoms of Pseudocercospora leaf spot. *Photo:* Florent P. Trouillas.

Neofabraea leaf and shoot lesions and fruit rot

Neofabraea diseases are caused by the fungal pathogens *Neofabraea kienholzii* (Seifert, Spotts & Lévesque) Spotts, Lévesque & Seifert (syn. *Cryptosporiopsis kienholzii* Seifert, Spotts & Lévesque) and *Phlyctema vagabunda* Desm. [syn. *Neofabraea alba* (E. J. Guthrie) Verkley and *N. vagabunda* (Desm.) P. R. Johnst.]. *P. vagabunda* has long been known in Italy and

Spain as the causal agent of leprosy fruit rot of olive, although that designation and associated symptoms often have been confused with olive anthracnose caused by *Colletotrichum* species (Romero et al. 2018).

P. *vagabunda* was first reported in 2016 as an emerging pathogen of olive in mechanized orchard systems in Spain, causing branch cankers and twig dieback as well as small circular, necrotic lesions on leaves on Arbequina and Picual varieties (Romero et al. 2016, 2018). In California, Neofabraea leaf and shoot lesions are common in SHD orchards of the northern San Joaquin and Sacramento Valleys; however, they mainly affect leaves and twigs of Arbosana and, to a lesser extent, Arbequina, and have not been detected on Koroneiki (Trouillas et al. 2019). Fruit rot rarely occurs in SHD orchards in California, although fruit spots have been found occasionally on Arbequina olive fruit left on the tree in SHD orchards in the San Joaquin Valley (Trouillas et al. 2019). Fruit spots caused by *P. vagabunda* have also been reported on Coratina and Picholine varieties in Sonoma County (Rooney-Latham et al. 2013).

Symptoms

In California, symptoms develop primarily on leaves and shoots in SHD olive orchards. Leaf, shoot, and twig lesions appear on trees usually around mid-February but are most visible during early spring (March and April). Lesions on leaves are necrotic and circular to elongated, usually occur singly, and range from 0.2 to 0.4 inch (0.5 to 1 cm) in diameter (Trouillas et al. 2019) (fig. 12.7). Leaf lesions occur at sites of injuries caused by mechanical harvesters and also at abrasion sites where leaves rub against each other from strong air movement during harvest. In severely affected Arbosana orchards, defoliation of trees may occur (Trouillas et al. 2019).

Reddish-brown lesions on shoots and twigs develop at wounds caused by mechanical harvesters (Trouillas et al. 2019) (fig. 12.8). Lesions may develop around the circumference of a shoot or twig, causing dieback. Occasionally, cankers form on larger branches and appear as sunken, reddish lesions on the bark, elongating from the site of injury.

In Spain and Italy, *P. vagabunda* has been reported to cause fruit rot that is characterized by

Figure 12.7

A Neofabraea lesion has developed on this leaf. *Photo:* Florent P. Trouillas.

Figure 12.8

Neofabraea lesions may appear on shoots. *Photo:* Florent P. Trouillas.

Figure 12.9

Neofabraea fruit spots damage olives. *Photo:* Paul Vossen.

small circular, dark brown, necrotic lesions that are slightly depressed and surrounded by a chlorotic halo (Romero et al. 2018). Fruit spots have been detected occasionally in SHD orchards in California, mainly on Arbequina and on fruit that remain attached to the tree after harvest. Fruit spots on Coratina and Picholine are scattered, slightly sunken, and brown surrounded by a green halo (Rooney-Latham et al. 2013) (fig. 12.9).

Comments on the disease

The mechanical harvest of oil olives combined with rainfall and mild temperatures following harvest are the main factors contributing to the occurrence of Neofabraea leaf and shoot lesions. In California, oil olives are generally harvested in the fall, which corresponds with the onset of the rainy season. As shoots, twigs, and leaves suffer injuries from mechanical harvesting and the first rains occur, wounds become exposed to infection by the pathogens *N. kienholzii* and *P. vagabunda*. Conidia of these pathogens are released from rain-splashed fruiting bodies on diseased leaves that remained on the trees after the previous year's infections.

Field observations of seasonal symptom progression suggest that infections of leaves and shoots remain latent during the early winter months. Lesions become apparent mainly in February and March. The optimal mycelial growth temperature for *N. kienholzii* is 68°F (20°C) and, for *P. vagabunda,* 59°F (15°C). *N. kienholzii* and *P. vagabunda* are common pathogens of apples and pears in the Pacific Northwest, causing bull's-eye rot. In California, bull's-eye rot disease has been found sporadically on pear and apple trees (Henriquez et al. 2004; Rooney-Latham and Soriano 2016); however, the role of these hosts as sources of inoculum for Neofabraea leaf and shoot lesions of olive has not been determined.

Management

Fungicide applications are effective in reducing the incidence of Neofabraea diseases (see table 12.1). One application of ziram or the premixture formulation of difenoconazole and cyprodinil immediately after harvest has been shown to significantly reduce the disease. These fungicides are currently pending registration on olive in California. A second application of these fungicides 2 to 3 weeks after the initial application may be considered if rain persists through December. Alternating fungicides with different modes of action is recommended to prevent fungicide resistance. Copper applications are ineffective against this disease.

Phytophthora root and crown rot

Phytophthora species are soil-inhabiting pathogens commonly referred to as water molds. Many perennial crops worldwide suffer losses due to Phytophthora root and crown rot. The disease occurs in olive trees but is of relatively minor importance in California. Risks for the disease increase, however, when planting orchards in marginal soils with poor water infiltration and drainage, including soils where a hardpan is present or heavy clay soils. Prolonged periods of saturated soil and wetness are typically required for the disease to occur. Several species, including *P. citricola* and *P. drechsleri*, have been reported in olive orchards in California, although additional species may occur.

Symptoms

Phytophthora-infected trees exhibit reduced growth, yellowing leaves, and defoliation, and may eventually die. Infected roots become dark brown to black inside and outside (fig. 12.10), whereas healthy roots remain white inside. In

Figure 12.10

Root of an olive tree affected by Phytophthora root rot; infection is characterized by the brown discoloration of the affected root. *Photo:* Paul Vossen.

contrast to the pathogen causing Armillaria root rot, *Phytophthora* species do not form visible mycelium in infected tissues. Infection may progress to the root crown, causing crown rot and a canker that tends to spread into aboveground portions of the tree trunk. The affected bark turns dark brown, eventually leading to rapid death of the tree as the pathogen girdles the trunk, preventing water and nutrient transport from roots to foliage.

Comments on the disease

Phytophthora root and crown rot is commonly associated with heavy soils and prolonged periods of high soil moisture. The disease usually occurs in spring and early summer in low-lying areas of an orchard, but affected trees can also be randomly distributed. *Phytophthora* species can survive in the soil during dry periods as oospores or chlamydospores (thick-walled resting spores).

In moist soils at moderate temperatures, resting spores produce reproductive structures called sporangia, from which motile zoospores are released when the soil moisture level reaches saturation. Zoospores require water in a flooded soil to swim to the root surface and invade root tissues. Each species has its own requirements of temperature, moisture, and nutrients. Furthermore, pathogenicity and virulence vary with host variety. Many *Phytophthora* species are present in canal and river water, and can be introduced in orchards irrigated from these sources. To date, *Phytophthora* species have not been found in well water.

Management

Proper water management, particularly avoiding overirrigation, is the basis of management of Phytophthora root and crown rot (see table 12.1). Sites with poor surface drainage that cannot be corrected, or very heavy clay soils, are not appropriate for most olive orchards. Cultural practices that avoid prolonged saturation of the soil, such as planting on berms, shortening irrigation time, and improving water penetration, lessen root rot. Sufficient irrigation should be provided to meet the trees' water demand, but the soil should not be saturated for longer than 24 hours, especially near the root crown. When roots have grown outside the root ball, and after active tree growth has begun, drip emitters should be moved away from the trunk to avoid excessive wetting of the crown.

Foliar applications of potassium phosphonate, also referred to as phosphite or phosphorous acid, are commonly used as a preventive treatment against Phytophthora root and crown rot in other tree crops in California. Potassium phosphonate directly inhibits the pathogen and is thought to enhance the plant's natural defense mechanisms. The efficacy of phosphite treatments on olive, however, has not been evaluated in California. Resistant olive varieties have not been identified.

Preplant soil fumigation can lower *Phytophthora* populations to minimal levels and enhance both orchard establishment and tree growth. However, the pathogen may later be reintroduced into the orchard if contaminated surface water is used to irrigate trees.

Armillaria root rot

Although Armillaria root rot is a serious fungal disease of many tree species, it has not been reported as a major disease of olive in California. The pathogen, *Armillaria mellea* (Vahl.) Quel., is endemic to the Central Valley, where it occurs naturally on oak trees (*Quercus* spp.). Because of its occurrence on oaks, it is also known as oak root fungus. Typically, trees growing along riparian ecosystems are infected.

Symptoms

Infected trees have slowly thinning canopies and are weakened by dying root systems. Symptoms often are observed first on one side of the tree; after several years, the whole tree declines and eventually dies. The roots, including the wood of the roots, as well as the bark and underlying wood of the tree crown, become discolored. Typically, a distinctive white, fan-shaped mycelial plaque forms between the bark and the cambium (generative tissue under the bark) that kills the phloem and cambium of the tree.

The disease commonly spreads tree to tree by fungal bootlace-like rhizomorphs that develop along the surface of roots. These rootlike structures contain highly differentiated tissues, including a whitish central core (medulla) of thin-walled, elongated hyphae embedded in mucilage surrounded by a cortex and rind of smaller, compact, thick-walled cells. The rind cells are darkly pigmented by melanin. In the final stages of disease, the fungus degrades the wood, causing a white rot. The decay is bleached-white, soft, and punky (spongy) from the breakdown of the major components: cellulose, hemicellulose, and lignin.

Comments on the disease

The pathogen survives as mycelium, on decaying wood of dead roots and tree crowns, and as rhizomorphs on the surface of diseased roots. The fungus can persist for many years and even decades in large dead roots deep in moist soil. If the roots and crown become desiccated, the fungus will not survive.

Rhizomorphs grow along the surface of roots and spread from tree to tree through root-to-root contact or natural root grafts. They may also grow short distances away from source roots through the soil and infect healthy adjacent roots. The circular pattern of infected trees (known as an infection center) slowly expands over several years; it reflects the main dissemination process of rhizomorphs infecting and rotting roots, and explains the name *shoestring root rot*.

In fall and winter, *A. mellea* forms honey-colored mushrooms (basidiomes), also known as honey mushrooms (fig. 12.11). The basidiospores are sexually produced on basidia on the white gills, or lamellae, of the mushroom. These spores are not considered important in the dissemination of the pathogen.

Management

Armillaria root rot is a difficult disease to manage. Fungal eradication from areas where the disease occurs has not been achieved. Practices that slow the spread of the disease include sanitation and fumigation (see table 12.1). Prior to planting, roots remaining in the soil from infected trees should be removed and destroyed, followed by deep fumigation of the site. Still, roots of newly planted trees eventually encounter infected roots deep in the soil outside the fumigation zone. The pathogen invades these roots, grows to the tree crown while infecting adjacent roots, and the cycle starts again.

No rootstocks of olive are resistant, and infected trees cannot be cured. On other crops, exposing the crown to desiccation or injecting the lower trunk with propiconazole has slowed the decline of the tree by inhibiting the pathogen from reaching and girdling the tree crown.

Verticillium wilt

Verticillium wilt, caused by the soilborne fungus *Verticillium dahliae* Kleb., is a global disease of

Figure 12.11

Armillaria mellea produces honey-colored mushrooms in fall and winter. *Photo:* Florent P. Trouillas.

Figure 12.12

The leaves on a single branch, or sector, of an olive tree affected by Verticillium wilt have died and browned. *Photo:* Florent P. Trouillas.

olive. *V. dahliae* is found in diverse agricultural soils worldwide and is a pathogen of many perennial and annual crops grown in California. It occurs sporadically in olive orchards of the Central Valley, in locations where highly susceptible crops, such as cotton, tomato, and peppers, were grown previously and inoculum in the soil has accumulated to high levels. The disease can affect young and mature olive trees each year, and it may be a serious economic disease in some locations.

Symptoms

The primary symptom of Verticillium wilt is a sudden browning of the leaves on a single branch, with leaves remaining attached (fig. 12.12). Symptoms often appear with the first warm weather of summer. One or multiple branches may collapse, and trees can die after repeated infections over several years. Infections increase with tree age as root systems enlarge and explore larger volumes of contaminated soil, but young trees also can be affected. Darkening of xylem tissues and vascular streaking of affected branches or stems are not readily apparent in olive as they are in many other susceptible plants.

Comments on the disease

The fungus persists in the soil for many years as small black, thick-walled resting structures called microsclerotia. They germinate following stimulation by root exudates of a susceptible host plant, allowing the fungus to penetrate the root and colonize the cortex. Hyphae of the pathogen then invade the xylem vessels, where conidia are produced and moved up the plant along with water. The presence of conidia in the water-conducting tissues, the current year's xylem vessels, leads to plugging of the vascular system, preventing water movement to the upper parts of the plant.

Leaves and stems deprived of water soon begin to wilt and show foliar chlorosis, followed by sudden wilting and death of limbs or the entire tree. As the infected tissues decay, the fungus produces microsclerotia, which are released into the soil with the decomposition of plant tissues left on-site. This disease has killed entire orchards, when trees were planted in heavily infested soil. Of the several strains of the fungus, the most virulent is the cotton-defoliating strain, named for its symptoms on cotton.

Most *V. dahliae* infections occur in cool, moist soil during late winter and spring, before high temperatures prevail. As temperatures rise, infections decrease until soils cool down again in early fall, and then another round of infections can occur. During the hot summer months, the fungus dies out in the upper parts of the plant and becomes difficult to isolate in culture, although disease is apparent. When cool and moderate spring temperatures persist or summers are mild, Verticillium wilt is very common.

Management

No reliable methods have been developed for control of Verticillium wilt. Site selection for new plantings should be based on crop history and an assessment of the inoculum level (number of microsclerotia) in the soil. Land previously planted with crops susceptible to Verticillium wilt is likely to harbor high counts of microsclerotia. Inoculum levels can be determined by soil analysis available through private laboratories.

Preplant fumigation may reduce microsclerotium populations in the soil but will not eradicate the fungus. Solarization, which involves heating the soil by covering with clear plastic for 4 to 6 weeks during a period of high temperatures lethal to soilborne pathogens, has provided inconsistent results. Flooding fallow fields and growing grass crops for several seasons can reduce inoculum (see table 12.1). No resistant variety is available, although some tolerance has been reported for Ascolano. Interplanting crops that are susceptible to *V. dahliae* should be avoided.

Anthracnose

One of the most important fungal diseases of olive fruit worldwide, anthracnose is caused by several *Colletotrichum* species, including *C. acutatum* J. H. Simmonds and *C. gloeosporioides* (Penz.) Penz. & Sacc. (Cacciola et al. 2012; Moral and Trapero 2012). The disease can cause high yield losses and impacts olive oil quality, particularly in humid production areas or during wet conditions, and where susceptible varieties are grown (Cacciola et al. 2012; Moral and Trapero 2012).

In California, anthracnose has been observed sporadically in winter on the fruit of Gordal Sevillana grown as an ornamental and left unharvested, as well as in an experimental research plot of the University of California. However, olive anthracnose has not been detected in commercial olive orchards, and it is likely that the disease will not thrive in olive under the environmental conditions of California's Central Valley. Olive varieties Arbosana, Arbequina, and Koroneiki range from resistant to moderately susceptible to anthracnose, and the disease generally can be avoided when harvest occurs while the olive fruit is still green (Moral and Trapero 2009).

Symptoms

Classic symptoms of olive anthracnose include fruit rot and mummification. Fruit rot is characterized by soft, dark brown fruit with orange masses of conidia on the surface. Affected fruit falls prematurely to the ground, and only a few mummies remain attached to the tree. The disease mostly affects mature fruit, but under favorable environmental conditions, green fruit of susceptible varieties can also be diseased. In varieties with oblong drupes, rot often starts from the apical end, but in varieties with large, nearly spherical drupes, symptoms generally manifest as circular sunken lesions with acervuli (small, cushionlike, spore-producing structures) forming in concentric rings (Cacciola et al. 2012) (fig. 12.13).

Under favorable environmental conditions, olive anthracnose can also cause chlorosis and necrosis of leaves, defoliation, and dieback of twigs and branches. In spring, infected leaves may exhibit yellowing of the leaf blade. In late

Figure 12.13

Circular sunken lesions are a typical symptom for olive anthracnose on fruit. *Photo:* Florent P. Trouillas.

spring and early summer, infected leaves fall to the ground and very susceptible trees become defoliated (Cacciola et al. 2012).

Comments on the disease

Colletotrichum species form acervuli that develop on the olive mummies in the tree canopy. Acervuli produce conidia that serve as the main inoculum for infection. Infection can occur at all stages of fruit development, from the emerging flower to the ripening fruit. However, early infection of young fruit generally remains latent until the fruit ripens in fall and early winter, causing the typical fruit rot and the production of conidial masses (Moral and Trapero 2012). At this stage, conidia can be rain splashed to other fruit, causing secondary infections before harvest. As temperatures rise in the spring, the rotten fruit becomes mummified (Moral and Trapero 2012). Acervuli may also develop on leaves and shoots.

Management

The disease is not serious enough in California to warrant specific control measures (see table 12.1). Generally, it can be avoided by harvesting the fruit before it ripens.

Botryosphaeriaceae cankers

Canker diseases caused by fungal pathogens in the Botryosphaeriaceae family occur sporadically in low-density olive orchards of Ascolano, Mission, Manzanillo, and Sevillano in the Sacramento and San Joaquin Valleys, as well as on the Central and North Coasts. *Neofusicoccum mediterraneum* Crous, M. J. Wingf. & A. J. L. Phillips, *Diplodia mutila* (Fr.) Mont., and *Botryosphaeria dothidea* (Moug.) Ces. & De Not. are the main pathogens associated with Botryosphaeriaceae cankers of olive in California (Úrbez-Torres et al. 2013). These canker diseases do not have significant economic impact in oil olive orchards in California and generally can be avoided if pruning is done outside the rainy season.

Symptoms

Symptoms associated with Botryosphaeriaceae cankers include twig and branch dieback. Cankers are characterized by vascular discoloration of the wood, which is generally wedge-shaped in cross section and often results in the death of cambial and bark tissue (fig. 12.14). Infections of branches can sometime move to the trunk, eventually causing tree death (Úrbez-Torres et al. 2013).

Comments on the disease

Cankers develop at pruning wounds made in the tree crown; large pruning cuts are most likely to become infected by Botryosphaeriaceae fungi. The disease also can develop at sunburn lesions and in areas affected by olive knot. Botryosphaeriaceae fungi can invade aging olive knots and progress into the twig to form a canker. The canker then girdles and kills small shoots and branches, exacerbating the damage caused by olive knot.

Botryosphaeriaceae pathogens have a wide host range, including forest trees, trees and bushes in riparian areas, and ornamental trees, as well as many woody crops, such as almond, avocado, citrus, grapevine, pistachio, and walnut. Botryosphaeriaceae fungi commonly produce pycnidia (asexual reproductive structures) on the diseased bark of susceptible host plants; these release rain-splashed conidia to fresh wounds, causing infection. Perithecia (flask-shaped fruiting bodies) also can be found on diseased hosts; during rain, these produce airborne ascospores (spores produced sexually) that likely function in long-distance dissemination of the pathogens.

Management

The main recommendation for controlling Botryosphaeriaceae cankers is to avoid pruning during and near rainfall events (see table 12.1).

Cytospora canker

Cytospora species are common canker pathogens of perennial tree crops worldwide. Cytospora canker of olive, caused by *C. oleicola* and *C. olivarum*, occurs sporadically in low-density olive orchards in California (Úrbez-Torres et al. 2020). The disease has also been reported in SHD orchards in the San Joaquin Valley, although its extent in these orchards has not been fully investigated.

Symptoms

The disease appears as brown, sunken lesions on the bark with discolorations of the wood; these lesions are known as cankers (fig. 12.15). As the disease develops, twig and branch dieback can occur, and eventually the entire tree may die. Cytospora canker typically develops at pruning wounds and produces cankers somewhat similar to those caused by Botryosphaeriaceae fungi.

Figure 12.14

This branch canker was caused by fungal species in the Botryosphaeriaceae family. *Photo:* José Ramón Úrbez-Torres.

Figure 12.15

Brown lesions on the bark of twigs are symptomatic of Cytospora canker. *Photo:* Florent P. Trouillas.

Comments on the disease

Mechanized orchard practices can cause severe damage to the bark and wood of olive tree trunks and branches. In particular, fresh wounds resulting from mechanical pruning potentially serve as entry sites for infection by *Cytospora* species and other fungal canker pathogens. *Cytospora* species produce numerous pycnidia on dead or diseased wood that serve as the primary inoculum for Cytospora canker. The spores can be disseminated by rain splash and wind to nearby healthy trees.

Management

There is no recommended method for controlling Cytospora canker, and the main recommendation is to avoid pruning during and near rain events (see table 12.1).

Pleurostoma decline

Pleurostoma decline of olive trees is caused by the fungal pathogen *Pleurostoma richardsiae* (Nannf.) Reblova & Jaklitsch [syn. *Pleurostomophora richardsiae* (Nannf.) L. Mostert, W. Gams & Crous, and *Phialophora richardsiae* (Nannf.) Conant]. In California, the disease has been found sporadically in SHD olive orchards of Arbosana and Koroneiki in San Joaquin County and in a low-density orchard in Glenn County (Lawrence et al. 2021). The disease has also been reported on olive trees in Italy, Brazil, and Croatia (Canale et al. 2019; Carlucci et al. 2013; Ivic et al. 2018). In grapevine, *P. richardsiae* is considered a global vascular pathogen involved in the Petri and esca disease complex (Canale et al. 2019; Carlucci et al. 2013).

Symptoms

Symptoms of Pleurostoma decline of olive include foliar yellowing and browning, leaf drop, and wilting and dieback of twigs and branches (fig. 12.16). Cross sections of the trunk of declining trees reveal irregular, brown to dark discoloration of the wood (fig. 12.17). Severely affected trees may die. *P. richardsiae* also has been isolated from olive trees with collar rot symptoms.

Comments on the disease

Little is known about the biology of Pleurostoma decline. Field observations suggest that infections

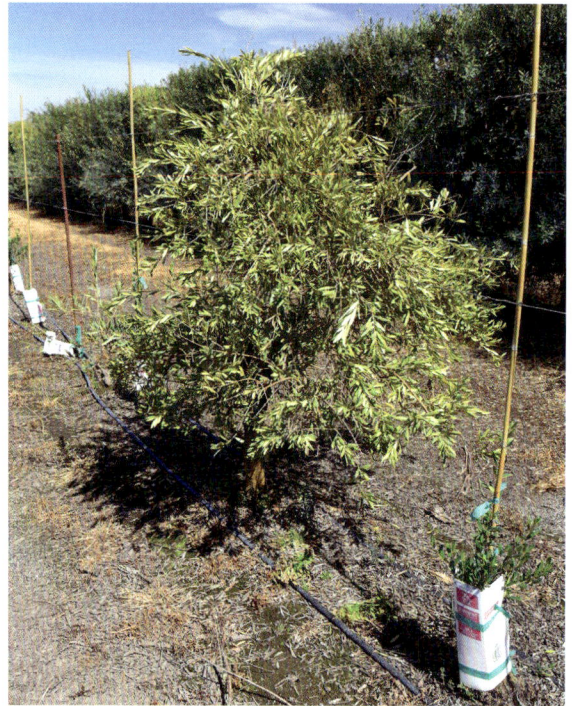

Figure 12.16

Pleurostoma decline caused leaf yellowing of this olive tree. *Photo:* Florent P. Trouillas.

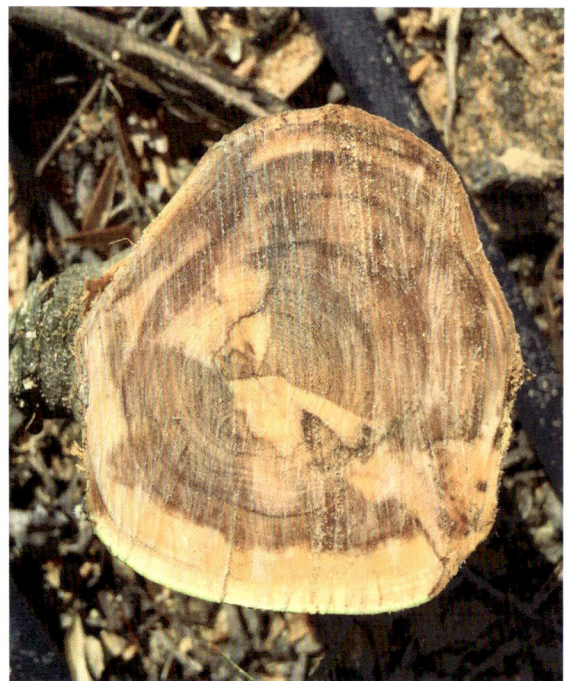

Figure 12.17

Wood discoloration caused by Pleurostoma decline is evident in this trunk. *Photo:* Florent P. Trouillas.

by *P. richardsiae* initiate at wounds in trunks and branches caused by field equipment; by large insect borers, including the Pacific flatheaded borer (*Chrysobothris mali*); or by any other injuries that may expose the tree's vascular tissues. In California SHD orchards, trees with severe injuries on trunks from field equipment, such as mechanical harvesters, have been most affected by the disease. *P. richardsiae* is likely to produce fruiting bodies on infected plant parts, including trunk and branches of susceptible hosts, and these can serve as inoculum sources for Pleurostoma decline. Spores may be disseminated by rain splash and wind to healthy trees, where fresh wounds serve as entry sites for new infections.

Management

The disease is not serious enough in California to warrant specific control measures (see table 12.1). However, infections may be avoided by preventing bark and trunk injuries from mechanical equipment. Proper orchard sanitation is also recommended to avoid population buildup of the Pacific flatheaded borer. This includes removal of weakened and dead branches in the orchard that can serve as overwintering sites for mature borer larvae during late fall and winter in California.

Olive leaf scorch and quick decline syndrome

Xylella fastidiosa is a xylem-limited bacterium and a serious pathogen of many ornamental and agricultural plants, inflicting severe economic damage worldwide. Primarily transmitted from plant to plant by insect vectors, it causes a variety of leaf scorch symptoms, including Pierce's disease of grapevine, oleander leaf scorch, and almond leaf scorch. On olive, the pathogen can cause leaf scorch. In 2013, a strain of the bacterium (*X. fastidiosa* ssp. *pauca*) was detected for the first time in Italy; it causes olive quick decline syndrome (OQDS).

Symptoms

The symptoms of olive leaf scorch consist of leaf scorching and dieback of twigs and branches. Scorch symptoms include yellowing and browning of leaf tips and margins or of entire leaves (fig. 12.18). Because water transpiration is blocked or limited by bacterial growth in the xylem, leaves show scorch or burn symptoms during warm weather in late spring, summer, and early fall. Symptoms have been reported in California; however, the less virulent *X. fastidiosa* ssp. *multiplex* is infrequently isolated, and inoculations on olive has caused infections that were asymptomatic (Krugner et al. 2014). In Italy, OQDS is a devastating disease. It causes dieback of twigs and branches (fig. 12.19) so severe that trees are no longer capable of producing a crop.

Figure 12.18

Scorched leaves with browning initiating at leaf tips are symptoms of olive leaf scorch. *Photo:* S. Pollastro.

Figure 12.19

Olive quick decline syndrome is a serious disease in the Apulia region of Italy. Symptoms include twig and branch dieback. *Photo:* S. Pollastro.

Comments on the disease

X. fastidiosa is a gram-negative, xylem-limited, slow-growing bacterium that is transmitted by numerous xylem-feeding insect vectors. This pathogen is widely distributed throughout North and South America, where it infects a large number of hosts, including ornamental and tree crop species. *X. fastidiosa* isolates from olive in California belong to subspecies *multiplex*. Inoculation of olive and grapevine plants with these isolates did not cause disease. The isolates, however, caused leaf scorch disease in almond (Krugner et al. 2014). *X. fastidiosa* ssp. *fastidiosa*, which causes Pierce's disease of grapevine, infected olive in greenhouse experiments but was never found to infect olives in the field and cause disease (Krugner et al. 2014). In California, the bacterial pathogen is transmitted by xylem sap–feeding leafhoppers, including the blue-green sharpshooter (*Graphocephala atropunctata* Signoret), the green sharpshooter (*Draeculacephala minerva* Ball), and the glassy-winged sharpshooter [*Homalodisca vitripennis* (Germar)].

In Italy, OQDS is caused by *X. fastidiosa* ssp. *pauca*. In addition to infecting olive, the Salentian (from Salento peninsula of Italy) strains of *X. fastidiosa* in nature infect a number of woody (almond, cherry) and shrubby (*Nerium oleander, Acacia saligna, Polygala myrtifolia, Westringia fruticosa, Rosmarinus officinalis, Rhamnus alaternus, Myrtus communis*) hosts with no evidence of grapevines or citrus being hosts (Martelli et al. 2016). *Philaenus spumarius* (meadow spittlebug), a common froghopper that thrives on olive in Italy, was identified as the main vector (Martelli et al. 2016).

Management

In California, leaf scorch disease is not serious enough to warrant specific control measures. Because OQDS has caused devastating losses to the olive industry in the Apulia region of southern Italy, every effort should be made to prevent this highly virulent strain of *X. fastidiosa* from being introduced to California or other olive-growing countries. Monitoring of orchards should be done routinely, and affected trees should be destroyed immediately.

References

Cacciola, S. O., R. Faedda, S. Fulvia, G. Agosteo, L. Schena, S. Frisullo, and G. Magnano di San Lio. 2012. Olive anthracnose. Journal of Plant Pathology 94(1):29–44.

Canale, M. C., C. Nunes Nesi, B. Falkenbach, C. Hunhoff da Silva, and E. Brugnara. 2019. *Pleurostomophora richardsiae* associated with olive tree and grapevine decline in southern Brazil. Phytopathologia Mediterranea 58(1): 201–205. https://doi.org/10.14601/Phytopathol_Mediterr-23357

Carlucci, A., M. L. Raimondo, F. Cibelli, A. J. Lander Phillips, and F. Lops. 2013. *Pleurostomophora richardsiae, Neofusicoccum parvum* and *Phaeoacremonium aleophilum* associated with a decline of olives in southern Italy. Phytopathologia Mediterranea 52(3): 517–527. https://doi.org/10.14601/Phytopathol_Mediterr-13526

Comai, L., and T. Kosuge. 1980. Involvement of plasmid deoxyribonucleic acid in indoleacetic acid synthesis in *Pseudomonas savastanoi*. Journal of Bacteriology 143:950–957. https://doi.org/10.1128/jb.143.2.950-957.1980

———. 1982. Cloning characterization of iaaM, a virulence determinant of *Pseudomonas savastanoi*. Journal of Bacteriology 149:40–46. https://doi.org/10.1128/jb.149.1.40-46.1982

Henriquez, J. L., D. Sugar, and R. A. Spotts. 2004. Etiology of bull's eye rot of pear caused by *Neofabraea* spp. in Oregon, Washington, and California. Plant Disease 88(10):1134–1138. https://doi.org/10.1094/PDIS.2004.88.10.1134

Ivic, D., Z. Tomic, and S. Godena. 2018. First report of *Pleurostomophora richardsiae* causing branch dieback and collar rot of olive in Istria, Croatia. Plant Disease 102:2648. https://doi.org/10.1094/PDIS-04-18-0669-PDN

Krugner, R., M. S. Sisterson, J. Chen, D. C. Stenger, and M. W. Johnson. 2014. Evaluation of olive as a host of *Xylella fastidiosa* and associated sharpshooter vectors. Plant Disease 98(9):1186–1193. https://doi.org/10.1094/PDIS-01-14-0014-RE

Lawrence, D. P., M. T. Nouri, and F. P. Trouillas. 2021. Pleurostoma decline of olive trees caused by *Pleurostoma richardsiae* in California. Plant Disease 105(8):2149–2159. https://doi.org/10.1094/PDIS-08-20-1771-RE

Martelli, G. P., D. Boscia, F. Porcelli, and M. Saponari. 2016. The olive quick decline syndrome in south-east Italy: A threatening phytosanitary emergency. European Journal of Plant Pathology 144:235–243. https://doi.org/10.1007/s10658-015-0784-7

Moral, J., and A. Trapero. 2009. Assessing the susceptibility of olive cultivars to anthracnose caused by *Colletotrichum acutatum*. Plant Disease 93:1028–1036. https://doi.org/10.1094/PDIS-93-10-1028

———. 2012. Mummified fruit as a source of inoculum and disease dynamics of olive anthracnose caused by *Colletotrichum* spp. Phytopathology 102(10):982–989. https://doi.org/10.1094/PHYTO-12-11-0344

Nguyen, K. A., H. Förster, and J. E. Adaskaveg. 2017. Quaternary ammonium compounds as new sanitizers for reducing the spread of the olive knot pathogen on orchard equipment. Plant Disease 101(7):1188–1193. https://doi.org/10.1094/PDIS-11-16-1578-RE

———. 2018a. Efficacy of copper and new bactericides for managing olive knot in California. Plant Disease 102(5):892–898. https://doi.org/10.1094/PDIS-08-17-1162-RE

——— 2018b. Genetic diversity of *Pseudomonas savastanoi* pv. *savastanoi* in California and characterization of epidemiological factors for olive knot development. Plant Disease 102(9):1718–1724. https://doi.org/10.1094/PDIS-11-17-1709-RE

Romero, J., M. C. Raya, L. F. Roca, C. Agustí-Brisach, J. Moral, and A. Trapero. 2018. Phenotypic, molecular and pathogenic characterization of *Phlyctema vagabunda*, causal agent of olive leprosy. Plant Pathology 67(2):277–294. https://doi.org/10.1111/ppa.12748

Romero, J., M. C. Raya, L. F. Roca, J. Moral, and A. Trapero. 2016. First report of *Neofabraea vagabunda* causing branch cankers on olives in Spain. Plant Disease 100(2):527–527. https://doi.org/10.1094/PDIS-06-15-0657-PDN

Rooney-Latham, S., and M. C. Soriano, 2016. First report of *Neofabraea alba* causing branch canker dieback of apple in California. Plant Disease 100(5): 1011. https://doi.org/10.1094/PDIS-08-15-0949-PDN

Rooney-Latham, S., L. L. Gallegos, P. M. Vossen, and W. D. Gubler. 2013. First report of *Neofabraea alba* causing fruit spot on olive in North America. Plant Disease 97(10):1384. https://doi.org/10.1094/PDIS-04-13-0394-PDN

Surico, G., N. S. Iacobellis, and A. Sisto. 1985. Studies on the role of indole-3-acetic acid and cytokinins in the formation of knots on olive and oleander plants by *Pseudomonas syringae* pv. *savastanoi*. Physiological Plant Pathology 26:309–320. https://doi.org/10.1016/0048-4059(85)90006-2

Trouillas, F. P., M. T. Nouri, D. P. Lawrence, J. Moral, R. Travadon, B. J. Aegerter, and D. Lightle. 2019. Identification and characterization of *Neofabraea kienholzii* and *Phlyctema vagabunda* causing leaf and shoot lesions of olive in California. Plant Disease 103:3018–3030. https://doi.org/10.1094/PDIS-02-19-0277-RE

Úrbez-Torres, J. R., D. P. Lawrence, F. P. Hand, and F. P. Trouillas. 2020. Olive twig and branch dieback in California caused by *Cytospora oleicola* and the newly described species *Cytospora olivarum* sp. nov. Plant Disease 104(7):1908–1917. https://doi.org/10.1094/PDIS-09-19-1979-RE

Úrbez-Torres, J. R., F. Peduto, P. M. Vossen, W.
H. Krueger, and W. D. Gubler. 2013. Olive
twig and branch dieback: Etiology, incidence,
and distribution in California. Plant Disease
97(2): 231–244. https://doi.org/10.1094/
PDIS-04-12-0390-RE

Viruega, J. R., and A. Trapero. 1999.
Epidemiology of leaf spot of olive tree caused
by *Spilocaea oleagina* in southern Spain.
Acta Horticulturae 474:531–534. https://doi.
org/10.17660/ActaHortic.1999.474.109

Viruega, J. R., L.F. Roca, J. Moral, and A.
Trapero. 2011. Factors affecting infection and
disease development on olive leaves
inoculated with *Fusicladium oleagineum*.
Plant Disease 95(9):1139–1146.

Wilson, E. E., and H. N. Miller. 1949. Olive leaf
spot and its control with fungicides. Hilgardia
19(1):1–24.

13 Weed Management in Olive

Bradley D. Hanson,
UC Cooperative
Extension (UCCE)
Weed Science
Specialist,
Department of Plant
Sciences, UC Davis

John A. Roncoroni,
UCCE Farm Advisor
Emeritus, Napa
County

Highlights

- Weed control is an ongoing management concern in oil olive orchards.

- Weeds directly impact establishment and productivity of the trees through competition for resources.

- In addition to direct competition, weeds can interfere with irrigation, harvest, and other crop management practices; weeds can also interfere with the crop indirectly by serving as an alternate host for insects and pathogens and by providing habitat for vertebrate pests.

- Weed management typically involves a range of cultural, mechanical/physical, and chemical practices based on the weed challenges present in each orchard; there is no one-size-fits-all approach that will work in all olive orchards.

- Although weed control is critical to olive orchard productivity, care must be taken to use weed management practices appropriately to maximize efficacy, minimize crop injury, and reduce impacts on non-target organisms and the environment.

Weed control in olive orchards enhances the development of newly planted trees and improves the growth and yield of established trees. As it does for producers of other orchard and vineyard crops, weed control constitutes a significant production cost and management concern for oil olive producers from orchard establishment through maturity. Although growers have many tools available to achieve their weed management objectives, the best strategy for employing these tools will vary from year to year and from orchard to orchard, depending on local conditions.

Why weeds are controlled

The primary reason to manage weeds in olive orchards is to reduce competition for limited resources. Weeds can reduce tree growth and yields by competing for water, nutrients, sunlight, and rooting space. This competition is most severe during the first 4 to 5 years of the tree's life or where root growth is limited; however, competition can also impact the productivity of older trees and reduce overall production efficiency.

The effect of competition from weeds is somewhat lessened after the trees get large enough to shade a significant portion of the orchard floor and have larger and more expansive root systems. This is true not only in larger trees but also in the high- and super-high-density planting arrangements common in oil olive production systems.

While direct completion is less of a problem in established olive orchards, weed management can affect other orchard management practices. Vegetated orchard floors in frost-prone areas can result in colder orchard conditions and slightly increase frost hazards, which increases the potential for olive knot. Weeds may also increase humidity, making trees more susceptible to infection by peacock spot fungus and black scale infestation.

The type and density of vegetation on the orchard floor can influence other diseases and pests, such as insects, rodents, and nematodes, which may in turn impact olive productivity. An especially obvious example is the link between weeds and vertebrate pests. Weeds, particularly grasses that form a dense cover around the tree trunk, are highly competitive with young trees. These weeds also provide a good habitat for field mice and voles, which can girdle and kill young trees.

Gophers are most prevalent in nontilled orchards and where broadleaf weeds such as field bindweed and perennial clovers predominate; excessive gopher feeding on the roots can weaken or kill young trees. To reduce vertebrate pest damage, give extra weed control attention to the area within 3 feet (1 m) of a young tree's trunk.

Excessive weed growth can interfere with cultural practices and harvest, reducing the efficiency of these operations. For example, weeds can disrupt the water application pattern from sprinklers and low-volume spray emitters, and dense or large weeds can physically interfere with harvest machinery.

While it is important to control weeds, remember that poorly implemented weed control practices can negatively impact olive orchard productivity. Herbicide injury can occur from particle drift, misapplications, overly aggressive spray programs, or unusual or unexpected conditions in the field. Trunk or root damage from cultivation can reduce tree vigor and lead to excessive suckering; these wounds can also serve as an entry point for pathogens, leading to crown gall, olive knot, and other diseases. No matter the weed control practice, take care to minimize the risks of crop injury.

Identification and monitoring

The first step in developing a weed management program is to identify the weed species and understand the biology of weeds infesting the orchard or planting site. Various tools are available to identify weeds. They include *Weeds of California and Other Western States* (DiTomaso et al. 2007); the UC Statewide IPM Program website, www2.ipm.ucanr.edu/agriculture/; electronic plant identification keys; and, increasingly, web apps for phones and tablets. At a minimum, a familiarity with weed identification and an understanding of the basic biology, life cycle, and reproductive habits of groups of similar weeds are crucial for choosing an effective management strategy.

Many species of summer and winter annual and perennial weeds are found in California olive orchards. Weeds vary from year to year, among regions, and even within orchards. Therefore, weed surveys should be conducted at least twice each year, once in late winter and again in late spring or summer, to describe the spectrum of weeds present and determine which have escaped or survived the current weed management program. Weed surveys can be informal or formal; however, a written record of dates, weed species observed, location, and density can serve as a basis for decisions about weed management practices and how to refine the plan over time.

Weed control tactics

Orchard floor management decisions are influenced by regional location, climatic conditions, soils, irrigation practices, topography, and grower or market preferences. Weed management intensity may vary within and between the tree rows, and often several weed management techniques are used in combination or sequence to provide year-round control. In the orchard shown in figure 13.1, cultural weed management practices include precise water and fertilizer placement via drip irrigation system and a winter cereal cover crop. Mechanical practices include tillage to control weeds and incorporate the cover crop between rows. Chemical practices include a winter-applied preemergence herbicide applied in strips under the tree row (which is poorly controlling summer-growing yellow nutsedge at

Figure 13.1

In this young high-density olive orchard, several weed management techniques are used in combination and sequence. *Photo:* Bradley Hanson.

each dripper) and application of a postemergence herbicide using an over-the-row herbicide sprayer to retreat the narrow tree row strips. Most weed control tactics used in oil olive orchards are cultural, mechanical/physical, or chemical strategies (table 13.1).

Cultural practices

Many cultural practices in olive orchards can affect weeds; the purpose of some is to manage weeds while weed control is a secondary effect of others. Site preparation practices, such as grading, can reduce poorly drained areas, which commonly have greater weed problems. Orchard planting density and arrangement can impact how quickly the crop canopy develops and subsequently shades weeds.

The type of irrigation system used in the orchard affects the amount of wetted area, which can impact weed germination and growth as well as herbicide performance. Irrigation and fertilization practices, especially placement and timing, can sometimes be manipulated to favor the crop and disfavor weed access to these resources. Pruning practices, especially of the lower portions of the tree, can affect the growth of weeds by shading them or exposing them to light—and unpruned branches may affect weed control by interfering with spraying or cultivation

Table 13.1

General categories and examples of weed control tactics used in olive orchards

Cultural	Mechanical/physical	Chemical
Site selection and grading	Cultivation	Preemergence herbicides
Planting density and arrangement	Mowing	Postemergence herbicides
Irrigation and fertilizer management	Thermal (flame or steam)	Organic postemergence herbicides
Pruning practices	Hand labor	
Cover crops	String trimmers	
Mulches		
Animals		
Others*		

*Many production practices can be manipulated or applied in a way that offers at least a slight advantage to the crop and disadvantage to the noncrop plants in the orchard.

operations. These and many more cultural practices can collectively make a significant impact on weed management in olive orchards.

Cover crops

Cover crops are planted in some orchards to enhance soil quality and water infiltration, and to improve orchard access during wet winter conditions. They can also be used to suppress resident weedy vegetation on the orchard floor through direct competition. Cover crops are usually grown between tree rows while other practices are used to manage weeds within the tree row (fig. 13.2). The cover crop species selected and the manner

Figure 13.2

A cover crop is used to suppress weeds between rows of olive trees. In the row, mulch and chemical controls are used to keep the strip free of weeds. *Photo:* Paul Vossen.

in which it is managed will differ from one region to another.

Cover cropping has many benefits but may also create some challenges, particularly when the management needs of the orchard and cover crop conflict. To minimize competition with trees for water, plan any additional water needs of the cover crop or time of cover crop termination. Newly established cover crops may not tolerate the damage from fall and winter orchard traffic during pruning, brush removal, chipping and shredding, and spraying. In orchards where these operations are planned, cover crops can be seeded in alternate row middles and the operations carried out in the nonseeded row middles.

Common cover crops in California include fall-seeded cereal crops (for example, wheat, oat, cereal rye, barley), annual reseeding grass covers (e.g., Blando bromegrass, Zorro fescue), and subterranean or rose clovers. These cover crops are seeded into a prepared seedbed between tree rows in late September through mid-November. Cereal crops will likely need to be seeded each year, but others may reseed themselves if mowed in January or early February and then allowed to regrow in April and May. Mowing after the seed mature ensures seed for the next season.

Larger-seeded cover crops, such as bell bean, purple or common vetch, and crimson clover, are often planted in orchards and tilled in as green manure. Perennial grasses (e.g., tall fescue, Berber orchardgrass, perennial ryegrass) can be grown, but they require summer irrigation and may compete with tree growth. Avoid species such as white clover, strawberry clover, and bermudagrass because they can be very difficult to control once established.

Consider how cover crops will affect weed management and other critical olive orchard practices. For more information on cover crops, consult Grant et al. 2006, Ingels et al. 1998, and Miller et al. 1989.

Mulches

Mulches are used to reduce weed density and competition in the tree row (fig. 13.3). They prevent weed germination and growth by blocking light and serve as a physical barrier that keeps seedlings from reaching the surface. Mulches are most effective when applied to a relatively weed-free soil surface.

Mulches can be organic materials from off-site, such as cereal straw, green waste, composted wood chips, and sawdust, or materials from the orchard, such as cover crop residue or shredded tree prunings. For effective weed control, organic mulches should be maintained at a depth of 4 inches (10 cm). These materials naturally degrade, and their depth can shrink by as much as 60 percent after 1 year unless replenished. As organic mulches degrade, they can become a growth medium for weed species, such as common groundsel, prickly lettuce, common sowthistle, and panicle-leaf willowherb.

Figure 13.3

Whether synthetic (as shown here) or organic materials, mulches block light to reduce weed germination, emergence, and establishment. *Photo:* Paul Vossen.

Synthetic mulch materials, including polyethylene, polypropylene, and polyester, are installed around individual trees or to entire rows. Some woven fabric mulches offer excellent weed control for several years; however, the initial purchase and installation costs can be quite high.

Mulches create more uniform soil conditions, which can promote young tree growth; however, in some cases mulches can reduce soil temperature, which can slow tree growth in spring. Mulches may also provide a good habitat for gophers, voles, field mice, and snakes; organic mulches can potentially introduce new weed seed. Mulches do not generally control perennial weed growth unless all light can be excluded. As they emerge, some species, such as yellow nutsedge, can penetrate plastic and woven mulches. Weeds that grow in, on, or adjacent to the edge of a mulch can be controlled with spot treatments of herbicides or physically removed to extend the useful life of the mulch.

Animal weed control

Animals are occasionally used for selective grazing of weeds in olive orchards. Most common are sheep and goats, which reduce weed competition by lowering the height of weeds. Weed control success with animals is heavily dependent on a sufficient stocking density and following somewhat intense rotational grazing strategies. Additionally, browsing animals, such as goats, must be closely managed to avoid damage to trees. Some examples exist of using weeder geese, which prefer grasses, including difficult-to-manage perennial grasses, such as bermudagrass and johnsongrass; however, geese are not currently widely used for this purpose in orchards.

Although animals can suppress weeds, generally they are removed before achieving full control of weeds due to animal health and welfare concerns as feed becomes limited. During the time they are used for weed control, they require water and protection from predators. Additionally, due to food safety concerns related to animal manure, regulations may require the removal of animals up to 120 days prior to harvest. Contact the California Department of Food and Agriculture, the local county agricultural commissioner's office, or county health officials for more specific food safety regulations that may apply.

Mechanical and physical practices

Mechanical and physical weed control includes practices that uproot, bury, chop, cut, or otherwise damage weedy vegetation. Cultivation and mowing are the most common mechanical weed control practices in orchard production systems, especially between tree rows. Physical weed control includes pulling, hoeing, and chopping tasks commonly done by hand labor. Although some mechanical and physical weed control approaches can be used within the tree row or around individual trees, care must be taken to minimize damage to trees and irrigation equipment.

Cultivation

Cultivation is usually most effective on annual weeds and seedlings of biennial and perennial weeds. The equipment should be set to cultivate shallowly, usually around 2 inches (5 cm) or less, with the goal of uprooting small seedlings; these shallow operations also minimize damage to tree roots. Established plants and perennial weeds are more likely to survive or to regrow or reroot (reestablish); they may have already produced seed by the time they are controlled. Additionally, larger plants may clog equipment.

Tillage equipment can spread rhizomes, tubers, and stolons from perennial plants, such as field bindweed, bermudagrass, and johnsongrass. These species should be controlled before they are 3 weeks old, at which point they may have already formed perennial structures. In irrigated orchards, after perennial weeds are cultivated, water should be withheld as long as feasible to increase desiccation (drying out) of the damaged tissues and reduce reestablishment.

Weed seed can be moved from site to site in soil or plant debris clinging to tillage equipment. To avoid spreading weed seed, clean cultivation equipment before moving it to new fields, particularly if using it in areas with uncommon or difficult-to-control weeds. Generally, avoid deep tillage as it can bring deeply buried weed seed to the surface, where they are more likely to successfully germinate and establish. Deep tillage

is unnecessary for weed control and is more likely to damage tree roots, which can reduce the ability of the tree to take up nutrients and allow soil pathogens access to the tree.

Common tillage equipment used between tree rows includes disks, powered rototillers, and scrapers. Mechanical cultivators available for use in the tree row include weed knives, spyder cultivators, and rotary tillers. Rotary tillers, such as a Weed Badger, Kimco in-row tiller, or Clemens hoe, are most effective if used on loose soil with few rocks.

When developing a new orchard and planning cultivation as the primary means of weed control, consider how orchard and irrigation system layout will affect weed control operations. Relatively minor changes at this stage may make later weed control operations much more effective and efficient.

Mowing

A variety of tractor-mounted rotary mowers or flail-type mowers can be operated between the tree rows. Mowing is commonly used to maintain plant cover, whether a cover crop or weedy vegetation, in a low-growing state. However, depending on the weed species and timing of the operation, mowing may provide weed control or only suppression.

While upright-growing broadleaf weeds can be managed relatively well with mowing, grasses tend to be tolerant because their growing points are belowground. Similarly, low-growing and prostrate plants, such as field bindweed, common purslane, and prostrate knotweed, usually are poorly controlled with mowing. Consequently, over time, mowing often leads to a shift toward grasses and low-growing species. Some oil olive growers purposefully use the timing of mowing operations to reduce seed production of undesirable species and allow more desirable plants, such as cover crops, or less problematic resident vegetation to reseed.

Cross mowing between the trees within the row can be accomplished in widely spaced plantings; however, this may not be an option where aboveground drip or microsprinklers are used. Specialized tractor-mounted mowers and trimmers that are compatible with some irrigation systems allow mowing within the tree row.

Similarly, handheld string trimmers, such as Weed Eater, may be used within the row, but care must be taken because of the risk of injury to trees or damage to irrigation equipment.

Flame and steam weeding

Flaming is used to control very young weed seedlings in established orchards. It is not intended to burn the weeds but rather to damage the cells of tiny seedlings with heat. The plants then die of desiccation. Foliage that retains a thumbprint when pressure is applied between thumb and finger is considered to have been adequately flamed.

Species and growth stage are the most important factors affecting the successful control of weeds. Annual broadleaf weeds, especially those with fewer than two true leaves, can be controlled with flaming. Grasses are usually more tolerant and are likely to regrow because their growing points are belowground.

For safety, flaming should be done in the cooler months of the year, ideally after the rains have started in late fall through early spring. Extreme caution must be used when flaming around young trees to avoid damaging thin, green bark or girdling the tree and causing death or a setback of several years. It may be best to wait until trees are mature. Flaming may also damage plastic drip tubing and sprinkler heads or ignite mulches in the orchard. A burning permit may be required in some areas to protect air quality.

Propane-fueled flamers are most commonly used in agricultural flame weeding. A single flame directed to the base of the tree or several burners on a boom can be used to flame the weeds between tree rows. The specific flaming angle, flaming pattern, and flame length vary with the manufacturer's recommendations. Equipment speed should be adjusted for desired weed injury without damaging the tree trunks. Typically, flaming can be done at 3 to 5 miles (5 to 8 km) per hour through orchards, although speed depends on the heat output of the unit being used. Flaming should be done in windless conditions to allow the flame's heat to reach the target. Observe the flame for adjustment in the early morning or evening, when it is easier to see.

Hot-steam weeding is a thermal weed control practice that eliminates the fire danger of open-flame applications. Superheated water is delivered from a boom or spray nozzle attached to a diesel-fired boiler. After treatment, the leaves change color and the plant withers. Steam typically provides less consistent control than flaming.

Flame and steam weeding can be used for chemical-free postemergence weed control, but limitations include irregular weed control efficacy, relatively slow operational speeds, and the need for repeated applications. Both flaming and steaming are also energy-intensive due to the relatively large amount of propane or diesel used by the burners or boiler in addition to the fuel for the tractor.

Chemical practices

Herbicides registered for use in olive can provide high levels of weed control in many cases. Depending on the weeds present and site-specific conditions, tank-mix combinations or sequential applications of herbicides may be required for year-round effective control. Herbicide use is regulated in California, and a licensed applicator may be required in commercial agricultural situations. See table 13.2 for a list of herbicides currently registered for use in olive orchards in California. Before using any herbicide, identify the weed species to be controlled; carefully follow the product label directions to maximize performance and minimize risks to the applicator, the crop, and the environment.

Herbicides are traditionally described as belonging to two groups: preemergence (active against germinating weed seed and very small seedlings) and postemergence (active on emerged plants). Some herbicides have both pre- and postemergence activity. In olive orchards, herbicides are commonly applied on a strip 2 to 10 feet (0.6 to 3 m) wide (depending on orchard row spacing and tree size) centered on the tree row (fig. 13.4).

Preemergence herbicides

Preemergence herbicides primarily affect only newly germinating seed and very small seedlings; most have very limited activity on emerged weeds. For best results, preemergence herbicides should be applied to the soil before most weeds have germinated and emerged, just prior to an irrigation or rainfall of 0.25 to 0.5 inch (6 to 13 mm), which can incorporate the herbicide into

Table 13.2

Herbicides registered for use in olive in California*

Active ingredient	Example trade names	Limitations
Preemergence		
Diuron	Karmex, Direx	
Flumioxazin	Chateau, Tuscany	
Indaziflam	Alion	
Isoxaben	Trellis	Nonbearing only
Oryzalin	Surflan	
Oxyfluorfen	Goal, GoalTender	
Pendimethalin	Prowl H20, Satellite Hydrocap	
Penoxsulam/oxyfluorfen	Pindar GT	
Simazine	Princep	
Postemergence		
Carfentrazone	Shark	
Clethodim	Select Max	Nonbearing only
Diquat	Diquat	Nonbearing only
Fluazifop	Fusilade	Nonbearing only
Glyphosate	Roundup, many others	
Glufosinate	Rely 280, Lifeline, many others	
Paraquat	Gramoxone SL	
Pelargonic acid	Scythe	
Pyraflufen	Venue	
Saflufenacil	Treevix	
Sethoxydim	Poast	Nonbearing only
Postemergence (organic)		
Ammonium nonanoate	Axxe	
Caprylic/capric acid	Suppress	

*For details and the most up-to-date registration information, refer to the herbicide label or to the UC IPM Olive Pest Management Guidelines herbicide treatment table at www2.ipm.ucanr.edu/agriculture/olive/Herbicide-Treatment-Table/.

the shallow soil layers, where most weed seed germinate. However, applications should not be made if a large amount (several inches) of precipitation is expected in a short period, as runoff or excessive leaching (especially in coarse soils) of the herbicide may occur.

Preemergence herbicides provide residual weed control that can last for several months up to a year, depending on the soil type, solubility of the material, adsorption of the material to soil, sensitivity of the weed species present, and the dosage applied. They can be applied alone, in combinations of herbicides in fall after harvest, split into two applications (fall and spring), or in winter with a postemergence (foliar) herbicide. In some cases, it may be beneficial to delay the

Figure 13.4

In this orchard, preemergence and postemergence herbicides are sprayed in a strip centered on the tree row while mowing is used to manage weeds in the middles. *Photo:* Bradley Hanson.

preemergence application in winter until most weeds have germinated and then tank-mix a preemergence herbicide and a postemergence herbicide. This increases the likelihood that sufficient rainfall will occur to incorporate the herbicide and allow longer weed control into the summer.

If allowed to remain on the soil surface without being activated by rain or irrigation, some preemergence herbicides can degrade from sunlight, and subsequent weed control can be reduced. Most annual weeds germinate in the shallow soil layers and can be controlled with preemergence herbicides; however, some large weed seed may germinate in the soil below the herbicide zone and not be adequately controlled by the treatment. Because leaves or other debris covering the tree row can reduce preemergence herbicide contact with the soil, performance may be enhanced in some cases if the rows are blown or swept clean right before application.

Preemergence herbicide labels should be carefully followed to increase performance and reduce the risk of crop injury or environmental harm. In particular, the rate recommendations of some herbicides are based on soil type or orchard management practices. Some products can be used in young orchards but not bearing orchards, and others cannot be used in young orchards for a specified number of years after planting.

Postemergence herbicides

Postemergence herbicides are applied to the foliage of weeds that have already emerged. They can be combined with preemergence herbicides early in the season, applied alone as a broadcast treatment, or applied as spot treatments during the growing season.

Postemergence contact herbicides, such as paraquat, damage only those parts of the plant that are directly treated, making good coverage and wetting essential. A single treatment can kill small susceptible annual weeds, but retreatment may be necessary if perennial weeds regrow from roots or other underground structures, if weed size or density resulted in incomplete coverage, or if new germination of annual weeds occurs after the initial application.

Translocated postemergence herbicides, such as glyphosate, are absorbed by leaves and stems and moved to other above- and belowground portions of the plant, where they affect sites of action associated with the plant's growing points. These herbicides are often more effective on larger annual or perennial plants than contact herbicides (although glyphosate does not translocate well into mature nutsedge tubers). With translocated herbicides, complete coverage is not as essential as it is with contact herbicides; however, with both types of herbicides, better coverage often results in better weed control efficacy.

Postemergence herbicides should be applied when weeds are small and not under moisture stress. If the weed population is sparse or patchy, the amount of herbicide needed can be reduced by making spot applications or by using a visual weed-seeking sprayer. Some weeds, like spotted spurge, set seed soon after emergence, so they must be treated frequently to provide adequate control if a postemergence-only strategy is used.

Many postemergence herbicides used in olive orchards can injure trees if allowed to contact the foliage or green bark on young trees. Trunk protectors, lower limb pruning, timely removal of suckers, and appropriate sprayer setup should be considered to minimize the risk of tree injury.

Postemergence herbicides usually require the addition of an adjuvant (often either a nonionic surfactant or a nonphytotoxic oil) to be effective. Ammonium sulfate is often added to the spray water first, before adding the herbicide or combination of herbicides, to condition the water and help improve herbicide uptake by weeds, particularly where the water is high in calcium, sodium, magnesium, or iron. Many factors affect the performance of postemergence herbicides, especially glyphosate, including dust, spray volume, and hard water. For more information on the effective use of postemergence herbicides, see Miller et al. (2013).

Organic postemergence herbicides

Some organic herbicides are approved for use in olive orchards. Organic herbicides typically work as contact herbicides, damaging only the treated tissue; as with any contact herbicide, good coverage is essential to maximize performance. In some cases, a spray volume of up to 100 gallons per acre (935 L per ha) is required when using these products. Additionally, the inclusion of an organically approved surfactant may improve performance. Efficacy is generally greatest on seedlings or very small weeds. Repeated applications may be necessary to manage subsequently emerging weeds. Many organic herbicides are relatively expensive on a broadcast basis but may be useful for spot treatments in conjunction with mulches or other nonchemical weed management approaches.

Herbicide application equipment

Herbicide application equipment must be accurately calibrated and set up to apply the proper amount of herbicide to the soil and young growing weeds. For safe application and to minimize drift, the sprayer should be equipped with a short boom that has nozzles designed to minimize the amount of very small spray particles generated. Nozzle technology has advanced significantly in recent years, and many manufacturers have developed nozzles or attachments to decrease the proportion of very small droplets in the spray pattern. Many small-acreage growers utilize handheld wands to apply postemergence herbicides or to spot-treat individual weeds; this equipment is generally not suitable for application of preemergence herbicides because of the risk of overtreatment and missed areas. The importance of operator training cannot be overstated for ensuring safe and effective herbicide applications in orchards.

Herbicides and irrigation

In California, orchards can be irrigated by several methods, including low-volume drip, microsprinklers, misters, solid-set sprinklers, furrow irrigation, and basin flood irrigation. High-efficiency systems are becoming more common across the industry; most high-density and super-high-density olive production systems use drip irrigation systems. The irrigation method and practices can affect herbicide performance and should be considered when making decisions about herbicides and application rates. Low-volume irrigation is common in California olive orchards because it applies moisture more uniformly and efficiently than other methods. However, low-volume irrigation water applied too frequently can increase the chance of leaching and accelerate herbicide degradation, which can leave the areas around the emitters with vigorously growing weeds. It is important to monitor these areas closely and, when necessary, spot-treat with postemergence herbicides.

Weed management before planting

Ideally, weed management should be part of the planning process before the orchard is planted. See chapter 3, Site Selection and Preparation, for more information on how to reduce weed pressure in the season or two before planting.

Newly planted orchards

Olive trees are most sensitive to weed or cover crop competition during the first few years of growth. Weedy orchards may take more years than weed-free orchards to become economically productive. Unfortunately, young trees are also more sensitive to injury from many weed control practices. Regardless of the method used to control weeds, take care to avoid injuring trees with chemicals or mechanically damaging the trunk or roots.

Some growers prefer to manage weeds without herbicides for the first year or two after planting. This usually requires hoeing, cultivating, or using weed knives less than 2 inches (5 cm) deep around trees several times during spring and summer as well as cultivating or mowing between tree rows. Success is likelier when weeds are still in the seedling stage; control becomes more difficult when weeds are larger. Hand tools are generally used close to the tree to minimize injury from mechanical cultivators, particularly when the trees are young.

A limited number of herbicides are registered for control of weeds in new plantings. Preemergence herbicides (for example, flumioxazin, isoxaben, oryzalin, oxyfluorfen) can be used in the tree row after soil has settled around the transplants. Some herbicides can be used to control weeds after they emerge. These can be selective herbicides for annual grass control and suppression of perennial grasses (for example, clethodim, fluazifop, and sethoxydim). Some nonselective herbicides can also be used, but usually cartons or wraps are required to protect the green trunks and lower branches of the young, small trees. Translocated herbicides, such as glyphosate, are not typically used in young olive orchards due to concerns about crop injury.

Mowing or cultivation is used between the tree rows. In many cases, cover crops are not used for the first few years of an olive planting to minimize the competitive effects on the small trees.

Organic production systems

Organic systems cannot use synthetic herbicides for weed control; however, all the previously described cultural and mechanical/physical weed management approaches are applicable and even more important. Organic olive growers typically combine several techniques to reduce weed pressure. Cultivation and mowing are common weed management approaches between tree rows; physical barriers, flaming, in-row cultivation, and hand labor are common within tree rows. Organic producers should be especially diligent about minimizing weed seed production and reducing the development and spread of vegetative propagules.

A limited number of organic herbicides are registered for use in olive (table 13.2). Most organic herbicides work as contact herbicides and are typically applied at relatively high rates at high spray volume to ensure adequate coverage. Organic herbicides do not control established perennial weeds, are weak on annual grasses, and may not be cost-effective for commercial production. Because these products have no residual activity, several applications may be necessary.

Organic herbicides can damage leaves and young stems of olive trees but generally are safe for woody stems and trunks.

Approval of herbicides should be verified with an organic certifier before use as not all alternative herbicides are approved by all agencies and rules can change. One approach used by some growers is to start a new olive orchard using conventional weed control practices and transition later into organic. This approach allows the use of conventional herbicides to eliminate as many weeds as possible during the first years of a new orchard, minimizes weed competition during the most critical period of tree establishment, and reduces weed seed in the soil seed bank. A few years after establishment, the transition to organic can be initiated and the orchard certified by the time it reaches significant production levels.

Perennial weeds

Some of the most difficult weeds to manage in orchard crops, including olive, are perennial species. These plants can live for several years and spread via both seed and vegetative structures. While new plants arising from seed are readily controlled with tactics that work on seedlings of annual species, plants arising from spreading or creeping roots, rhizomes, or stolons can be extremely difficult to manage with mechanical/physical treatments or either preemergence herbicides or postemergence contact herbicides. Among the most challenging of these are field bindweed, bermudagrass, johnsongrass, and yellow nutsedge.

Field bindweed

Field bindweed is a vigorous perennial weed that reproduces from seed (which can survive for up to 30 years in the soil) or from rhizomes or the extensive root system. Because of the seed's longevity in the soil, controlling plants before they produce seed is critical. The plants may spread from rhizome or root sections that are cut and moved during cultivation; however, cultivation when the soil is dry controls seedlings. Spot treatments with high label rates of glyphosate can be effective, especially when the bindweed is actively growing and at the early stages of flowering.

Bermudagrass

Bermudagrass is a vigorous spring- and summer-growing perennial grass. It reproduces from seed but, because of its extensive system of rhizomes and stolons, can also be spread during cultivation. It frequently becomes a problem in mowed orchards because mowing increases the amount of light that the stolons receive, thus stimulating their growth. This grass is very competitive with the trees for moisture and nutrients. Preemergence herbicides can reduce seedling recruitment but are not effective on established plants. If bermudagrass develops in localized areas, spot-treat with a postemergence herbicide, such as glyphosate.

Johnsongrass

One of the most troublesome of perennial grasses, johnsongrass reproduces from underground stems and from seed. The mature plant grows during spring and summer in spreading patches that may be as tall as 6 to 7 feet (1.8 to 2 m). Before the orchard is planted, repeated tillage during the dry summer months can be used to control johnsongrass, but the soil must be fairly dry or rhizome buds may sprout. After the orchard is planted, tillage can suppress johnsongrass but may also spread rhizomes. Repeated applications of selective postemergence grass herbicides, such as fluazifop or sethoxydim, or nonselective herbicides, such as glyphosate, often are required for control of johnsongrass.

Yellow nutsedge

Yellow nutsedge is a difficult-to-control perennial weed and a particular problem in young orchards and around replacement trees because it does best in full sun. It reproduces from underground tubers that survive for 2 to 5 years in the soil. The tubers are easily spread by cultivation equipment. Each tuber contains several buds capable of producing plants; one or two sprout to form new plants. If the plant is removed with cultivation or an herbicide, then another bud in the tuber may be activated and a new plant initiated. In established orchards, patches of nutsedge can be spot-treated with glyphosate; treat at or before the five-leaf stage to prevent new tubers from forming. As with seedling bindweed, young nutsedge can be controlled by cultivating when the soil is dry.

Herbicide-resistant weeds

Herbicide resistance is the inherited ability of weeds to survive and grow at herbicide dosages many times greater than usually needed for control of that species. The risk for the development of herbicide resistance is greatest when the same herbicide or herbicides with the same mode of action are used repeatedly, as is often done in orchards. To prevent herbicide resistance, use a variety of weed control strategies, including cultural practices and alternating herbicides with different modes of action. Failure to do this can result in the rapid loss of herbicides as an effective pest management tool, although cultivation remains an option.

To avoid spreading resistant weeds from one field to another, clean equipment before moving it out of a field with known herbicide-resistant weeds. Schedule fields with known resistance problems as the last ones for field operations. Some populations of annual bluegrass, horseweed, annual or Italian ryegrass, junglerice, and hairy fleabane have developed resistance to glyphosate in California.

The first step in preventing herbicide resistance is early detection. Monitor the growth patterns of weeds that may indicate resistance. These could include a patch of dense weeds with less-dense populations radiating out from the central patch, or a mix of weeds that have escaped control scattered among controlled weeds of the same species in no particular pattern throughout the field.

One of the most important control strategies in managing resistant populations of weeds is to deter the plants from producing seed. To help prevent the development of resistance to herbicides in orchards, use the following tactics:

- Rotate herbicides that have different sites of action and the Weed Science Society of America (WSSA) group numbers to reduce selection pressure for resistant biotypes.
- Monitor for weed survival after an herbicide application.
- Include nonchemical weed control methods, such as cultivation or hand-weeding.
- Clean equipment after working in weed-contaminated orchards to prevent the spread of weed seed.
- Control weeds suspected of herbicide resistance before they can produce seed.

For more information on herbicide resistance, see Hanson et al. (2013) and Peachey et al. (2013).

An ongoing process

Managing weeds is an ongoing concern and expense in oil olive orchards. Poorly controlled weeds can impact establishment and productivity of the orchard, contribute to other orchard pest problems, and physically interfere with irrigation, harvest, and other orchard operations. While there is no one-size-fits-all solution to weed management in olive orchards, growers should consider a variety of cultural, mechanical, and physical practices to reduce weed pressure and supplement with carefully considered chemical approaches as appropriate.

Good weed management programs are typically based on an understanding of the weeds present in the orchard as they can vary greatly in different regions, within an orchard, or even during the lifespan of the orchard. Development and refinement of an orchard weed management program should be considered an ongoing and iterative process. Regularly monitor orchards to determine what weed control practices are working and where they fail; then use this information to adjust subsequent management operations.

We want to thank C. L. Elmore, W. T. Lanini, D. W. Cudney, and P. M. Vossen, whose writing on weed management provided the starting point for this chapter.

References

DiTomaso, J. M., and E. A. Healy. 2007. Weeds of California and other western states. 2 vols. Oakland: UC Agriculture and Natural Resources Publication 3488.

Grant, J., K. K. Anderson, T. Pritchard, J. Hasey, R. L. Bugg, F. Thomas, and T. Johnson. 2006. Cover crops for walnut orchards. Oakland: UC Agriculture and Natural Resources Publication 21627.

Hanson, B. D., A. Fischer, A. Shrestha, M. Jasieniuk, E. Peachey, R. Boydston, T. Miller, and K. Al-Khatib. 2013. Selection pressure, shifting populations, and herbicide resistance and tolerance. Oakland: UC Agriculture and Natural Resources Publication 8493.

Ingels, C. A., R. L. Bugg, G. T. McGourty, and L. P. Christensen, eds. 1998. Cover cropping in vineyards: A grower's handbook. Oakland: UC Agriculture and Natural Resources Publication 3338.

Miller, P. R., W. L. Graves, W. A. Williams, and B. A. Madson. 1989. Covercrops for California agriculture. Oakland: UC Agriculture and Natural Resources Publication 21471.

Miller, T., B. Hanson, E. Peachey, R. Boydston, and K. Al-Khatib. 2013. Glyphosate stewardship: Maintaining the effectiveness of a widely used herbicide. Oakland: UC Agriculture and Natural Resources Publication 8492.

Peachey, E., R. Boydston, B. Hanson, T. Miller, and K. Al-Khatib. 2013. Preventing and managing glyphosate-resistant weeds in orchards and vineyards. Oakland: UC Agriculture and Natural Resources Publication 8501.

Suggested reading

Elmore, C. L., J. J. Stapleton, C. E. Bell, and J. E. Devay. 1997. Soil solarization: A nonpesticidal method for controlling diseases, nematodes, and weeds. Oakland: UC Agriculture and Natural Resources Publication 21377.

Ferguson L., M. W. Johnson, W. H. Krueger, W. T. Lanini, M. V. McKenry, R. A. Van Steenwyk, P. M. Vossen, and F. G. Zalom. Revised regularly. UC IPM Pest Management Guidelines: Olive. Oakland: UC Agriculture and Natural Resources Publication 3452.

Jarvis-Shean, K., A. Fulton, D. Doll, B. Lampinen, B. Hanson, R. Baldwin, D. Lightle, and B. Vinsonhaler. 2018. Young orchard handbook. UC Cooperative Extension Capitol Corridor.

14

Spray Application Principles and Techniques for Olive

Peter Ako Larbi,
Assistant UC
Cooperative
Extension Specialist,
Kearney Agricultural
Research and
Extension Center

Highlights

- The goal of applying a protective spray is uniform coverage of the tree canopy from top to bottom.

- Air-assisted sprayers are commonly used for spray application in olive and other tree crops grown in California and around the world.

- A pruning program that opens up the canopy is essential for good spray coverage and will more likely lead to successful chemical spray pest control.

- For properly pruned trees, the prerequisites for successful high-volume or low-volume spray coverage are the same: evaluation of tree size and spacing, calibration of equipment, proper calculation of the materials needed, and proper timing of the application.

- High-volume applications require frequent refill and pesticide handling; the greater volume of spray does not necessarily mean that droplets reach higher or deeper into the foliage.

- Low-volume applications, which require less refill and pesticide handling, can be as effective as high-volume sprays in reducing pest populations to an acceptable level.

- The level of spray coverage, deposition, and pest control in each case is a result of the spraying technique, not necessarily the total gallons of spray per acre. A high volume of spray does not compensate for an excessive speed of travel, poor nozzle configuration, or badly directed air pattern.

- Chemical failure or insect resistance is too often blamed when the real problem is poor coverage.

- With improperly calibrated sprayers, radially applied spray, no matter the volume of the application, oversprays the bottom half of large mature tree canopies. Medium and small canopies of younger trees present potential for overspraying an empty space above the canopy, which is wasteful and leads to increased drift potential.

- It is good practice for all sprayer operators to ensure through calibration the correct speed of travel, proper nozzle configuration, amount of spray needed per acre, and careful measurement of the chemical.

- Proper calibration is critical for effective pest control, minimal waste, reduced environmental impact, and compliance with applicable law as represented by the pesticide label.

live trees grow differently than deciduous fruit trees. The resulting tree canopy has heavy dense foliage, mostly on the periphery of the tree. Consistent coverage of the tree canopy from top to bottom is essential when spraying. However, the natural foliage density of olive canopies resists spray penetration and may result in poor coverage, or spray distribution, within the tree canopy and especially the top center. Thus, a pruning program that opens up the canopy (see chapter 9, Canopy Management) is essential for good spray coverage and will more likely lead to successful chemical spray pest control.

Spray application in orchards

Spraying is a primary means of applying pesticides to crops for pest control. In olive and other tree crops grown in California and around the world, air-assisted sprayers are commonly

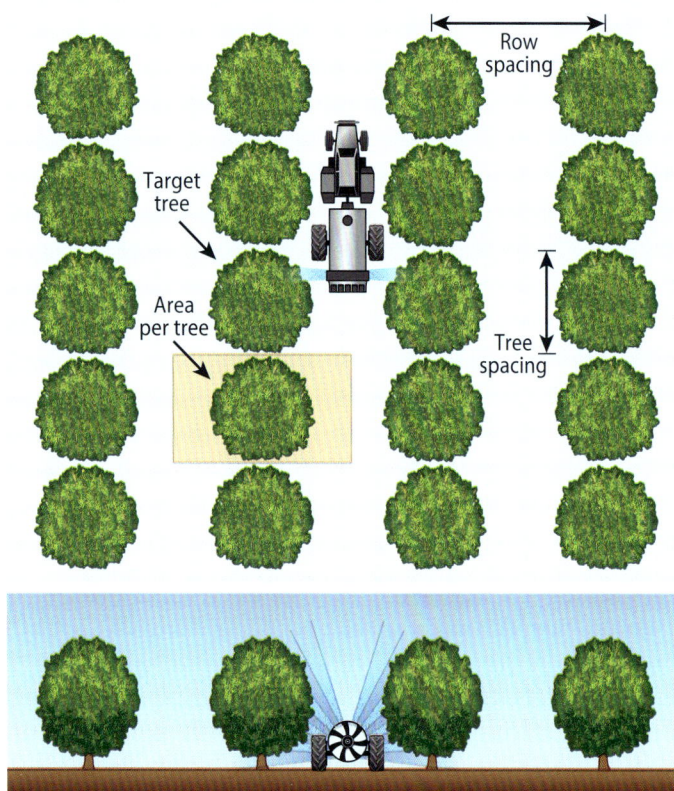

Figure 14.1

As an airblast sprayer proceeds through an orchard, it sprays half of each tree canopy on either side of it. Efficient spraying minimizes the amount of spray lost to the ground and air.

used. This type of sprayer uses a stream of air to transport the spray to the target tree canopy while opening up the canopy. In California, airblast sprayers are the most widely used sprayers, but other types of air-carrier sprayers, such as cannon-type low-volume sprayers, are sometimes used. They come in different designs and typically apply spray radially away from the sprayer.

Figure 14.1 depicts a view of a spray application in progress in an olive orchard. The ground area surrounding each tree equals the row spacing times the tree spacing within the row. The figure shows an operator driving a tractor connected to an airblast sprayer between tree rows while applying spray from both sides to cover half tree canopies on each side. At any point during the application, the immediate tree canopies adjacent to the sprayer are the target canopies. For the application to be considered complete, each row is sprayed once from each side. Only the tree side of the sprayer is sprayed when passing the edge rows.

The spray applied is intended to deposit on the target canopy. However, spray droplets can be lost to the ground within the application area, or they can be lost to the air and carried by the wind as drift to unplanned areas, including sensitive sites. As the spray outcomes equal the total spray applied, any gain in one outcome conversely results in a decrease in some other outcome (Salyani et al. 2007). Because deposition of the spray in the canopy is the desirable outcome, maximizing it is indicative of an efficient application. However, achieving this level of efficiency does not occur without effort.

Spray coverage

The type of spray deposit is a result of the spraying technique. The total number of gallons of spray per acre is not the only factor in obtaining proper spray coverage and pest control. For properly pruned trees, the prerequisites for successful spray coverage are evaluation of tree size and spacing, calibration of equipment, proper calculation of the materials needed, and proper timing of the application. When these prerequisites are met, the pesticide can be applied satisfactorily so that it effectively hits the target pest. Chemical failure or insect resistance is too often blamed when the real problem is poor coverage and inadequate deposition on the target canopy.

Drift problems

The drift of minute chemical droplets through the air to unintended areas outside the application site can be hazardous to people, animals, and wildlife and can contaminate adjacent crops. A certain percentage of fine droplets, produced by most types of nozzles and application techniques, tend to remain airborne and drift downwind. In general, large droplets help counteract the effects of the wind.

As spray application rate comes down in spray gallonage per acre, it must come down in droplet size to produce the droplet surface area required for good coverage. Potentially, low-volume application poses a greater drift hazard due to its lighter spray (see the section "Spray volume," below). Therefore, spray equipment design must include features that mitigate drift. In addition, spray applications must be done under appropriate weather conditions (wind speed < 10 mph) to reduce the chances of drift.

Most damage by pesticide drift is done within the specific area of application. However, more extensive damage can result from the transport of chemicals through direct water contact, from soil water runoff from treated fields, and from excessive airborne drift. Pesticide drift can also produce unnecessary residue on nearby crops, rendering them unfit for the retail market or for animal fodder. Confining chemicals to their

intended target field and reducing losses from chemical drift onto nontarget areas are important objectives to minimize environmental contamination and human exposure, as well as prevent further restrictions on agricultural chemicals.

Safety

The dangers and precautions involved in spraying are similar for all application techniques. Operators should keep in mind that the chemical is most concentrated in the container and that the material may still be highly concentrated in the tank. Guided by the label and law, and depending on the compound used, operators must wear protective clothing for their greatest protection, including a spray-protecting hat, footwear, gloves, and an approved respirator, during and between spray runs and especially when loading.

Lower-volume spraying requires the operator to handle the actual pesticide or containers less frequently than when spraying at higher volumes, a factor that favors lower-volume spraying. Depending on gallons per acre (L per ha) and tank capacity, handling can occur at least 5 to 10 times less frequently at low volumes than at high volumes. Furthermore, an operator using an open station tractor is not subjected to drenching by the spray, nor is the cover crop or ground saturated with runoff.

Considerations for proper spray application

To be effective, the spray application should result in sufficient spray deposit on the target. The operation and functionality of the sprayer as well as the dynamics of the spray should agree with the target canopy characteristics. Hence, the sprayer should be well maintained, well calibrated, and well adjusted to the target canopy. Additionally, the sprayer must be operated by a trained and knowledgeable spray team (that is, operator and crew) that is familiar with and adheres to best practices (Deveau 2015). The team must also be attentive and able to quickly respond to changing situations as necessary.

For maximum coverage, correct nozzle calibration is necessary. It is good practice for all sprayer operators to ensure proper nozzle configuration, correct travel speed, accurate calculation of the amount of spray needed per acre, and careful measurement of pesticides.

Proper timing of spray application in terms of the target pest and weather conditions is crucial, as is an accurately applied spray that is well directed to the target canopy and adequately penetrates the canopy. Otherwise, less-than-desired pest control will result. Moreover, the spray formulation should be optimized for on-target canopy coverage and deposition, for example, using spray additives that enhance the ability of the droplets to stick to the target.

Spray application parameters

Spray application parameters refer to the sprayer settings that determine the application rate. These include the operating pressure and the nozzle configuration and size, which determine the spray volume; the ground speed, which meters the volume applied per area covered; and the volume of air available to carry the spray to the target.

Spray volume

Typical airblast sprayers use hydraulic nozzles. The volume of spray delivered by the sprayer per unit of time (sprayer output) is determined by the operating pressure as well as the number of open nozzles and their capacities (flow rates). Sprayer output increases with increasing pressure. As the sprayer travels forward, the spray volume is distributed over the orchard area covered. For a given sprayer output, the ground speed of the sprayer determines the spray application rate, that is, the volume of spray applied per area of orchard. This is commonly stated in gallons per acre (gpa) or liters per hectare (L per ha).

Spray application rates are categorized: as high volume, or dilute (350 to 800 gpa [3,274 to 7,483 L per ha]); midvolume (100 to 350 gpa [935 to 3,274 L per ha]); low volume, or concentrate (25 to 100 gpa [234 to 935 L per ha]); and extralow volume, or high concentrate (< 25 gpa [< 234 L per ha]). High-volume spray applications usually result in spray runoff from leaves nearer to the sprayer. Conversely, low-volume applications resulting in a mist will minimize runoff or drip.

For a given spray formulation, the application rate is a measure of the treatment dose used to control the target pest, but only to an extent. The true treatment dose is the mass of active

ingredient in the pesticide that actually deposits on the target tree canopy. This is also termed *on-target spray deposition,* defined as the mass of active ingredient per unit of target area, for example, leaf area. Spray deposition is commonly measured in micrograms per square centimeter (μg per cm^2).

Nozzle configuration

All noncalibrated or improperly calibrated, radially applied, air-assisted ground spray applications, from high- to extralow volume, will most likely overspray the bottom half of large mature tree canopies (fig. 14.2). Such spray chemical deposits in the bottom half of the canopy can

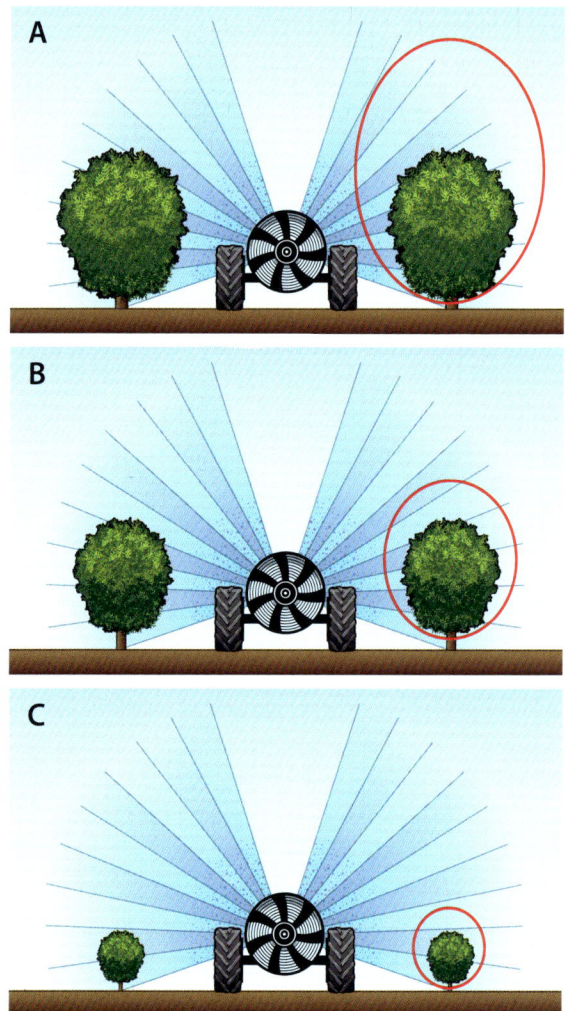

Figure 14.2

Airblast spray applications with all nozzles open have a different effect on different size tree canopies: spray imbalance in a large canopy (A); and spray waste in medium (B) and small (C) canopies.

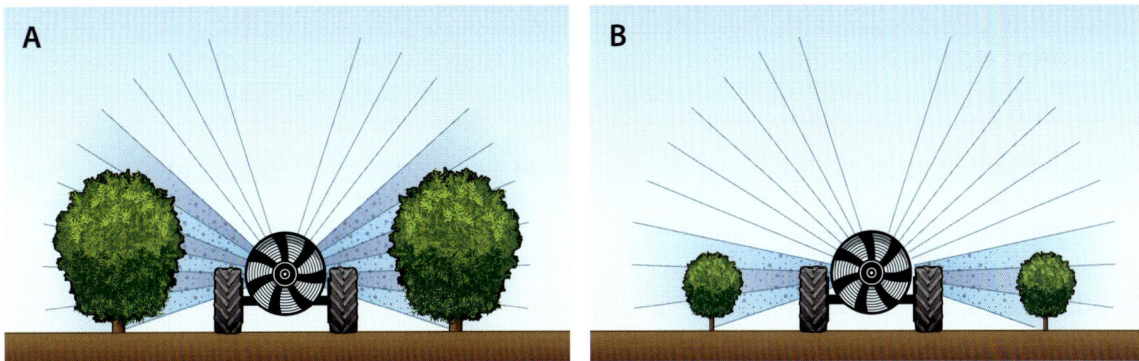

Figure 14.3

Airblast sprayer nozzles can be shut off to match a medium canopy (A) and a small canopy (B).

be 4 to 5 times greater than in the top half. This overspraying is due in part to the height and location of the sprayer in relation to the tree's height and configuration. The empty space above the canopy of younger trees with medium or small canopies can potentially be oversprayed, which is wasteful and leads to increased drift potential.

The waste (see figs. 14.2B and 14.2C) can be easily eliminated by matching the number of nozzles to the canopy size. This can be accomplished by shutting off nozzles at positions where the spray discharged is directed away from (above or beneath) the target canopy (fig. 14.3). A different nozzle arrangement is needed to properly adjust the gallons (L) per minute discharge of the spray to suit the canopy characteristics.

The imbalance in large trees (see fig. 14.2A) can be largely corrected through proper nozzle configuration to deliver a nonuniform spray to compensate for gravity and other factors that reduce coverage in the upper canopy. This can be accomplished by adjusting spray distribution via nozzle placement, for example, so that approximately two-thirds of the spray is emitted from the top half of the nozzle manifold. Such an adjustment can be made on most sprayers by putting large nozzle tips at positions 2, 3, and 4 on the spray manifold, medium tips at 5, 6, and 7, and small tips at 1, 8, and 9 (fig. 14.4). It should be noted that the success of this correction depends on the sprayer specifications and canopy characteristics.

Air discharge

By carrying pesticide to the target tissue, air from the sprayer's fan plays a very important role in the proper distribution of the pesticide inside the canopy. It is sufficient and most efficient for the air to carry spray droplets to the target and not beyond (Derksen et al. 2006). The reason is that spray droplets carried beyond the target tend to be lost either as ground fallout or potentially as drift when they miss canopies in the subsequent rows. Hence, the air velocity and air volume should be enough to cause the spray to penetrate the target canopy (Derksen et al. 2006).

The sprayer's ground speed and air volume should be matched to the canopy size and foliage density. If properly matched to the canopy, air discharge can aid coverage. Fine-tuning of air discharge is done quite easily if the sprayer has adjustable air vanes in the fan housing or high and

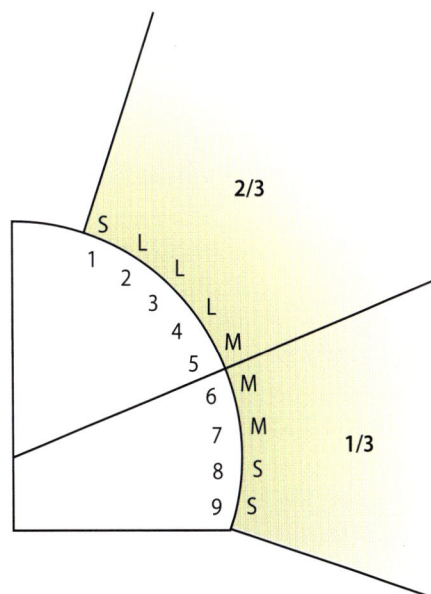

Figure 14.4

This arrangement of large (L), medium (M), and small (S) nozzle tips on the spray manifold suits radially applied airblast sprays to large mature trees.

low settings for the fan. If not, reconfiguration of the air discharge opening may be necessary by installing air vanes, narrowing or widening the fan inlet or air discharge outlet, or moving nozzles and plumbing out of the airstream area, if possible.

Air velocity measured at the fan housing discharge (for example, with a commercially available air velocity transmitter) can give air velocity differences or deficiencies from top to bottom of the air discharge outlet in miles (km) per hour. Air volume and air velocity are very difficult to control and measure once the air leaves the sprayer. Therefore, it is best to evaluate the air's effect on spray droplets by using water-sensitive cards to assess spray coverage during sprayer calibration. The cards can be placed high and low in the tree canopy, and inside as well as on the periphery. Alternatively, a food dye can be added to the tank and white target cards used (the amount of dye should be sufficient for the dye deposit to be visible).

Placement of cards to assess spray coverage is dictated by the pest and where the coverage is needed. For instance, for scale insect control, cards should be stapled to interior branches as that is where scale insects cause the most damage. For foliar pest control, cards can be stapled on both sides of paper or plastic tags (for example, livestock ear tags and shipping tags) for placement within the branches.

Ground speed

The ground travel speed of sprayers is an important variable. At a constant sprayer output and air volume rate, the ground speed serves as a metering variable. More spray is applied per unit of area at lower speeds and less spray at higher speeds. Unless the actual ground speed is known, the spray may be either underapplied (if the speed is greater than expected) or overapplied (if the speed is less than expected). Underapplication leads to poor coverage and deposition, limiting pest control effectiveness. Overapplication leads to waste and increased chemical loading on the environment.

Insufficient coverage is most easily seen when comparing the effectiveness of scale control on the interior and at the top of the canopy. Because the effective air volume rate needed to carry the spray into the target canopy decreases with increasing sprayer travel speed, the extent of penetration is limited. Therefore, the sprayer must move slowly enough, that is, remain adjacent to the tree long enough, to allow sprayer air and pesticide to penetrate the tree canopy.

Equipment need assessment

The general configuration and basic function of air-carrier sprayers are well known to growers. These sprayers are available in many styles and sizes, but they all perform in basically the same way. An important factor in considering sprayer need is the range of applications (high volume to extralow volume) that will be done. To determine the tank capacity needed, the maximum application rate is another consideration. Otherwise, the selected tank capacity may not allow for maximum productivity. Generally, the smaller the tank capacity, the greater the labor and fuel costs of spraying. The time spent driving the sprayer from the orchard to the filling station, refilling the tank and mixing the chemical, and driving back into the orchard to continue spraying increases operating costs and reduces productivity.

High-volume (dilute) spraying

High-volume spray application has been used by olive growers for many years. Although the advent of the air-carrier high-volume sprayer greatly reduced problems caused by the old high-pressure handgun sprayers, it has not solved all coverage problems. Both systems use large volumes of water. The handgun's poor coverage resulted mostly from human fatigue, while poor coverage with the air-carrier sprayer often results from improper nozzle configuration and travel speed. A high-volume application does not necessarily mean that spray reaches higher or deeper into the foliage.

Drawbacks of high-volume spraying include the frequent need to refill the tank with water and handle the pesticide. Many such sprayers also have a large number of nozzles, increasing the potential for plugging and wear. However, a good nozzle maintenance practice will minimize those potential problems.

Low-volume (concentrate) spraying

Low-volume spraying is a common practice in California and the Pacific Northwest deciduous tree fruit areas. Most low-volume spray applications made commercially are in the range of 25 to 100 gallons per acre (234 to 935 L per ha). On the plus side, this type of spray application requires a smaller number of refills, which increases productivity. On the negative side, the small droplet size desired for low-volume spraying implies a higher drift potential. Thus, appropriate measures need to be in place to limit drift; they may be incorporated into the sprayer design. The extralow volume, or high-concentrate, range of 5 to 25 gallons per acre (47 to 234 L per ha) requires the fewest fill trips and stops.

Low-volume advantages

On fruit trees, low-volume spraying can be as effective as high-volume spraying. Low-volume applications permit the use of equipment that is smaller, initially less costly, and potentially more easily and cheaply maintained; for example, power take-off (PTO) units are more economical than engine-powered units (Dibble 2005).

Savings in time per sprayer operation are notable with the low-volume method. There is less downtime because of fewer refills. Using a 400-gallon (1,514-L) low-volume sprayer calibrated for 40 gallons per acre (374 L per ha), 10 acres (4 ha) can be sprayed between fills (table 14.1). In a 12-foot (3.7-m) row spacing at a travel speed of 2 miles (3.2 km) per hour, coverage takes 0.34 hour per acre (51 min per ha) or 2.9 acres (1.2 ha) per hour, resulting in 3.44 hours of continuous spraying for each fill (table 14.2).

In addition, the total spray water used per acre in low-volume applications is reduced by 75 to 90 percent from that used in high-volume spraying. Transporting less water weight through the orchard means reduced soil compaction, which is even more desirable when the soil is wet.

Equipment selection

Air-carrier sprayers can be categorized based on such features as intended application type, design, source of power, and mode of mobility. The number of nozzles on the sprayer can vary widely, although not all may be kept open, depending on the target canopy characteristics. With the wide range of nozzles on the market, some sprayers are able to apply both low-volume and high-volume sprays. Some sprayers are specially designed for certain terrains, such as slopes.

Air-carrier sprayers use either axial or centrifugal fans. Axial fans move the air parallel to the axis of the fan shaft and have deflectors that push the air out equally from both sides. Centrifugal fans move the air at a right angle to the axis of the fan. Some sprayers have a leaf guard near the air inlet to prevent debris from getting into the fan assembly, and some have adjustable air vanes for versatility and better spray coverage.

Table 14.1

Orchard area coverage by an air-carrier sprayer

| Application rate (gpa) | Area coverage (ac) | | | | | | |
| | Tank capacity (gal) | | | | | | |
	100	200	300	400	500	600	1,000
15	6.7	13.3	20.0	26.7	33.3	40.0	66.7
30	3.3	6.7	10.0	13.3	16.7	20.0	33.3
40	2.5	5.0	7.5	10.0	12.5	15.0	25.0
50	2.0	4.0	6.0	8.0	10.0	12.0	20.0
100	1.0	2.0	3.0	4.0	5.0	6.0	10.0
200	0.5	1.0	1.5	2.0	2.5	3.0	5.0
400	0.3	0.5	0.8	1.0	1.3	1.5	2.5
600	0.2	0.3	0.5	0.7	0.8	1.0	1.7

Table 14.2

Air-carrier sprayer coverage time

Application rate (gpa)	Coverage time (hr)*						
	Tank capacity (gal)						
	100	200	300	400	500	600	1,000
15	2.29	4.58	6.88	9.17	11.46	13.75	22.92
30	1.15	2.29	3.44	4.58	5.73	6.88	11.46
40	0.86	1.72	2.58	3.44	4.30	5.16	8.59
50	0.69	1.38	2.06	2.75	3.44	4.13	6.88
100	0.34	0.69	1.03	1.38	1.72	2.06	3.44
200	0.17	0.34	0.52	0.69	0.86	1.03	1.72
400	0.09	0.17	0.26	0.34	0.43	0.52	0.86
600	0.06	0.11	0.17	0.23	0.29	0.34	0.57

*For a sprayer traveling at 2 mph in an orchard with 12-foot rows.

Growers should always check the performance of any sprayer system to determine if it meets their particular needs and request a demonstration in their orchard. Depending on the severity of pest problems, growers may want a larger unit with a bigger fan or greater tractor horsepower to operate a PTO sprayer. This would allow the equipment to direct greater amounts of air into particularly dense or tall trees. After affordability, growers should also consider the convenience and complexity of the equipment's operation and maintenance.

Intended application

Although air-carrier sprayers may be intended for use in tree cultivation (arboriculture) or vine cultivation (viticulture) based on their size and tank capacity, some sprayers can be used for both purposes. Nonetheless, the intended application of a given sprayer is subject to design limitations.

Design

Sprayers can be grouped into several design categories: radially discharging, tower, above-row, and multirow. Radially discharging sprayers have nozzles arranged at different angles to aim at different sections of the tree canopy (fig. 14.5). Tower sprayers are suitable for trees grown in high walls; the spray nozzles tend to be closer to the canopy, requiring less air to carry droplets

Figure 14.5

A radially discharging airblast sprayer has nozzles at many angles. *Photo:* Peter Larbi.

into the target canopy. Some radially discharging sprayers come with tower attachment options.

The height of the target tree determines the steepness of the top boundary of the spray cloud. The steepness affects the reach as gravity acts on the spray droplets. More air force is needed to carry the spray high to a greater reach. Some radially discharging sprayers can reach up to 16 feet (5 m). Others with greater air power can reach higher. Some tower sprayers can reach up to 26 feet (8 m) high, while others can reach just up to 10 feet (3 m) high. Designs that reduce the steepness of the top spray angle minimize the potential for drift as less spray goes above the canopy to become susceptible to drift.

Above-row sprayers have a tower design with wings that extend sideways above the target canopy to increase spray coverage at the top of the canopy (fig. 14.6). Spray and air volume are split between the tower and the side of the sprayer, reducing the spray and air volume from the side of the sprayer, thus limiting drift potential. The use of this type of sprayer is limited to trees no taller than the top spray reach.

Multirow sprayers have folding arms on both sides that go above the tree rows and drop on the other side. Spray and air volume are provided to the three sides of each tree row: sideways away from the sprayer side, downward from the top arm, and sideways toward the sprayer. Figure 14.7 shows a two-row sprayer that can treat two complete rows at a time. There are other multirow sprayers with longer folding arms and additional dropping arms to treat more rows at a time. Some multirow sprayers have antidrift tunnels or protective screens that partially enclose the target canopy. The use of multirow sprayers is limited to tree canopies shorter than the top arm and no wider than the distance between the sprayer and the dropping arm.

Power and mobility

Air-carrier sprayers may be powered by a tractor's PTO or by an engine. The biggest difference between the two is the air power generated: engine-powered airblast sprayers are generally more powerful (depending on engine capacity) than PTO-powered sprayers and can create very large air volumes and speeds to carry spray higher to the top of tall tree canopies and much farther into dense canopies. However, engine-powered sprayers tend to be pricier than PTO-powered ones, which also provide adequate power for the spray application in most cases. These factors make PTO-powered sprayers more popular than engine-powered ones.

The mode of mobility of sprayers varies. Some sprayers are designed to be mounted on a tractor's 3-point hitch (fig. 14.8). Others are made to be trailed behind the tractor as shown in figures 14.5, 14.6, and 14.7. Trailed sprayers have a shaft in front to connect to the tractor's PTO and

Figure 14.6

An above-row sprayer limits drift potential. *Photo:* Peter Larbi.

Figure 14.7

This custom-built two-row research sprayer features three tanks for different spray formulations. Each fan has eight nozzles evenly placed around the outlet. *Photo:* Peter Larbi.

Figure 14.8

A mounted sprayer is held up by the tractor's 3-point hitch; this one has slightly inclined tower outlets. *Photo:* Peter Larbi.

a set of tires on an axle midway on the tank-fan assembly to provide both support and mobility. They tend to have a larger tank capacity than mounted sprayers held up by the tractor's 3-point hitch. Mounted sprayers are also powered by the tractor's PTO.

Some sprayers are self-powered, providing their own power for operation and mobility. This type of sprayer has an enclosed cabin in front for the operator. With advancements in technology, some newer sprayers are unmanned and autonomous (fig. 14.9), eliminating the need for direct operation. Instead, a remote operator programs, controls, and monitors the sprayer from a control station. This allows for management of multiple sprayers from a single location.

Other features

Some groups of sprayers (mainly low-volume types) share other distinctive characteristics. For instance, some sprayers use compressed air to produce finer spray, known as fog. Some use low-pressure systems to move the spray liquid to special nozzles where a high-velocity (160 to 220 mph [257 to 354 km per hour]) air discharge shears the spray liquid into small droplets. Some also use electrostatic charging of the microdroplets to increase attraction between the spray and the vegetation, aiming to increase deposition (fig. 14.10).

Another group of low-volume sprayers delivers jets of mist or fog from multiple cannons. The air from the fan is divided among the cannons, which are directed at different parts of the target canopy. Figure 14.11 shows two cannon-type sprayers, one delivering the spray radially and the other with a multirow design.

Sprayer calibration

Air-carrier sprayers need to be calibrated, that is, adjusted based on measurements taken, to ensure that the correct amount of spray material is applied. Accuracy is essential for effective pest

Figure 14.9

An autonomous orchard sprayer is operated remotely. *Photo:* Peter Larbi.

Figure 14.10

An electrostatic sprayer increases attraction between the spray and foliage. *Photo:* Peter Larbi.

Figure 14.11

Cannon-type low-volume sprayers deliver jets of spray. *Photos*: Peter Larbi.

control, minimal waste, reduced environmental impact, and compliance with applicable law as represented by the pesticide label. Before calibrating, ensure that the following components are clean and in good working condition: tank, filter screens and strainers, pressure gauge, and nozzles.

Ideally, sprayer calibration should be done at the beginning of the growing season and whenever there is a significant change in conditions, such as changes in ground condition (for example, soil type, soil wetness, ground cover) and changes in target condition (e.g., crop type, tree size, canopy density). The principles involved in proper calibration are relatively simple and have already been described generally. A major objective is to optimize the application. There are different ways to calibrate, with the same or similar outcomes. Larbi (2020) suggests the following steps: (1) determine travel speed, (2) assess air profile to determine number of nozzles, (3) select nozzles, (4) measure sprayer output, (5) adjust sprayer output, and (6) evaluate spray coverage.

Figure 14.12 is an orchard sprayer calibration worksheet that can be used in the field; it provides detailed step-by-step procedures with useful formulas for calibration. To demonstrate use of the worksheet, below are three calibration worked examples, for hybrid, high-density, and super-high-density orchard systems. Table 14.3, a typical chart from a nozzle manufacturer's catalog, is used for the nozzle selection step in the worked examples. Tables 14.4 and 14.5 are useful general references.

Example 1: Hybrid orchard system
Known:

1. Desired application rate = 400 gpa
2. Intended operating pressure = 150 psi
3. Intended ground speed = 2 mph
4. Number of nozzles per side = 9
5. Orchard with large mature trees
6. Row spacing = 12 feet
7. Tree spacing in row = 10 feet

Figure 14.13 shows the completed calibration worksheet. The measured travel speed was very close to the intended ground speed of 2 mph. All nine nozzles per side were identified to be used for spraying. After deciding a nonuniform spray would be used, the nozzle sizes were selected with reference to table 14.3; the arrangement on

the manifold is shown in figure 14.14. When the sprayer output was measured, the application rate was 399.3 gpa instead of the desired 400 gpa, but no further adjustment was performed.

Example 2: High-density orchard system
Known:

1. Desired application rate = 300 gpa
2. Intended operating pressure = 150 psi
3. Intended ground speed = 2 mph
4. Number of nozzles per side = 9
5. Orchard with medium-size trees
6. Row spacing = 22 feet
7. Tree spacing in row = 10 feet

Figure 14.15 shows the completed calibration worksheet. The measured travel speed was close enough to the intended ground speed of 2 mph. Six nozzles per side were identified to be used for spraying. After deciding a uniform spray would be used, the nozzle size was selected with reference to table 14.3. When the sprayer output was measured, the application rate was 295.88 gpa. To correct it to the desired 300 gpa, the travel speed was adjusted to 1.97 mph.

Example 3: Super-high-density orchard system
Known:

1. Desired application rate = 50 gpa
2. Intended operating pressure = 150 psi
3. Intended ground speed = 2.5 mph
4. Number of nozzles per side = 9
5. Orchard with small trees
6. Row spacing = 12 feet
7. Tree spacing in row = 5 feet

Figure 14.16 shows the completed calibration worksheet. The measured travel speed was close enough to the intended ground speed of 2.5 mph. Three nozzles per side were identified to be used for spraying. After deciding a uniform spray would be used, the nozzle size was selected with reference to table 14.3. When the sprayer output was measured, the application rate was 57.78 gpa. It was corrected to the desired 50 gpa by adjusting the operating pressure from 150 psi to 113 psi.

Figure 14.12

Orchard sprayer calibration worksheet.

Orchard sprayer calibration worksheet

Step 1: Determine travel speed

i   In the target orchard, measure with a measuring tape a known distance, D, (typically 100 or 200 ft)
and mark out with marking flags. $D =$ _____ feet

Alt   With tree spacing known, TS, mark out a selected number of trees, NT, along a row with marking flags.
$D = TS \times NT =$ _____ $\times$ _____ $=$ _____ feet

ii   Using a stopwatch, measure the time, T, it takes for the half-filled sprayer to travel the marked distance at a preselected gear setting. Repeat the measurement and determine the average.
$T =$ _____ $+$ _____ $+$ _____ $/3 =$ _____ seconds

iii   Calculate speed, S_2, as:
$$S_2 \text{ (mph)} = 0.68 \frac{D \text{ (ft)}}{T \text{ (s)}} = 0.68 \underline{\qquad} = \underline{\qquad} \text{ mph}$$

Step 2: Determine air profile/number of nozzles

i   Attach about 2 to 4 feet of flagging tape to each nozzle on the sprayer manifolds.

ii   Start the fan and observe the direction of the flagging tapes from behind the sprayer. *Optional:* Take a photo of the scene with a camera for reviewing.

iii   Determine the number of nozzles, N, and their positions, NP, that are well directed on the target canopy.
$N =$ _____ ; $NP =$ _____ , _____ , _____ , _____ , _____ , _____ , _____ , _____ , _____ , _____

iv   Shut nozzles at positions that miss the target canopy on both sides.

Step 3: Select nozzles

i   Knowing the desired application rate, AR_1 and row spacing, RS, determine the total sprayer output per side, SO_1, from all open nozzles as:
$$SO_1 \text{ (gpm)} = \frac{AR_1 \text{ (gpa)} \times S_2 \text{ (mph)} \times RS \text{ (ft)}}{990}$$
$$= \frac{\underline{\quad} \times \underline{\quad} \times \underline{\quad}}{990} = \underline{\qquad} \text{ gpm}$$

ii   From a nozzle catalog, select a mix of tips to give a total flow rate equal to SO_1 at the intended operating pressure.

a. For uniform spray per side, $FR_1 \text{ (gpm)} = \dfrac{SO_1}{N}$

b. For nonuniform spray per side
(upper flow $= 2/3\ SO_1$; lower flow $= 1/3\ SO_1$).
- For even flow per section,
$FR_{1.nozzle} = \dfrac{\text{Section flow}}{\text{No. of section nozzles}}$
- For noneven flow per section, use the figure on the right as a guide.

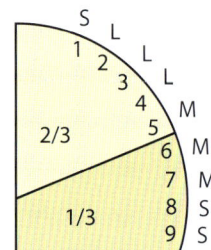

Orchard sprayer calibration worksheet

Step 4: Measure sprayer output

i   Connect hoses to the open nozzles to catch flow and run sprayer in stationary mode.

ii   Confirm the accuracy of pressure gauge reading, P_2, using a tool.

$$P_2 = \underline{\hspace{2cm}} \text{ psi}$$

iii   Use a measuring pitcher and a stopwatch to collect spray from each nozzle for 1 minute and calculate flow rate (gpm) as volume (in oz) divided by 128. Add them up to get the actual sprayer output per side, SO_2 (gpm).

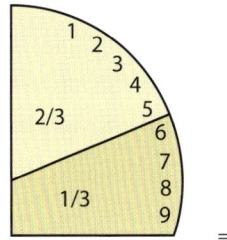

Alt   Use an automatic nozzle calibrator to measure flow rate from each nozzle.

iv   Calculate the actual application rate, AR_2, as:

$$AR_2 \text{ (gpa)} = \frac{990 \times SO_2 \text{ (gpm)}}{S \text{ (mph)} \times RS \text{ (ft)}} = \frac{990 \times \underline{\hspace{1.5cm}}}{\underline{\hspace{0.5cm}} \times \underline{\hspace{0.5cm}}} = \underline{\hspace{1.5cm}} \text{ gpa}$$

v   Compare AR_2 from step 4(iv) with the desired application rate, AR_1. If they differ greatly or if AR_2 is greater than the maximum label rate, do step 5.

Step 5 (optional): Adjust sprayer output

i   Adjust the speed:

$$S_{adj} \text{ (mph)} = \frac{AR_2 \text{ (gpa)} \times S_2 \text{ (mph)}}{AR_1 \text{ (gpa)}} = \frac{\underline{\hspace{0.5cm}} \times \underline{\hspace{0.5cm}}}{\underline{\hspace{1.5cm}}} = \underline{\hspace{1.5cm}} \text{ mph}$$

Or

ii   Adjust the operating pressure:

$$P_{adj} \text{ (psi)} = \frac{SO_1^2 \text{ (gpm}^2) \times P_2 \text{ (psi)}}{SO_2^2 \text{ (gpm}^2)} = \frac{\underline{\hspace{0.5cm}} \times \underline{\hspace{0.5cm}}}{\underline{\hspace{1.5cm}}} = \underline{\hspace{1.5cm}} \text{ psi}$$

Step 6: Verify spray coverage

i   Attach water-sensitive cards (yellow cards that turn blue when moist) to different locations in the target canopy.

ii   Run the sprayer and apply water as calibrated for the intended application.

iii   Evaluate spray coverage to ensure that it is suitable.

iv   Make necessary adjustments to obtain a suitable coverage.

Note: Subscript "1" refers to desired, target, or initial value; subscript "2" refers to actual or measured value.

Table 14.3

Sample chart from a nozzle manufacturer's catalog showing gallons per minute at different pressures

Nozzle size*	Gallons per minute (gpm)					
	50 psi	*100 psi*	*150 psi*	*200 psi*	*250 psi*	*300 psi*
01	0.112	0.158	0.194	0.224	0.250	0.274
02	0.224	0.316	0.387	0.447	0.500	0.548
03	0.335	0.474	0.581	0.671	0.750	0.822
04	0.447	0.632	0.775	0.894	1.000	1.095
05	0.559	0.791	0.968	1.118	1.250	1.369
06	0.671	0.949	1.162	1.342	1.500	1.643
07	0.783	1.107	1.356	1.565	1.750	1.917
08	0.894	1.265	1.549	1.789	2.000	2.191

*01 to 03 are small (S); 04 to 06 are medium (M); 07 and 08 are large (L).

Table 14.4

Orchard sprayer calibration chart for sprayer traveling at 2 mph

Spray application rate (gpa)	Discharge necessary from each side of sprayer (gpm)						
	Row spacing (ft)						
	8	*10*	*11*	*12*	*13*	*14*	*15*
50	0.81	1.01	1.11	1.21	1.31	1.41	1.52
100	1.62	2.02	2.22	2.42	2.63	2.83	3.03
150	2.42	3.03	3.33	3.64	3.94	4.24	4.55
200	3.23	4.04	4.44	4.85	5.25	5.66	6.06
250	4.04	5.05	5.56	6.06	6.57	7.07	7.58
300	4.85	6.06	6.67	7.27	7.88	8.48	9.09
350	5.66	7.07	7.78	8.48	9.19	9.90	10.61
400	6.46	8.08	8.89	9.70	10.51	11.31	12.12
450	7.27	9.09	10.00	10.91	11.82	12.73	13.64
500	8.08	10.10	11.11	12.12	13.13	14.14	15.15
550	8.89	11.11	12.22	13.33	14.44	15.56	16.67
600	9.70	12.12	13.33	14.55	15.76	16.97	18.18

Table 14.5

Orchard sprayer calibration chart for three different spray rates per acre

Ground speed (mph)	Discharge necessary from each side of sprayer (gpm)						
	Row spacing (ft)						
	8	10	11	12	13	14	15
100 gpa							
1.00	0.81	1.01	1.11	1.21	1.31	1.41	1.52
1.25	1.01	1.26	1.39	1.52	1.64	1.77	1.89
1.50	1.21	1.52	1.67	1.82	1.97	2.12	2.27
1.75	1.41	1.77	1.94	2.12	2.30	2.47	2.65
2.00	1.62	2.02	2.22	2.42	2.63	2.83	3.03
400 gpa							
1.00	3.23	4.04	4.44	4.85	5.25	5.66	6.06
1.25	4.04	5.05	5.56	6.06	6.57	7.07	7.58
1.50	4.85	6.06	6.67	7.27	7.88	8.48	9.09
1.75	5.66	7.07	7.78	8.48	9.19	9.90	10.61
2.00	6.46	8.08	8.89	9.70	10.51	11.31	12.12
2.25	7.27	9.09	10.00	10.91	11.82	12.73	13.64
2.50	8.08	10.10	11.11	12.12	13.13	14.14	15.15
500 gpa							
1.00	4.04	5.05	5.56	6.06	6.57	7.07	7.58
1.25	5.05	6.31	6.94	7.58	8.21	8.84	9.47
1.50	6.06	7.58	8.33	9.09	9.85	10.61	11.36
1.75	7.07	8.84	9.72	10.61	11.49	12.37	13.26
2.00	8.08	10.10	11.11	12.12	13.13	14.14	15.15
2.25	9.09	11.36	12.50	13.64	14.77	15.91	17.05
2.50	10.10	12.63	13.89	15.15	16.41	17.68	18.94

Figure 14.13

Completed calibration worksheet for worked example 1, hybrid orchard system.

Orchard sprayer calibration worksheet

Step 1: Determine travel speed

i In the target orchard, measure with a measuring tape a known distance, **D,** (typically 100 or 200 ft)
and mark out with marking flags. **D** = ____ feet

Alt With tree spacing known, **TS,** mark out a selected number of trees, **NT,** along a row with marking flags.
D = **TS** × **NT** = _10_ × _10_ = _100_ feet

ii Using a stopwatch, measure the time, **T,** it takes for the half-filled sprayer to travel the marked distance at a preselected gear setting. Repeat the measurement and determine the average.
T = _34.2_ + _34.6_ + _33.7_ /3 = _34.17_ seconds

iii Calculate speed, S_2, as:

$$S_2 \text{ (mph)} = 0.68 \frac{D \text{ (ft)}}{T \text{ (s)}} = 0.68 \frac{100}{34.17} = \frac{1.99 \approx 2}{} \text{ mph}$$

Step 2: Determine air profile/number of nozzles

i Attach about 2 to 4 feet of flagging tape to each nozzle on the sprayer manifolds.

ii Start the fan and observe the direction of the flagging tapes from behind the sprayer.
Optional: Take a photo of the scene with a camera for reviewing.

iii Determine the number of nozzles, **N,** and their positions, **NP,** that are well directed on the target canopy.
N = _9_ ; **NP** = _1_ , _2_ , _3_ , _4_ , _5_ , _6_ , _7_ , _8_ , _9_ , ____

iv Shut nozzles at positions that miss the target canopy on both sides.

Step 3: Select nozzles

i Knowing the desired application rate, AR_1 and row spacing, **RS,** determine the total sprayer output per side, SO_1, from all open nozzles as:

$$SO_1 \text{ (gpm)} = \frac{AR_1 \text{ (gpa)} \times S_2 \text{ (mph)} \times RS \text{ (ft)}}{990}$$

$$= \frac{400 \times 2 \times 12}{990} = \frac{9.70}{} \text{ gpm}$$

ii From a nozzle catalog, select a mix of tips to give a total flow rate equal to SO_1 at the intended operating pressure.

a. For uniform spray per side, $FR_1 \text{ (gpm)} = \dfrac{SO_1}{N}$

b. For nonuniform spray per side
(upper flow = 2/3 SO_1; lower flow = 1/3 SO_1).

   • For even flow per section,
   $FR_{1.nozzle} = \dfrac{\text{Section flow}}{\text{No. of section nozzles}}$

   • For noneven flow per section, use the figure on the right as a guide.

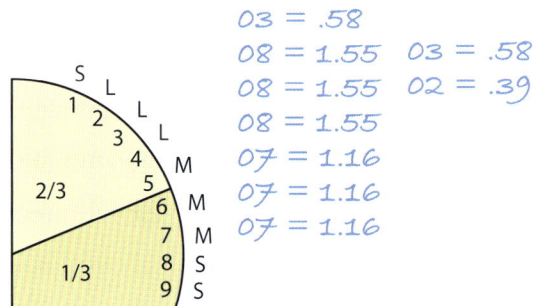

03 = .58
08 = 1.55   03 = .58
08 = 1.55   02 = .39
08 = 1.55
07 = 1.16
07 = 1.16
07 = 1.16

Figure 14.13, continued. *Page 2 of 2*

Step 4: Measure sprayer output

i Connect hoses to the open nozzles to catch flow and run sprayer in stationary mode.

ii Confirm the accuracy of pressure gauge reading, P_2, using a tool.

$$P_2 = \underline{\quad 150 \quad} \text{ psi}$$

iii Use a measuring pitcher and a stopwatch to collect spray from each nozzle for 1 minute and calculate flow rate (gpm) as volume (in oz) divided by 128. Add them up to get the actual sprayer output per side, SO_2 (gpm). ✓

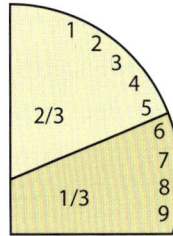

Alt Use an automatic nozzle calibrator to measure flow rate from each nozzle.

$$
\begin{aligned}
03 &= .58 \\
08 &= 1.54 & 03 &= .59 \\
08 &= 1.55 & 02 &= .39 \\
08 &= 1.56 \\
07 &= 1.15 & \text{TOTAL} \\
07 &= 1.16 & &= 9.68 \\
07 &= 1.15
\end{aligned}
$$

iv Calculate the actual application rate, AR_2, as:

$$AR_2 \text{ (gpa)} = \frac{990 \times SO_2 \text{ (gpm)}}{S \text{ (mph)} \times RS \text{ (ft)}} = \frac{990 \times\ 9.68}{2\ \times\ 12} = \underline{\quad 399.3 \quad} \text{gpa}$$

v Compare AR_2 from step 4(iv) with the desired application rate, AR_1. If they differ greatly or if AR_2 is greater than the maximum label rate, do step 5.

Step 5 (optional): Adjust sprayer output

i Adjust the speed:

$$SR_{adj} \text{ (mph)} = \frac{AR_2 \text{ (gpa)} \times S_2 \text{ (mph)}}{AR_1 \text{ (gpa)}} = \frac{\quad\times\quad}{} = \underline{\quad\quad} \text{ mph}$$

Or

ii Adjust the operating pressure:

$$P_{adj} \text{ (psi)} = \frac{SO_1{}^2 \text{ (gpm}^2) \times P_2 \text{ (psi)}}{SO_2{}^2 \text{ (gpm}^2)} = \frac{\quad\times\quad}{} = \underline{\quad\quad} \text{ psi}$$

Step 6: Verify spray coverage

i Attach water-sensitive cards (yellow cards that turn blue when moist) to different locations in the target canopy.

ii Run the sprayer and apply water as calibrated for the intended application.

iii Evaluate spray coverage to ensure that it is suitable.

iv Make necessary adjustments to obtain a suitable coverage.

Note: Subscript "1" refers to desired, target, or initial value; subscript "2" refers to actual or measured value.

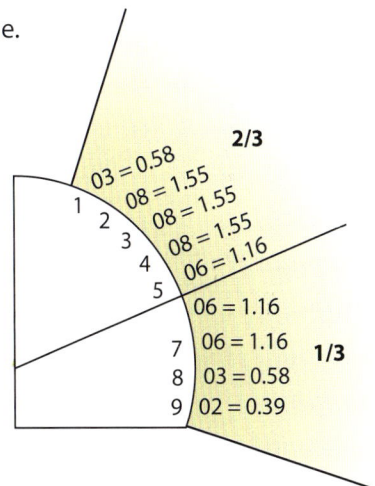

Figure 14.14.

This arrangement of nozzle tips was used for worksheet example 1. It discharges a total of 9.7 gpm.

Figure 14.15

Completed calibration worksheet for worked example 2, high-density orchard system.

Orchard sprayer calibration worksheet

Step 1: Determine travel speed

i   In the target orchard, measure with a measuring tape a known distance, **D,** (typically 100 or 200 ft) and mark out with marking flags. **D** = _100_ feet

Alt   With tree spacing known, **TS,** mark out a selected number of trees, **NT,** along a row with marking flags.
D = **TS** × **NT** = _____ × _____ = _____ feet

ii   Using a stopwatch, measure the time, **T,** it takes for the half-filled sprayer to travel the marked distance at a preselected gear setting. Repeat the measurement and determine the average.
$T = \frac{34.0}{} + \frac{33.8}{} + \frac{34.5}{}/3 = \frac{34.1}{}$ seconds

iii   Calculate speed, S_2, as:
$$S_2 \text{ (mph)} = 0.68 \frac{D \text{ (ft)}}{T \text{ (s)}} = 0.68 \frac{100}{34.1} = \frac{1.99 \approx 2}{} \text{ mph}$$

Step 2: Determine air profile/number of nozzles

i   Attach about 2 to 4 feet of flagging tape to each nozzle on the sprayer manifolds.

ii   Start the fan and observe the direction of the flagging tapes from behind the sprayer. *Optional:* Take a photo of the scene with a camera for reviewing.

iii   Determine the number of nozzles, **N,** and their positions, **NP,** that are well directed on the target canopy.
N = _6_ ; **NP** = _4_ , _5_ , _6_ , _7_ , _8_ , _9_ , _____ , _____ , _____ , _____

iv   Shut nozzles at positions that miss the target canopy on both sides.   _1_ , _2_ , _3_

Step 3: Select nozzles

i   Knowing the desired application rate, AR1, and row spacing, RS, determine the total sprayer output per side, SO1, from all open nozzles as:
$$SO_1 \text{ (gpm)} = \frac{AR_1 \text{ (gpa)} \times S_2 \text{ (mph)} \times RS \text{ (ft)}}{990}$$
$$= \frac{300 \times 2 \times 22}{990} = \frac{13.3}{} \text{ gpm}$$

ii   From a nozzle catalog, select a mix of tips to give a total flow rate equal to SO_1 at the intended operating pressure.

a. For uniform spray per side, $FR_1 \text{ (gpm)} = \frac{SO_1}{N}$ ✓   13.3 / 6 = 2.22

b. For nonuniform spray per side (upper flow = 2/3 SO_1; lower flow = 1/3 SO_1).

  • For even flow per section,
  $FR_{1.nozzle} = \frac{\text{Section flow}}{\text{No. of section nozzles}}$

  • For noneven flow per section, use the figure on the right as a guide.

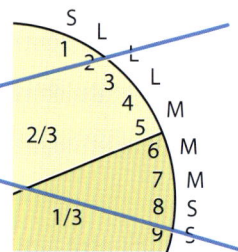

150 psi not possible

Pressure = 300 psi ✓

S
1 L
2 L
3 L
4 M
5 M
6 M
7 M
8 S
9 S

2/3

1/3

Figure 14.15, continued. *Page 2 of 2*

Step 4: Measure sprayer output

i   Connect hoses to the open nozzles to catch flow and run sprayer in stationary mode.

ii   Confirm the accuracy of pressure gauge reading, P_2, using a tool.

$$P_2 = \underline{\quad 300 \quad} \text{ psi}$$

iii   Use a measuring pitcher and a stopwatch to collect spray from each nozzle for 1 minute and calculate flow rate (gpm) as volume (in oz) divided by 128. Add them up to get the actual sprayer output per side, SO_2 (gpm).

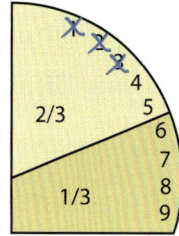

Alt   Use an automatic nozzle calibrator to measure flow rate from each nozzle. ✓

08 tips

4 = 2.19
5 = 1.94 check 2.19
6 = 2.21
7 = 2.19
8 = 2.18   *TOTAL*
9 = 2.19   *= 13.15*

iv   Calculate the actual application rate, AR_2, as:

$$AR_2 \text{ (gpa)} = \frac{990 \times SO_2 \text{ (gpm)}}{S \text{ (mph)} \times RS \text{ (ft)}} = \frac{990 \times 13.15}{2 \times 22} = \underline{\quad 295.88 \quad} \text{ gpa}$$

v   Compare AR_2 from step 4(iv) with the desired application rate, AR_1. If they differ greatly or if AR_2 is greater than the maximum label rate, do step 5.

Step 5 (optional): Adjust sprayer output

i   Adjust the speed:

$$\text{✓} \quad SR_{adj} \text{ (mph)} = \frac{AR_2 \text{ (gpa)} \times S_2 \text{ (mph)}}{AR_1 \text{ (gpa)}} = \frac{295.88 \times 2}{300} = \underline{\quad 1.97 \quad} \text{ mph}$$

Or

ii   Adjust the operating pressure:

$$P_{adj} \text{ (psi)} = \frac{SO_1{}^2 \text{ (gpm}^2\text{)} \times P_2 \text{ (psi)}}{SO_2{}^2 \text{ (gpm}^2\text{)}} = \frac{\quad \times \quad}{\quad} = \underline{\quad\quad} \text{ psi}$$

Step 6: Verify spray coverage

i   Attach water-sensitive cards (yellow cards that turn blue when moist) to different locations in the target canopy.

ii   Run the sprayer and apply water as calibrated for the intended application.

iii   Evaluate spray coverage to ensure that it is suitable.

iv   Make necessary adjustments to obtain a suitable coverage.

Note: Subscript "1" refers to desired, target, or initial value; subscript "2" refers to actual or measured value.

Figure 14.16

Completed calibration worksheet for worked example 3, super-high-density orchard system.

Orchard sprayer calibration worksheet

Step 1: Determine travel speed

i   In the target orchard, measure with a measuring tape a known distance, **D,** (typically 100 or 200 ft)
and mark out with marking flags. $D = $ _____ feet

Alt   With tree spacing known, **TS,** mark out a selected number of trees, **NT,** along a row with marking flags.
$D = TS \times NT = $ _5_ $\times$ _20_ $=$ _200_ feet

ii   Using a stopwatch, measure the time, **T,** it takes for the half-filled sprayer to travel the marked distance at a preselected gear setting. Repeat the measurement and determine the average.
$T = $ _55.2_ $+$ _55.0_ $+$ _54.7_ $/3 = $ _54.97_ seconds

iii   Calculate speed, S_2, as:
$$S_2 \text{ (mph)} = 0.68 \frac{D \text{ (ft)}}{T \text{ (s)}} = 0.68 \frac{200}{54.97} = \underline{2.47} \text{ mph}$$

Step 2: Determine air profile/number of nozzles

i   Attach about 2 to 4 feet of flagging tape to each nozzle on the sprayer manifolds.

ii   Start the fan and observe the direction of the flagging tapes from behind the sprayer. *Optional:* Take a photo of the scene with a camera for reviewing.

iii   Determine the number of nozzles, **N,** and their positions, **NP,** that are well directed on the target canopy.
$N = $ _3_ ; $NP = $ _7_ , _8_ , _9_ , ____ , ____ , ____ , ____ , ____ , ____ , ____

iv   Shut nozzles at positions that miss the target canopy on both sides.

Step 3: Select nozzles

i   Knowing the desired application rate, AR1, and row spacing, RS, determine the total sprayer output per side, SO1, from all open nozzles as:

$$SO_1 \text{ (gpm)} = \frac{AR_1 \text{ (gpa)} \times S_2 \text{ (mph)} \times RS \text{ (ft)}}{990}$$

$$= \frac{50 \times 2.47 \times 12}{990} = \underline{1.5} \text{ gpm}$$

ii   From a nozzle catalog, select a mix of tips to give a total flow rate equal to SO_1 at the intended operating pressure.

   a. For uniform spray per side, $FR_1 \text{ (gpm)} = \dfrac{SO_1}{N}$ ✓ $1.5 / 3 = 0.5$

   b. For nonuniform spray per side
       (upper flow = 2/3 SO_1; lower flow = 1/3 SO_1).
       • For even flow per section,
       $FR_{1.nozzle} = \dfrac{\text{Section flow}}{\text{No. of section nozzles}}$
       • For noneven flow per section, use the figure on the right as a guide.

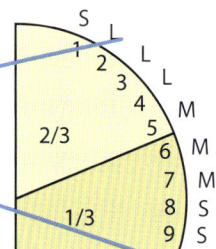

Figure 14.16, continued. *Page 2 of 2*

Step 4: Measure sprayer output

i   Connect hoses to the open nozzles to catch flow and run sprayer in stationary mode.

ii   Confirm the accuracy of pressure gauge reading, P_2, using a tool.

$$P_2 = \underline{\quad 150 \quad} \text{ psi}$$

iii Use a measuring pitcher and a stopwatch to collect spray from each nozzle for 1 minute and calculate flow rate (gpm) as volume (in oz) divided by 128. Add them up to get the actual sprayer output per side, SO_2 (gpm).

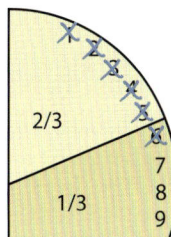

03 tips
7 = 0.58
8 = 0.58
9 = 0.57

TOTAL = 1.73

Alt Use an automatic nozzle calibrator to measure flow rate from each nozzle.

iv   Calculate the actual application rate, AR_2, as:

$$AR_2 \text{ (gpa)} = \frac{990 \times SO_2 \text{ (gpm)}}{S \text{ (mph)} \times RS \text{ (ft)}} = \frac{990 \times 1.75}{2.47 \times 12} = \underline{57.78} \text{ gpa}$$

v   Compare AR_2 from step 4(iv) with the desired application rate, AR_1. If they differ greatly or if AR_2 is greater than the maximum label rate, do step 5.

Step 5 (optional): Adjust sprayer output

i   Adjust the speed:

$$S_{adj} \text{ (mph)} = \frac{AR_2 \text{ (gpa)} \times S_2 \text{ (mph)}}{AR_1 \text{ (gpa)}} = \frac{\quad \times \quad}{} = \underline{\quad\quad} \text{ mph}$$

Or

ii   Adjust the operating pressure:

$$P_{adj} \text{ (psi)} = \frac{SO_1^2 \text{ (gpm}^2) \times P_2 \text{ (psi)}}{SO_2^2 \text{ (gpm}^2)} = \frac{1.5^2 \times 150}{1.73^2} = \underline{113} \text{ psi}$$

Step 6: Verify spray coverage

i   Attach water-sensitive cards (yellow cards that turn blue when moist) to different locations in the target canopy.

ii   Run the sprayer and apply water as calibrated for the intended application.

iii Evaluate spray coverage to ensure that it is suitable.

iv   Make necessary adjustments to obtain a suitable coverage.

Note: Subscript "1" refers to desired, target, or initial value; subscript "2" refers to actual or measured value.

References

Derksen, R., C. R. Krause, R. D. Fox, R. D. Brazee, and R. Zondag. 2006. Effect of application variables on spray deposition, coverage, and ground losses in nursery tree applications. Journal of Environmental Horticulture 24(1):45–52. https://doi.org/10.24266/0738-2898-24.1.45

Deveau, J. 2015. Airblast 101: A handbook of best practices in airblast spraying. Ontario Ministry of Agriculture, Food and Rural Affairs.

Dibble, J. E. 2005. Spray application principles and techniques for olives. In G. S. Sibbett, L. Ferguson, J. L. Coviello, and M. Lindstrand, eds. Olive production manual. 2nd ed. Oakland: UC Agriculture and Natural Resources Publication 3353.

Larbi, P. A. 2020. Spray calibration and coverage: The basics of spray application. Progressive Crop Consultant.

Salyani, M., M. Farooq, and R. D. Sweeb. 2007. Spray deposition and mass balance in citrus orchard applications. Transactions of the American Society of Agricultural and Biological Engineers (ASABE) 50(6):1963–1969. https://doi.org/10.13031/2013.24092

15

The Olive Harvest

Leandro Ravetti,
Agricultural Engineer,
CEO, Cobram Estate
Olives

Highlights

- Optimal harvesting time is generally determined by visual and physical factors that correlate with maximum oil yields and optimal product quality. An olive is considered mature when it reaches its maximum oil content.

- Color change and the amount of oil in the olive are not directly linked. The color of mature olives varies considerably, depending on the variety, crop load, and environmental conditions.

- Oil and moisture levels in the fruit can significantly influence oil extractability and yields. Oil accumulation starts immediately after pit hardening. The oil accumulation process concludes around 28 weeks after full bloom.

- The time of harvest has a great influence on oil quality. Oils from early harvested fruit show consistently greater fruitiness, greener tones, pungency, and bitterness than the sweeter oils produced from more mature fruit. The biophenols level and shelf life also drop with the ripening of the fruit.

- Manual harvesting costs are prohibitive for many growers, but small growers selling at a premium retail price often find the costs acceptable.

- Trunk shakers are often used for high-density (HD) olive orchards. Studies show that most olives drop in the first 3 seconds and that applying vibration beyond 5 seconds promotes tree damage and damage to the trunk shaker and tractor.

- Canopy contact harvesters are required for super-high-density (SHD) olive orchards and are suitable only for large HD operations.

- Continually checking the performance of the harvester is extremely important to maintain or improve harvesting efficiency.

- The time between harvest and processing is one of the most important factors affecting olive oil quality.

The modern, high-volume olive oil industry is a contemporary growing model pursuing maximum profit with sustainable production methods, and harvest is one of the most important aspects of that model (fig. 15.1). To achieve maximum profit, it is necessary to obtain high fruit and oil yields, acceptable prices for the final product, and minimal production costs. These yields need to be achieved with consideration of environmental conditions and sustainability, but the yields do not alone ensure profitability. Quality throughout production and processing has a strong influence on the final price. Furthermore, minimal production costs are likely to be achieved only if harvest mechanization is adopted in the orchards. Modern olive growing is significantly different from traditional olive cultivation, where horticultural limitations, such as a lack of irrigation and steep slopes, do not allow orchards to reach maximum and constant profitability.

Achieving high levels of harvesting efficiency has been an extremely important component in the industry's adoption of mechanical harvesting. Olive harvesting efficiency is defined as the percentage of the fruit on the trees that is effectively harvested and delivered to the mill. Mechanical harvesting is rarely 100 percent efficient because of the olive's high fruit retention force (FRF) and low fruit weight, and because of the structure and size of the trees. More than 90 percent efficiency is achieved only in homogenous trees that have been pruned to help with mechanical harvesting. An average of 80 to 90 percent efficiency is more commonly achieved during the oil olive harvest, but it could be much lower (as low as 60% to 65%) at the start of the harvest, when FRF is higher, and with high-FRF varieties.

All crops are harvested when they are judged to be horticulturally mature, a stage of development that meets an agreed-upon standard established in the marketplace. An oil olive is considered mature when it reaches its maximum oil content, which occurs about 2 months before physiological maturity. Harvesting at this time allows the olives to be processed into high-quality olive oil. The color range of olives at the harvest stage varies considerably depending on the variety, crop load, and environmental conditions.

Olive development and maturity index

Olive fruit growth and development takes about 6 to 7 months; the exact length of this period is influenced by variety, growing conditions, and fruit yield per tree. Olive development, like the development of other stone fruit, follows a double sigmoid curve, with a latent period at the beginning of the process and another one at the end.

Figure 15.1

Mechanical harvest is important to minimizing production costs in the modern high-volume olive oil industry. *Photo: Corto Olive Co.*

After pollination, in the first phase of development, cell division occurs relatively quickly, and during this fast growth period almost all the cell division processes are completed. During the second phase, growth is slow, but the pit-hardening process takes place, until both pit and embryo reach their final size (it will not change in the following months). The final phase, again one of fast growth, determines the size of the fruit, as the cells in the flesh enlarge and the processes of oil and moisture accumulation occur.

Olives are initially green and turn yellowish in response to a sharp reduction in chlorophyll in the cells. Then anthocyanin starts to accumulate in the cells, determining color intensity, which may range from reddish to intense purple and black. In most varieties, skin coloring begins at the apex of the fruit and continues toward the other end, next to the stem. Then the mesocarp begins to color, from the outermost part inward, until the purple hue reaches the pit.

Color change and the amount of oil in the olive are not directly linked. Consequently, a maturity index based on color is not the best method to determine when the oil accumulation has stopped and when the harvesting period should start. However, a maturity index based on color provides useful information regarding the style of oil that can be produced from the olives and any adjustments that will be needed during the oil extraction process. The index suggested by Ferreira is the most commonly used. An olive sample of approximately 4.5 pounds (2 kg) of fruit is taken at the height of the harvest operator and from the four sides of the tree. Once the sample has been homogenized, 100 olives are separated, and each olive is classified into one of the eight classes according to the color of its skin and flesh (table 15.1). The maturity index (MI) is the sum of the multiplication of the number of olives in each class by the numerical value of each class, divided by 100. The formula uses A, B, C, D, E, F, G, and H as the number of fruit in classes 0, 1, 2, 3, 4, 5, 6, and 7, respectively:

$$MI = (A \times 0 + B \times 1 + C \times 2 + D \times 3 + E \times 4 + F \times 5 + G \times 6 + H \times 7) \, 100$$

Table 15.1

Maturity index classes

Class	Description
0	Deep green skin
1	Yellowish-green skin
2	Green skin with reddish patches over less than half of the fruit; onset of ripening
3	Reddish or purple skin over more than half the fruit; end of onset of ripening
4	Black skin and white flesh
5	Black skin and less than half of the flesh purple
6	Black skin and more than half of the flesh purple but not reaching the pit
7	Black skin and purple flesh to the pit

Oil and moisture content

Oil and moisture levels in the fruit can significantly influence oil extractability and yields obtained in the mill. The oil accumulation process in the fruit starts immediately after pit hardening, with the leaves and the fruit as the carbon source for oil synthesis. During the first phase of the process (9 to 11 weeks after full bloom), oil accumulation is slow and mainly constituted of structural lipids; oil content during this phase reaches only 4 percent of the final oil accumulation. The second phase is the most important in terms of oil accumulation, accounting for almost 90 percent of the total oil; the process reaches its peak approximately 18 weeks after full bloom. During the third phase, oil accumulation slows down and usually concludes around 28 weeks after full bloom.

While the moisture content of olive fruit can range between 40 and 75 percent during the growing season, a range of about 50 to 56 percent at harvest will maximize the extractability of the oil. Different varieties planted in similar soils and managed similarly show significant variations in fruit moisture levels. This variation is genetically determined, so different management techniques are required for different varieties to arrive at an ideal moisture level at harvesting time without compromising vegetative growth or oil yields.

Large deviations from normal irrigation levels can reduce oil accumulation, biophenols, and shelf life.

When the fruit arrives at the processing plant with more than 56 to 58 percent moisture, it will most likely lead to emulsions in the paste. Emulsions reduce extraction efficiency and slow down processing, which increases crushing costs. If fruit moisture levels are very low, below 47 to 49 percent, the paste will have a high viscosity, forcing the need for water to be added during the crushing, malaxation, and/or separation steps. Adding water in these processing stages may reduce fruitiness intensity and biophenol content.

Varieties showing higher moisture levels, such as Leccino, Picual, and Manzanillo, are typically classified as varieties that produce more difficult pastes. Varieties like Frantoio and Koroneiki, which consistently tend to have lower moisture levels, are classified as easy paste varieties. Varieties that have intermediate moisture levels include Arbequina, Arbosana, and Barnea.

Harvest and rain

Once the harvest starts, as a rule, the sooner it finishes, the better the result. A short and compact harvest lessens the potential of pests and diseases to affect oil quality, reduces the chance of fruit drop, avoids the negative impact of excessive delays on next year's production, and keeps harvesting costs low. Significant rainfall during harvest is the most common reason picking is stopped; rainfall increases fruit moisture, which, as described earlier, can have a negative impact on oil extraction efficiency. Additionally, in certain environments and with certain varieties, harvesting with free water in the foliage can lead to a higher incidence of olive knot in the harvested area. Harvesters may also become bogged down in the saturated soil of rain-soaked orchards.

Fruit retention force and fruit drop

Regardless of the harvesting method, how easily the fruit separates from the tree (the ease with which it can be harvested) depends on how strongly the fruit is attached to the shoot. This strength of attachment is called the fruit retention force (FRF), or fruit attachment force. The FRF for olive ranges from 0 to 42 ounces (1,200 g); it can be measured with a small handheld dynamometer or spring scale.

Some researchers consider that more than the FRF itself, it is the ratio between the FRF and olive fruit weight (which can range from 0.02 to 0.4 ounce [0.5 to 10 g]) that better describes how easily a fruit will be harvested, particularly if trunk shakers will be used. The average ratio ranges from 200 to 400 with oil varieties, and 100 to 200 with table olive varieties. Olive is one of the most difficult fruit to harvest mechanically due to these relatively high ratios. Other mechanically harvested fruit have much lower ratios: for example, for peach the ratio is 40 to 60, plum 30 to 70, and cherry 50 to 100.

The FRF decreases as an olive fruit ripens. However, the decrease is not an even linear process; it is affected daily by weather conditions, orchard management, and other factors. As the FRF decreases, it has an increasing impact on natural fruit drop. The amount of natural fruit drop depends mainly on the variety, but it can be affected by weather conditions and tree health. Normally, there is a small fruit drop during the beginning of the harvesting period, which can increase as the harvest period extends, particularly with certain varieties, in warmer climates, and in light cropping years.

Impact of harvest time on next year's crop

The time of harvest has a significant influence on the next year's crop. When mature fruit remains on the tree for an extended period, flower differentiation in the next season is inhibited to a varying degree. Certainly, it is observed that, at equal crop loads, early harvested trees always crop more heavily the next year than those harvested later in the season, particularly in the case of heavy cropping trees (Dag et al. 2009; Lavee 1996). In most cases, a harvest delay of 2 to 3 weeks after the fruit reaches maximum oil content does not produce a significant damage to the next year's crop. Nonetheless, almost all trees harvested after that period will show some degree of loss in fruiting potential the following year.

Oil quality evolution

The time of harvest, the fruit quality, and the time between harvest and processing have a great influence on oil quality. Organoleptic analysis of oils produced from early harvested fruit shows consistently greater fruitiness, greener tones, pungency, and bitterness than the oils produced from more mature fruit, which tend to have milder and riper flavors. The biophenols level and shelf life also drop with the ripening of the fruit. The free fatty acid value (FFA) of the oil increases in relation to the level of damage to the fruit at harvest and to the time between harvest and processing. It increases slightly during the first 36 hours between harvest and processing; after 36 hours, it rises more sharply for both damaged and undamaged fruit, but the increase is more significant in damaged fruit and riper fruit.

Fruit-loosening agents

Fruit-loosening agents allow for an earlier harvest, with the consequent improvement in oil quality and reduction of alternate (biannual) bearing, and they increase the efficiency of mechanical harvesting. The most-evaluated fruit-loosening agents are ethephon and monopotassium phosphate (MKP), and combinations of them, with and without specific wetting agents. (At the date of the publication of this manual, ethephon is not a registered product for olive in California.)

Applied at correct rates, times, and conditions, most fruit-loosening agents cause a clear decrease in the FRF of olives 7 days after application compared to the FRF of untreated olives (Barranco et al. 2002; Bem-Tal and Wodner 1994; Martin 1994; Ravetti and McClelland 2008). The maximum effect of the fruit-loosening agents occurs generally 2 to 3 weeks after application. Fruit-loosening agents also show a positive effect on increasing harvesting efficiency in most years. This efficiency is reflected in larger percentages of fruit removal and faster harvesting, which reduces harvest costs and the risks associated with a late harvest.

Fruit-loosening agents are especially helpful in a high crop year, or when greener fruit is harvested early in the season, or to increase harvesting efficiency on varieties with a high FRF,

for example, Frantoio, Arbosana, Koroneiki, and Arbequina. Being a hormonal product, a fruit-loosening agent carries a risk of causing undesirable fruit drop or tree defoliation, particularly when applied to stressed trees. Education on the use of fruit-loosening agents is essential to avoid those negative effects.

Optimal harvesting time

To determine the optimal harvesting time of oil olives, evaluate all the following factors:

- The olives must have the maximum weight of oil.
- The quality of the oil must be optimal.
- Fruit and tree damage due to harvest must be minimal.
- Next year's crop must not be affected.
- Harvesting must be as inexpensive as possible.

The moment when the olives have the maximum amount of oil of the highest quality varies according to environmental conditions, variety characteristics, and the amount of fruit on the tree. Olive harvest in California has customarily begun in mid-October, peaked in early November, and finished in early December. The primary reason for delaying the start of the harvest is to accrue the value of the olives gaining more oil content and being easier to process. The main reason for accelerating harvest late in the season is to decrease losses due to overripeness, fruit drop, and cold damage.

It is important to recognize that once the fruit reaches the optimal time for harvesting, the oil quantity remains almost constant. The oil percentage per fresh fruit weight may still increase, but that's because of a loss of moisture from the fruit, not a higher amount of oil. Riper fruit is also usually easier to process, and the higher processing efficiency yields more oil.

The following process can be used to determine when to harvest based on maximum oil accumulation in the fruit:

1. Beginning in mid-August, take weekly samples from three olive trees (1 lb, [0.45 kg], of randomly selected olives), in each representative area of the orchard.

2. Use near-infrared (NIR), solvent extraction (Soxhlet), or another appropriate analytical method to determine the oil content in dry weight (g) of the weekly samples. If none of those methods are available, determine just the dry weight of the samples: simply dry a 100-olive sample in a fan-forced oven until constant weight.

3. When samples are showing an oil increase of less than 1 percent oil in dry weight or a total dry weight increase of less than 2 percent for 2 consecutive weeks, picking should begin within 1 week and last no longer than 3 weeks for that area of the orchard.

Growers eager to follow this process may be frustrated by the difficulty in completing the harvest of each variety within this timeframe. Having multiple varieties with sequential ripening schedules, adequate harvesting capacity, and compatible processing capacity are critical to being able to complete the harvest at the best possible time.

Manual harvesting

Manual harvesting in general is more expensive than mechanical harvesting due to the labor costs. Although small growers often find the costs acceptable when selling at a premium retail price, the cost is prohibitive for many olive growers. However, in the first few years after planting, when the trees are too small for mechanical harvesting, oil olives are hand-harvested. A picker at ground level or on a ladder removes the fruit by sliding a cupped, gloved hand down the shoot in a milking action. The shoot is positioned so that the fruit falls into a picking bucket or nets placed on the ground (fig. 15.2). Then, the fruit is dumped into standard orchard bins that hold 1,000 pounds (454 kg) or smaller crates holding 50 pounds (22 kg).

Harvesting with poles is the most common method used in oil olive orchards in traditional areas of the Mediterranean region. The poles range from 3 to 9 or even 12 feet (0.9 to 3.7 m), depending on the region. The operator uses the pole to hit the branches just above the fruit, taking care not to damage the fruit by striking it directly. The fruit drops onto nets that have been placed under the trees. Then the nets are folded and the contents poured into boxes, crates, or

bins. The drawback of pole harvesting is that it is labor intensive; and hitting the branches may damage the following year's crop.

The working time per pound of fruit manually harvested depends on yield and decreases as production increases. Despite this being a very common method in oil olive–growing countries around the world, either as the only method of harvest or as a complement to trunk shakers, its high cost and the impact on the following year's crop have seen it decrease in popularity. The harvest capacity of a worker picking by hand or with a pole depends on the worker's ability, the variety being picked, the size of the canopy, and the crop load. It can range between 250 and 1,000 pounds (115 and 455 kg) per worker per day.

Trunk-shaking harvesters

Most of the early harvesting machines designed for olives were based on shaking principles. A trunk-shaking harvester, used in traditional and high-density (HD) but not super-high-density (SHD) orchards, shakes the main trunk or each scaffold branch separately; the fruit falls onto a net or into a collecting device that is part of the harvester. Trunk-shaking harvesters show a wide range of harvesting efficiencies, ranging from 50 percent to above 90 percent, with the higher efficiencies linked to late harvest and light crop loads. There are three main types of vibration mechanisms: unidirectional, multidirectional, and orbital.

Figure 15.2

The cost of manual harvesting is prohibitive for many growers, although growers selling at a premium price often find the costs acceptable. *Photo:* Gregory Urquiaga.

Unidirectional shakers

A unidirectional, or pole-handle, shaker is mainly used to harvest fruit from secondary branches that are no more than 2 inches (5 cm) in diameter. It hangs from the shoulder of an operator and consists of a two-stroke engine with a speed reducer and a pole that ends in a U shape. The U is placed around the branch and transmits vibration to the branch at approximately 600 cycles per minute.

Unidirectional shakers have a high fruit detachment efficiency. They are very popular for small operations, early picking, and as an alternative or supplement to trunk shakers because of their low cost and because their use does not depend on the soil being dry enough to allow machinery access. However, their performance is slow, and they have high maintenance costs. Their biggest drawback is the impact on the operator. Daily use should not be more than 10 to 15 minutes per worker, which necessitates a large harvesting team taking turns to use the shaker—no single operator should be assigned to it.

There are also unidirectional trunk shakers pulled by or mounted on a tractor. These are used only to harvest young olive trees or the main limbs of older trees.

Multidirectional shakers

Multidirectional shakers are tractor mounted, with a cushioned shaker head that grips the trunk (fig. 15.3). Two circular, eccentric masses inside the shaker head rotate in opposite directions and at different speeds, driven by the action of one or two hydraulic motors. Most multidirectional shakers have a frontal grab and are suitable for single- and multitrunked trees because of the mobility of the tractor (those with reverse gears are particularly easy to maneuver). For single-trunk olive trees, tractor-pulled shakers can be used; the approximation and grabbing movement of the grab head on these shakers is done hydraulically, thus not imposing such a strain on the tractor.

Orbital shakers

An orbital shaker is a simplified version of a multidirectional shaker. A single mass creates a centrifugal force that produces a circular or orbital movement in the trunk. It has a harvesting efficiency similar to that of a multidirectional

Figure 15.3

Tractor-mounted multidirectional shaker harvesters have a cushioned head that grips the trunk. *Photo: Louise Ferguson.*

shaker. Self-propelled orbital shakers have a hydrostatic transmission that assists their mobility; the tricycle-type orbital shakers and the ones with both axes that rotate are particularly easy to maneuver. Maneuverability is the main advantage of the self-propelled orbital shakers over tractor-mounted orbital shakers. Their primary disadvantage is their high price, which is more than the price of a tractor-shaker set.

Fruit-catching devices

The most common trunk shakers have a fruit-catching device that opens like an umbrella (fig. 15.4). The fruit falls into it and travels into a hopper at its base. The drawback of the fruit-catching device is that it doesn't allow access to the canopy for supplementary pole harvesting to increase harvesting efficiency. More than one run through the orchard with a trunk-shaking harvester, to maximize harvesting efficiency, is likely cost prohibitive, and would increase the risk of tree damage.

Another fruit-catching method allows the olives to fall onto nets or shade cloths that are attached to the shaker; workers unfurl and gather the nets, and supplementary pole harvesting can be done at the same time. Fruit in the nets

Figure 15.4

As a trunk shaker sends vibrations through the tree, olives drop into an umbrella-like fruit-catching device. *Photo:* Gianni Sorrenti, Sicma SRL.

empties into the skirt of the shaker, and from there is off-loaded either with a boom to a trailer traveling with the harvester, or, the preferred method, into a wide loader bucket attached to the harvester, which can be tipped into a stationary trailer. Alternatively, a trailer traveling alongside the harvester can be fitted with nets that are gathered and spread out using a hydraulic motor or mechanical system. These trailers are long so that they can cover the entire fruit fall line.

Trunk shakers with a fruit-catching frame, which are used in plum and pistachio harvests, can also be used to harvest olives. These shakers can harvest only single-trunk trees, and optimally the trunk should be clear of branches to at least 3 feet (0.9 m). The harvester consists of two units, each with a catching frame tilted at 45 degrees, working in tandem on opposite sides of the tree, each unit maneuvered by its own driver. One of the units has a shaker clamp; the other unit has a conveyor that receives the olives that have rolled down the catching frames and moves them past a leaf blower and into a standard bin (fig. 15.5).

Best practices for trunk shakers

A key aspect of using any kind of trunk shaker is to apply the appropriate amount of vibration.

The vibration time provided by many operators is usually around 20 seconds, which is excessive. Studies show that most olives drop in the first 3 seconds and that applying vibration beyond 5 seconds promotes tree damage and damage to the shaker and tractor (Gil-Ribes 1986; Gil-Ribes and Gracia 1979). Two short vibrations rather than a single long one help prolong the life of the equipment. Fruit that fails to drop after trunk shaking often is concentrated in specific areas of

Figure 15.5

A trunk shaker with a fruit-catching frame has two units. This unit catches the olives and moves them into a bin. *Photo:* Louise Ferguson.

the tree; harvesting efficiency can be increased with supplemental pole harvesting.

If not operated well, a trunk shaker may damage the tree bark in the gripping area and result in some defoliation or the breaking of small branches and even large limbs (fig. 15.6). A trunk shaker generally causes minor damage to the aerial part of the tree, but on occasion it may pull off the trunk bark or separate the cambium. In the latter case, fungal growth may occur in the damaged area. To prevent these problems, the operator must work carefully and the grip needs to be well designed so that the contact area with the trunk is large enough and the vibration force causes only radial stress, not longitudinal stress. Longitudinal stress is prevented by placing the grip perpendicular to the trunk; sometimes this is not possible because of poor access.

Tree structure is an important factor in harvesting efficiency for many types of mechanically harvested fruit trees, and particularly for olive. The structure of a classically pruned olive tree, with many branches overhanging the perimeter of the tree, produces a strong buffer and a substantial inertia to the transmission of vibration. Studies on the efficiency of vibration transmission from the trunk to the fruit-bearing branches showed that the farther away the point of application of the vibration, the lower the vibration (Fridley et al. 1971; Gil-Ribes 1986; Gil-Ribes and Gracia 1979).

Ideally, there should be only three or four primary branches growing off the trunk, and all branches should be as close as possible to the vertical, particularly the primary and secondary branches. Ideally, also, the tree should have a single, 3-foot- (0.9-m) tall trunk and a canopy that is not too large. Remove crossover limbs that absorb energy, low-hanging branches that could interfere with the shaker clamp, and dead limbs. Thin out limbs with overly dense twigs (see chapter 9, Canopy Management).

Canopy-contact harvesters

The first canopy contact harvesters used in olive orchards were designed to harvest grapes grown on espaliers and trellises. This type of harvester, which has tines in the harvesting head that shake the canopy, is also used to harvest blueberry

Figure 15.6

A poorly operated trunk-shaking harvester can damage the tree bark. If the cambium is separated, fungal growth may occur. *Photo:* Louise Ferguson.

and coffee crops. Large-scale evaluation of these harvesters for olive began in 1996, and some manufacturers have modified their harvesters to address the specific requirements of modern olive production.

Super-high-density orchards

The most common harvester for SHD orchards drives over the tree row, with the trees passing through a tunnel where horizontally oriented bow rods compress and shake the canopy to detach the olives. Retractable catcher plates or buckets the operator positions close to the trunk catch the olives, and a conveyor belt carries them past leaf blowers and then into the machine hopper or hoppers (fig. 15.7). Each hopper can hold 2,500 to 3,500 pounds (1,135 to 1,590 kg) of fruit and has a conveyor on the back that moves the fruit into a bin trailer or gondola.

The harvesters for SHD systems are mostly self-propelled machines, with hydrostatic transmission based on a variable flow pump and power steering. They have a lifting system and a lateral compensation system with hydraulic cylinders on the wheels that help the machine work on slopes of up to 30 percent. They typically operate at speeds from 0.6 to 1.5 miles per hour (1 to 2.4 km per hr). The harvesting head is held by the chassis at four points with antivibratory anchoring.

Bow rods
compressing tree

Figure 15.7

In a canopy contact harvester, fruit from the compressed tree moves on to a conveyor that carries the fruit past a leaf blower into the hopper.

Harvesting efficiency typically ranges between 80 and 98 percent, with the highest efficiency observed in young and well-pruned orchards.

Trees should be topped at 8 to 10 feet (2.4 to 3 m), depending on the spacing between rows,

skirt-pruned so there are no branches lower than 2 feet (0.6 m) above the ground, and the canopy pruned so that it is no wider than 5 to 8 feet (1.5 to 2.5 m), depending on the row width. The pruning can be done with a mechanical pruning machine, by hand, or a combination of both.

After a few years, SHD olive trees can accumulate a lot of wood, particularly on the upper part of the canopy. This can seriously affect harvesting speed, efficiency, and ultimately yields if not properly addressed (see chapter 9, Canopy Management).

High-density orchards

A canopy contact harvester for HD olive orchards is larger than the SHD harvester due to the larger tree size in the HD system (fig. 15.8). The harvester consists of a single or multiple vertically mounted, cylindrical heads with radiating rods approximately 2.5 to 3.5 feet (0.8 to 1.1 m) long. The heads are mounted on a mobile unit above a catch frame. As the harvester moves forward, at 0.2 to 0.6 miles per hour (0.3 to 1 km/hr), a rotating counterweight at the top of each

Figure 15.8

A canopy contact harvester drives through a high-density orchard. The harvested fruit is off-loaded to an accompanying trailer. *Photo:* Cobram Estate Olives Ltd.

head produces a horizontal whipping motion in the distal tip of the radiating rods. It is this horizontal whip of the rods perpendicular to and against the hanging olive branches that removes the olives. From the catch frame, a sequence of conveyors and buckets takes the olives to a side chute that discharges them into an accompanying bin or trailer. Two to four fans remove leaves and other debris.

Harvesting efficiency is directly dependent on fruit accessibility. If the radiating rods contact the bearing canopy, the harvester can remove up to 96 percent of the fruit. Fruit on the top of the tree, on the rounded leading and trailing edges of the canopy, and in the tree skirt below 3 feet is the most difficult to remove. It's been demonstrated that canopy contact harvesters are more efficient and cost-effective than trunk shakers for HD orchards, particularly during early season harvests or under heavy cropping conditions (Ravetti and Robb 2010). The most important drawbacks of these machines are their cost, limited maneuverability, and difficulty to transport, which makes them highly suitable only for large operations.

Regarding pruning an HD orchard for a canopy contact harvester, hedgerowed trees appear to be the best shape, either with a straight (rectangular) or slightly sloped (pyramidal) canopy. These shapes present the maximum bearing surface to the harvester with minimal harvester head manipulation and maximum harvester speed. Ideally, the tree should be no more than 14 feet (4.3 m) tall, with the skirt 3 feet (0.9 m) above the ground, and a canopy width of 12 feet (3.7 m) or less, with no big structural branches growing toward the interrow and protruding more than 3 feet (0.9 m) from the center of the tree.

As with all tree crops, but particularly so with an evergreen tree that bears on 1-year-old shoots, hedging decreases the crop in the year of pruning. Fruit size in topped and hedged trees is improved in the year of pruning, but the larger fruit size is not enough to offset the yield-decreasing effect of the hedging. These effects dissipate in subsequent years if the hedging is not repeated. A canopy contact harvester is equally efficient at removing fruit from trees that were hedged in the current or in prior years.

The interior canopy must be hand-pruned to remove thicker wood that is growing perpendicular to the movement of the harvester, because it will be broken by the machine or compromise its efficient action. If the correct hand-pruning is conducted regularly, mechanical topping and light or no hedging in alternate years are enough for a canopy contact harvester. Data have not been collected, but it appears that skirt pruning is needed annually.

Cleaning and transport

During harvesting, many leaves and twigs may be collected with the fruit, to such a point that the standard fruit cleaning systems on the harvester or at the oil mill will not sufficiently remove the debris. An excessive amount of debris may result during mechanical harvesting when trees are not properly pruned or the orchard has a high incidence of disease, such as olive knot.

Depending on the harvesting method and the size of the operation, olives are transported to the mill in small, perforated crates of up to 50 pounds (22.7 kg), in vented orchard bins that hold 1,000 pounds (453.6 kg) (fig. 15.9), or in large trailers that can carry between 10,000 and 20,000 pounds (4.5 and 9.1 t). In California, large operations commonly use a standard hopper trailer with a capacity of 20,000 to 25,000 pounds (9.1 to 11.3 t).

Figure 15.9

These vented orchard bins are one option for transporting oil olives to the mill. Large operations use hopper trailers. *Photo: Dan Flynn.*

Storage

The time that elapses between harvest and processing is one of the most important factors affecting olive oil quality. Ideally, there would be no need for olive storage—the oil would be extracted from the fruit immediately upon delivery to the mill to preserve the most oil and the best characteristics of the fresh fruit. A perfect synchronization of harvesting and processing is very difficult to implement, so at least some of the fruit will often need to be stored.

To prevent alterations that could modify the quality of the oil, fruit storage needs to be short term so that the olives are processed within 12 to 24 hours of harvest. This timing is essential to get the best-quality oils from high-quality fruit, but it is also important for lower-quality fruit. Storing fruit for a long time is the main cause of oil quality spoilage; it results in severe alterations of the oil's organoleptic characteristics and also increases acidity due to microorganism activity and decreases oil stability.

When storage is needed, the fruit is stored at the mill to avoid significant increases in costs. The fruit is kept in the same crates, bins, or hopper trailers used to transport it, and those are placed in a cool, shaded, and ventilated area; or the fruit is emptied into specifically designed, clean stainless steel storage hoppers normally placed under a shade roof in the receiving area of the olive mill.

Improving/maintaining harvesting efficiency

A quality control process is extremely important to improve on or maintain harvesting efficiency levels. Growers should visually inspect how well each harvester is performing. For example, in large orchards, growers should drive past a minimum of 500 yards (457 m) of trees harvested during each shift or 8- to 12-hour period. The trees on both sides of the interrow should be carefully checked, with special attention paid to the following:
- tree damage (trees being pushed over or seriously ringbarked)
- canopy damage (number of broken branches)
- fruit losses from the catchment system
- fruit left on the trees

- top of the trees
- skirt of the trees
- front and back of the trees
- inside of the canopy
- outside of the canopy

Visual inspection should be complemented with sampling measurements to calculate the harvesting efficiency. The method used is to select two average trees per harvester and harvesting shift and set up shade cloth under the trees before the harvester passes through. After the harvester goes through, first the fruit left on the ground is collected and weighed. After that, the fruit left on the trees is picked and weighed. The two amounts are combined and added to the average actual yields obtained from those trees to obtain the total fruit before harvest. The ratio between the fruit harvested and the total amount of fruit (before harvest) provides the harvesting efficiency percentage.

When there is a difference of more than 20 percent between visual estimates of harvesting efficiency and sampling measurements of harvesting efficiency, the grower should double-check the sampling measurements and, if accurate, recalibrate future visual estimates to be more consistent with sampling measurements. Accurate measurements of harvesting efficiency should help growers to make improvements in other areas of orchard management, such as pruning for better fruit removal or delaying the time of harvest.

References

Barranco, D., C. C. de Toro, M. Oria, and H. F. Rapoport. 2002. Monopotassium phosphate (PO_4H_2K) for olive fruit abscission. Acta Horticulturae 586:263–266. https://doi.org/10.17660/ActaHortic.2002.586.50

Bem-Tal, Y., and M. Wodner. 1994. Chemical loosening of olive pedicels for mechanical harvesting. Acta Horticulturae 356:297–301. https://doi.org/10.17660/ActaHortic.1994.356.62

Dag, A., A. Bustam, A. Avni, I. Tzipori, S. Lavee, and J. Riov. 2009. Timing of fruit removal affects concurrent vegetative growth and subsequent return bloom and yield in olive. Scientia Horticulturae 123(4):469–472. https://doi.org/10.1016/j.scienta.2009.11.014

Fridley, R. B., H. T. Hartmann, J. J. Mehlschau, P. Chen, and J. Whisler. 1971. Olive harvest mechanization in California. California Agricultural Experiment Station Bulletin 855.

Gil-Ribes, J. 1986. Sistemas vibratorios de recoleccion de frutos. Fruticultura Profesional 2:9-13.

Gil-Ribes, J., and C. Gracia. 1979. Estudio de la eficiencia de transmisión de vibraciones en la estructura del olivo. Anales INIA 5:95–117.

Lavee, S. 1996. Biología y fisiología del olivo. In Enciclopedia mundial del olivo. Barcelona: International Olive Council. 61–110.

Martin, G. C. 1994. Mechanical olive harvest: Use of fruit loosening agents. Acta Horticulturae 356:284–291. https://doi.org/10.17660/ActaHortic.1994.356.60

Ravetti, L. M., and B. McClelland. 2008. Improving the efficiency of mechanical olive harvest. RIRDC Publication No. 08/052. Australian Government, Rural Industries Research and Development Corporation.

Ravetti, L. M., and S. Robb. 2010. Continuous mechanical harvesting in modern Australian olive growing systems. Advances in Horticultural Science 24(1):71–77.

Suggested reading

Barasona, J., M. L. Barasona, R. Rodríguez, and J. Cano. 1999. Rendimientos y costes de mecanización de la recolección de aceituna. Consejería de Agricultura y Pesca. Junta de Andalucía.

Barranco, D., R. Fernandez-Escobar, and L. Rallo, eds. 2010. Olive growing. Translated by Susan E. Hovell, William A. Hovell. Pendle Hill, NSW, Australia: RIRDC. Originally published as El cultivo del olivo. Madrid: Junta de Andalucia, Consejeria de Agricultura y Pesca, and Ediciones Mundi-Prensa, 2004.

Blanco, G. L., J. Agüera, J. Gil-Ribes, and F. Agrela. 2001. Mejora del comportamiento y diseño de los vibradores de troncos en olivar. VI Congreso Ibérico de Ciencias Hortícolas. SECH. Cáceres.

Ferguson, L., et al. 2008. Efficiency evaluation of picking head canopy harvester for 'Manzanillo' olives in traditional and hedgerow orchards. Proceedings of the 6th International Symposium on Olive Growing. Evora, Portugal.

Fontanazza, G. 1996. Olivicoltura intensiva meccanizzata. Edagricole. Bologna, Italy.

Gil, A., V. Osuna, and J. Gil-Ribes. 1998. Evolución de los parámetros coyunturales que condicionan la recolección de la aceituna. Mercacei 2:54–57.

Hartmann, H. T., W. Reed, J. E. Whisler, K. W. Opitz. 1975. Mechanical harvesting of olives. California Agriculture 29(6):4–6.

Lavee, S. 1989. Involvement of plant growth regulators and endogenous growth substances in the control of alternate bearing. Acta Horticulturae 239. https://doi.org/10.17660/ActaHortic.1989.239.50

Porras Piedra, A. et al. 1994. Recolección mecanizada de aceituna. Madrid: IOOC.

Rius, X, and J. Lacarte. 2010. La revolución del olivar: El cultivo en seto. Madrid: Ediciones Paraninfo S.a.

Tombesi, A., M. Boco, M. Pilli, and D. Farinelli. 2002. Influence of canopy density on efficiency of trunk shaker on olive mechanical harvesting. Proceedings of the 4th International Symposium on Olive Growing. ISHS Acta Horticulturae 586. https://doi.org/10.17660/ActaHortic.2002.586.56

Visco, T., M. Molfese, M. Cipolletti, R. Corradetti, and A. Tombesi. 2008. The influence of training system, variety and fruit ripening on the efficiency of mechanical harvesting of young olive trees in Abruzzo, Italy. Proceedings of the 5th International Symposium on Olive Growing. ISHS Acta Horticulturae 791. https://doi.org/10.17660/ActaHortic.2008.791.64

16

Processing Virgin Olive Oil

Dan Flynn, Executive Director Emeritus, UC Davis Olive Center

David Garci-Aguirre, Master Miller, Corto Olive

Pablo Canamasas, Olive Oil Production and Quality Consultant

Highlights

- Minimizing the delay between harvest and processing has a significant positive impact on the sensory and chemical quality of the oil.

- Olives that have a moisture content of 50 to 56 percent when delivered to the processor are more likely to achieve maximum oil yield than olives with lower or higher moisture content.

- Washing fruit prior to crushing can increase fruit moisture and decrease extraction, fruitiness, and shelf stability, so washing should be limited to fruit that carries substantial field soil and other impurities.

- Processing aids such as enzymes can facilitate the release of oil from early-maturity fruit, and talc can allow processors to manage emulsions in high-moisture fruit.

- A facility should strive to achieve > 85 percent extraction efficiency (percentage of the oil in the fruit that is extracted). Grid size and rotation speed in the crusher, and time and temperature in malaxation, are key variables in finding the balance between oil quality and extraction efficiency.

- Evaluating oil yields as well as oil losses in the pomace allows a processing facility to monitor their extraction efficiency.

- Freshly extracted oils contain moisture and sediment, which lead to quality problems if the oil is not frequently racked or filtered in due time.

- Ideal storage conditions for olive oil are stainless steel storage tanks regularly flushed with an inert gas like nitrogen or argon and maintained in the range of 59° to 64°F (15° to 18°C).

- Packaging on demand, while holding oil in bulk storage under ideal conditions, extends shelf life compared to packaging the entire stock soon after harvest.

Virgin olive oil is essentially a minimally processed fresh fruit juice that retains many of the nutritional qualities of fresh olive fruit. The processor uses mechanical means to crush olives to form a paste, then separates the oil from the paste without using heat or solvents (see chapter 1, California Olive Oil Industry and Standards, for a discussion of how the virgin olive oil process differs from the processes for refined olive oil and olive-pomace oil). Decisions made by the processor influence flavor, quality, oil clarity, extraction efficiency (percentage of the oil in the fruit that is extracted), and shelf stability. Knowing how processing works helps growers understand how to maximize oil quality and yield.

Oil quality and shelf life are predominantly dependent on fruit quality, transportation, processing, storage, and packaging. Fruit should be delivered to the processor within the shortest possible time after harvest, ideally within 4 hours (see chapter 15, The Olive Harvest). Minimizing the delay between harvest and processing has a significant positive impact on the sensorial and chemical quality of the oil.

Processing facility

Siting considerations for olive processing facilities include convenient road access, proximity to production areas, availability of staff, and access to water, electricity, and propane or natural gas. A typical processing line to produce virgin olive oil includes a crusher, malaxer, decanter, separator, and oil storage (fig. 16.1).

The entire processing line should be maintained at a high level of cleanliness and sanitation.

The fruit receiving area should be designed for ease in unloading and inspecting fruit and should have an excellent drainage system. The processing area should be easy to clean, be insulated, and have protected electric lighting and well-designed drainage and ventilation systems.

The storage area should be easy to clean, odor free, climate controlled, and insulated; have electric illumination rather than windows to control light exposure; and feature a well-designed drainage system. The by-product disposal area should be well connected to the processing area, easy to access, and kept clean.

Receiving

Processors should inspect the fruit upon receiving it to assess its condition and maturity index. Processors also should analyze oil and moisture content to decide on the parameters for paste preparation in the crusher and malaxer.

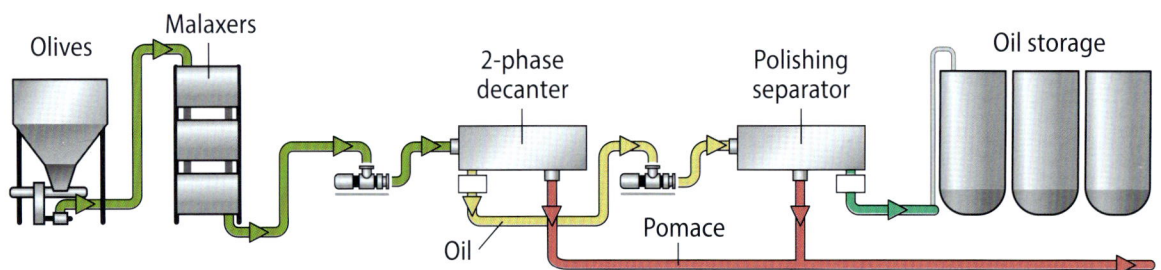

Figure 16.1

Typical components in a processing line for producing virgin olive oil.

Growers generally achieve better oil extraction by delivering fruit with moisture levels in the 50 to 56 percent range. Fruit with a moisture level below 50 percent produces a dry paste, which poses a significant operational challenge to the processor and may affect the oil's sensory quality by giving it "dry" flavor notes. Fruit with a moisture level above 56 percent produces a paste prone to emulsions, and the oil is more difficult to extract. To maximize oil yield at the processing facility, it's important to carefully implement a preharvest irrigation plan in the orchard.

Growers under contract with a processor may get paid a higher rate if the fruit is delivered in optimal condition—that is, meets the following criteria:

- Low percentage of material other than olives (MOO) because MOO creates problems for the washing equipment and adds astringency to the oil; optimal MOO percentage is less than 1 percent, good is less than 2 percent, increasing risk is 2 to 5 percent, and not acceptable is greater than 5 percent.
- Little or no damage from pests, diseases, or frost.
- Few or no mummified fruit, which accelerate oil fermentation and impart a sensory defect.
- No cross-contamination with hydraulic oils and free from pesticide residues.
- Low free fatty acidity, which is an indicator of initial fruit fermentation; oils from sound, healthy fruit that is processed in a timely manner should not exceed 0.35 percent free fatty acidity.
- Low peroxide value, indicating low fruit oxidation; oil from sound, healthy fruit that is properly processed should not exceed 10 meq O_2 per kilogram of oil.
- Bin temperature minimally elevated above air temperature, which is an indicator of minimal fruit fermentation; optimal is less than 9°F (5°C), and good is less than 18°F (10°C).

Fruit storage

Smaller processing facilities that work in batches tend to use bins for temporary storage, which allows them to classify fruit lots prior to crushing. The bins should be ventilated or perforated to limit fermentation during storage.

Processing facilities that run continuously typically hold fruit in small hoppers at specific points in the process. It is critical that fruit completely exit the hopper before the hopper is refilled. Olives that lodge in the hopper for an extended period ferment and cause a "fusty" sensory defect, increasing oxidation and reducing shelf life. Larger processors may have two hoppers to ensure continuous operation of the facility and allow for visual inspection of each hopper prior to refilling.

Deleafing and washing

It is important that harvest crews maximize the removal of mummified fruit, olive knot, dead wood, and other MOO. Keeping MOO in the delivered fruit to a minimum improves the ability of deleafing and washing equipment to eliminate it, which yields higher oil extraction and quality and reduces potential damage to processing equipment and downtime.

MOO removal equipment typically uses multiple processes to distinguish MOO from olives. Belted chain, vibratory conveyance, and sizing rollers remove anything smaller or larger than olives, such as branches, sticks, mummies, or underdeveloped fruit. Blowers and aspiration devices remove material that is lighter than olives, such as mummies, leaves, and twigs.

The washing unit traps heavier objects, such as rocks and metals; magnets on the incoming fruit belts supplement the removal of metals, although they do not attract nonferrous metals. To remove soil and residues, a line of nozzles sprays water over the olives as they are conveyed toward the crusher (fig. 16.2). Some processors augment the deleafing and washing equipment with optical sorting technology: cameras detect and reject MOO and olives that are damaged by fermentation, pests, or disease.

If processors do not have the capacity to dry the olives with an air knives system, they may bypass the washing step to avoid increasing fruit moisture, which decreases oil extractability, phenols, fruitiness, and shelf life. Washing may be necessary only for fruit that is carrying field soil and impurities. However, processors should weigh the value of bypassing the washing system with the risk that rocks or metals may enter and damage processing equipment.

Figure 16.2

Washing removes dirt but also adds moisture that can reduce oil extraction. *Photo:* Hector Amezcua.

If the facility is using a recirculating washer, the water needs to be changed at adequate intervals, as dirty wash water can impart negative flavors to the oil. An appropriate interval for changing the water is after the washer has washed 5 to 10 times its hourly capacity. For example, if a washer has a capacity of 1 ton per hour, change the wash water after it processes 5 to 10 tons of fruit. Heavily soiled fruit may require more frequent water changes.

Crushing

The objective in crushing is to break the cells of the olive and release the oil for extraction. Crushing tools include stones, discs, knives, rollers, and depitters, but the most common crusher is the hammer mill, due to its high output, simplicity, easy maintenance, and adjustability. The hammers in a hammer mill are stainless steel blocks that rotate at high speed and crush olives through a stainless steel grid perforated with holes ranging in diameter from about 0.2 to 0.25 inch (5 to 6 mm). California research has shown that the combination of rotation speed and grid size can impact extraction efficiency, phenol content, aromatic complexity, bitterness, and pungency (Polari et al. 2018b; Polari and Wang 2019).

The processor adjusts the grid size based on the variety, moisture content, size, and maturity of the fruit to obtain the processor's desired outcome. Generally, olives should be crushed so that the paste is as fine as possible. A small grid size is typically appropriate for small, low-maturity, and dry olives (less than 56% moisture), and a large grid size is appropriate for large ripe olives and olives with high moisture.

Fruit with high moisture may create emulsions, which are gel formations of oil and fat that can be spotted in the malaxer. Emulsions significantly reduce oil extraction. To avoid the risk of emulsions, growers should prepare for harvest by testing the fruit for moisture and adjust field irrigation schedules accordingly (see chapter 15, The Olive Harvest). The processor can reduce the risk of emulsions by increasing the crusher grid size and/or slowing the rotation speed of the crusher.

For firm, low-maturity fruit, the processor can add enzymes to the crusher to help release the oil. Enzymes (produced from the fungus *Aspergillus aculeatus* or *A. niger*) are biologically active proteins used extensively in wine, beer, and bread production. In oil extraction, they degrade the pectin and cellulose of the olive cell walls and vacuoles, helping to release the oil. Enzymes have no impact on oil quality and are not in the final product—their higher density makes them easy to remove during centrifugation. Processors often find that the cost of adding enzymes is more than offset by higher oil yields.

Crusher grids and hammer edges show wear with use, and the wear will eventually significantly reduce oil extraction efficiency. Replacing worn grids and hammers is a highly cost-effective routine maintenance procedure.

Malaxation

A malaxer is a jacketed tank with rotating blades that slowly mix the paste that emerges from the crusher. It initiates the process of separating the oil from the paste by increasing the proportion of large oil droplets, which result from the coalescence of smaller droplets (fig. 16.3). Malaxation is a critical stage of paste preparation. The processor's decisions in managing malaxation will have a significant impact on the quality and yield of oil. The impact of the decisions is conditioned by choices already made during crushing; for example, smaller grid sizes combined with longer malaxation times result in higher yields (Polari et al. 2018a). Many of the volatile compounds that compose the positive aroma of the oil are formed during malaxation.

Emulsions in the malaxer can be broken down

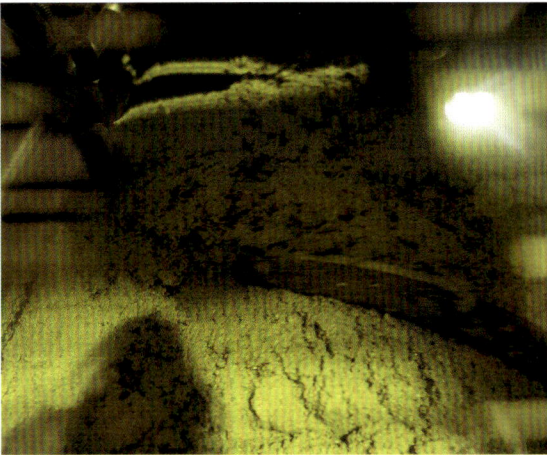

Figure 16.3

The malaxer kneads the paste to facilitate oil separation in a closed system. *Photo:* Hector Amezcua.

by adding talc (hydrated magnesium silicate). The processor may consider adding talc for high-moisture fruit (> 56% moisture), for varieties that are prone to producing emulsions (such as Arbequina, Picual, Hojiblanca, and Manzanillo), when using smaller crusher grids, and for fleshy varieties. The cost of adding talc likely will be offset by increased oil extraction and improved centrifugation throughput in the next stage of the process. Talc does not negatively impact oil quality and is not in the final product—its higher density makes it easy to remove during centrifugation. The most critical variables in malaxation are temperature and time.

Temperature

Malaxation temperature is influenced by the temperature of the paste emerging from the crusher and the temperature of the fluid (water and glycol) that regulates temperature in the malaxer jackets. High-quality olive oils require a temperature that is as low as possible but not less than 72°F (22°C) to facilitate adequate extraction efficiency and ensure the solubility of phenols and chlorophyll.

European regulations define "cold extraction" as extraction at temperatures below 80.6°F (27°C). In California, processors have found that 77° to 86°F (25° to 30°C) is an appropriate range for balancing quality with extraction efficiency. California standards recognize this flexibility by defining "cold pressed" and "cold extracted" as

extraction at a "temperature that does not lead to significant thermal alterations in the oil."

The processor should avoid temperatures that exceed this recommended range. While high paste temperatures can increase extraction efficiency, phenols, and chlorophylls (making for greener oil), they decrease shelf stability and fruity aroma, amplify oil bitterness, and produce a "cooked" sensory defect. Installation of a modern heat exchange system can rapidly condition paste temperature to the recommended range by reducing the paste temperature if fruit temperature was too high or increasing it if fruit temperature was too low.

Time

Malaxation requires sufficient time for the oil droplets to coalesce and be of sufficient size to be extracted by the decanter, but an extended malaxation time reduces the oil's bright fruity qualities and shelf life stability (because the paste is exposed to oxygen as it churns). A calculation of malaxation time takes into account the time to fill and empty the malaxer, as well as the time to mix the paste, as the malaxer blades are churning during filling and emptying. The calculation is different for malaxers that are working in continuous mode than those working in batch mode.

If the facility is working in continuous mode, a good malaxation time estimate can be made by dividing the holding capacity of the malaxer by the working capacity of the decanter. For example, if the malaxer has a holding capacity of 6,000 kilograms and the working capacity of the decanter is 4,000 kilograms per hour, then the malaxation time will be 1.5 hours (6,000 ÷ 4,000 = 1.5).

If the processing facility is working in batch mode, the filling and emptying time are divided by 2 and added to the mixing time. For example, if filling time is 30 minutes, and mixing time with the malaxer full is 30 minutes, and emptying time is 50 minutes, then the malaxation time will be 70 minutes (15 + 30 + 25 = 70).

Malaxation time significantly affects the oil's sensory profile and generally should be as short as possible. Typically, it is not less than 45 minutes, and up to 90 minutes or even longer may be necessary to achieve adequate separation of emulsified pastes. A processor may choose to malax for less than 45 minutes to achieve outstanding oil quality, although this benefit should be weighed against a probable reduction in oil yield, because extraction efficiency generally increases with greater malaxation time. Some research indicates that using inert gas in sealed malaxers (which reduce the paste's exposure to oxygen) may allow for longer malaxation times without compromising oil quality (Clodoveo 2012).

The paste provides visual clues that it has been properly malaxed. A large pool of oil floats on the paste, the rotating blades emerge clean as they rotate above the surface, the paste exhibits cracking (not a smooth surface), and it behaves more like a solid than a liquid. An emulsified paste appears excessively fluid, with little oil pooling and with paste sticking to the blades as they emerge above the surface.

New technologies to improve efficiency without compromising quality include pulsed electrical field, ultrasound, and microwave technologies. These technologies show promise but implementing them may not be cost-effective at this time.

Decanter

Modern processors largely have abandoned the centuries-old olive press method of separating oil from paste, which was labor intensive and could degrade oil quality, in favor of the horizontal decanter method. The decanter has three primary components: a feeding cane that delivers the paste pumped from the malaxer to the feed zone inside the decanter, a horizontal rotating bowl that creates the force necessary for separation,

and a scroll that moves the paste through the bowl (fig. 16.4).

The decanter bowl and scroll assembly spins at approximately 3,000 to 5,000 revolutions per minute, the centrifugal force separating the paste into three layers (known as phases): pomace, vegetable water, and oil. Pomace, the heaviest of the three phases, moves to the decanter's outside walls, the water collects next to it, and the oil, the lightest component, remains close to the center of the decanter.

Decanters generally are categorized as two-phase or three-phase, which describes how each type of decanter separates the phases. A two-phase decanter has two outlets: one for oil and one for wet pomace (55 to 75% moisture). A three-phase decanter has three outlets: one for oil, one for water, and one for a drier pomace (45 to 55% moisture). A recently developed three-phase system with water-saving technology, called a multiphase decanter, if properly adjusted, can deliver oil, a pâté-textured pomace, and dry pomace.

Each decanter is rated by the manufacturer for total capacity, such as 1 metric ton of paste per hour. Total capacity is based on decanter length, bowl diameter, feeding cane length, and the differential speed of the scroll and bowl. Generally, the decanter's most cost-effective working capacity is 65 to 75 percent of its total capacity—the processor slows down the pumping speed to get better oil extraction. Reaching good oil extraction efficiency rates is achieved by obtaining good separation between the pomace, water, and oil

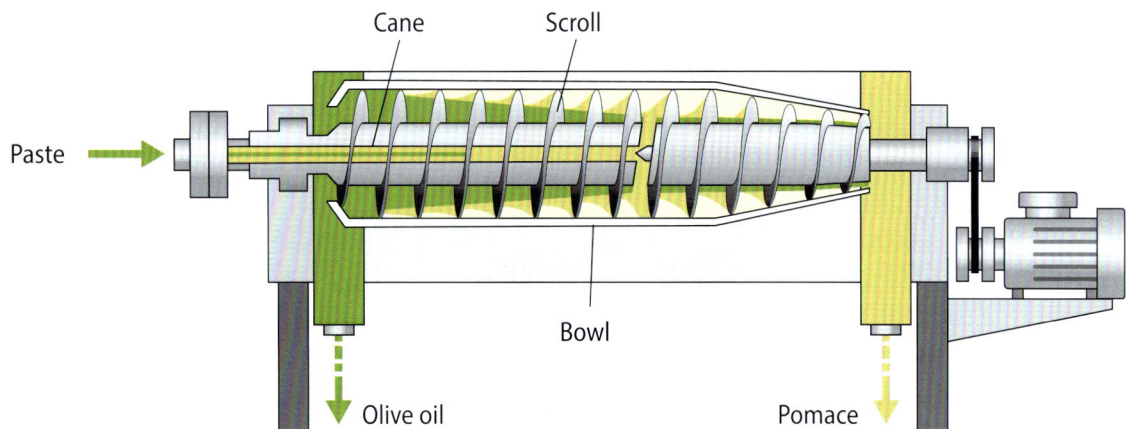

Figure 16.4

Paste enters a decanter via a feeding cane, and the oil is separated from the pomace as the bowl rotates at high speed.

layers in the decanter, which depends on the processor's ability to properly prepare the olive paste during crushing and malaxation.

The position of the weir plates, which control the flow of oil exiting the decanter, must be adjusted according to the fruit condition to allow for proper oil evacuation from the decanter. Typically, fruit with higher oil content requires a larger oil exit opening. The appropriate use of talc and enzymes increases the oil layer inside the decanter and therefore also requires a larger exit opening.

Injecting water into a decanter assists with proper separation of the pomace, oil, and water layers, which becomes very important when processing olives with moisture less than 55 percent. The injected water should be at about the same temperature as the paste because colder water can lead to emulsions and hotter water can have a negative impact on oil quality. Water injection rates can be up to 6 percent of the volume of the total paste throughput for two-phase decanters and range from 10 to 20 percent for three-phase decanters.

Although the extraction results are similar for two-phase and three-phase decanters, the lower water injection rates are a key factor in why two-phase decanters are more prevalent in modern processing facilities. Lower water injection rates result in less energy use and less wastewater to manage, while producing oils of greater intensity and phenolic content. However, because two-phase systems use less water, they demand greater vigilance in paste preparation because water injection can't be used to the same degree to make oil extraction easier with difficult pastes.

Processors should measure the oil content in the pomace to check on the facility's oil extraction efficiency (fig. 16.5). Facilities should strive to reach or exceed a benchmark of 85 percent extraction efficiency. A common practice is to take a sample of pomace from the decanter exit and use NIR (near-infrared) equipment to measure its oil and moisture content. Pomace samples can also be sent to a laboratory for evaluation using solvent extraction methods and oven dryers.

Vertical separator

Oil leaving a horizontal decanter has a significant amount of water and sediment due to the

Figure 16.5

Extraction efficiency can be assessed by checking the oil content in the pomace. *Photo:* Hector Amezcua.

intermixing that occurs inside the decanter. If not sufficiently removed, the sediment and water produce anaerobic fermentations that quickly result in defects in the oil. To ensure good shelf life, processors must remove sediments and lower the moisture content of the oil to less than 0.2 percent. That process usually begins with the vertical separator (although some processors forego this stage of processing); the oil is further clarified using settling, filtration, or racking, or some combination of these methods.

The vertical separator is a bowl-shaped centrifuge with a central vertical shaft that spins at a very high speed, two times higher than the decanter, to remove the remaining solids and separate the oil and water, which have partially intermixed in the decanter. Inside the bowl is a stacked set of conical discs, on the surface of which the oil and water separation occurs. The heavy phase (water) is flung to the outer edges of each disc and the lighter phase (oil) is pushed to the center of the disc, where it is pumped out of the separator. The inward/outward separation zone on the disc is regulated by the processor's choice of a water ring. With too large a water ring, there's a risk of losing oil in the water outlet; with too small a ring, there may be too much water in the oil (fig. 16.6).

The processor adds warm water to the separator, ideally no more than 20 percent by volume of the oil-water mixture emerging from the decanter. The temperature of the added water should be the same as or no more than 6°F (3°C)

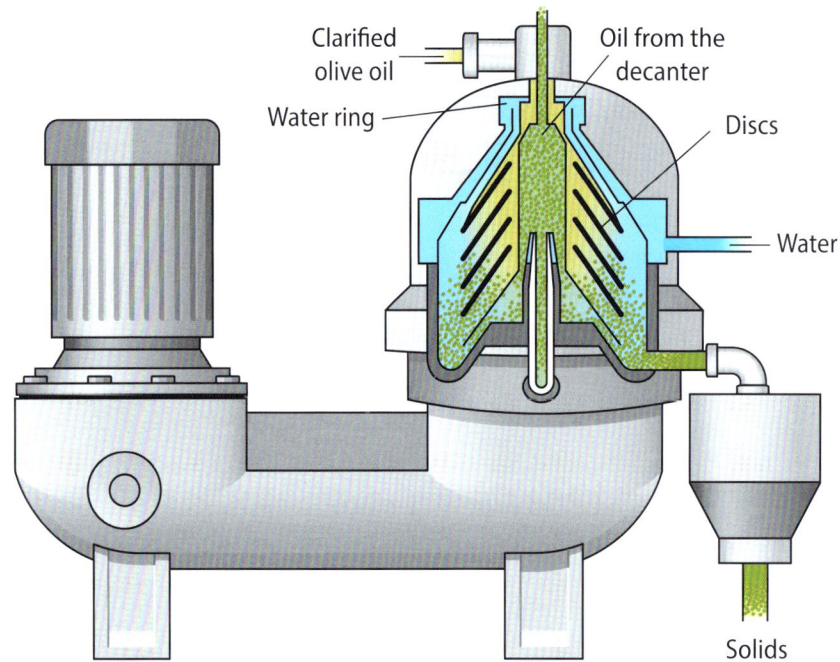

Figure 16.6

In a vertical separator, centrifugal separation of oil from water and solids occurs on conical discs.

higher than the temperature of the mixture coming from the decanter to ensure a proper separation of the oil, water, and fine particles. Do not let the added water temperature exceed 93°F (34°C) because such a high temperature can increase oil oxidation, decrease phenols, and impart a "burnt" sensory defect. Avoid adding water that is colder than the oil, which could produce emulsions.

Some modern separators are designed to minimize oil contact with oxygen by hermetically sealing the inlet to the separator, and by creating back pressure and submerging the centripetal pump for the oil phase in the oil, thereby ensuring the pump is not mixing oil and oxygen. Some modern separators are also jacketed for the circulation of a cooling fluid (typically water), which maintains cooler oil temperatures to improve the quality and shelf life of the oil.

To ensure oil quality and separator reliability, the processor must monitor the clarity of the oil and the composition of the solids (fig. 16.7) and make adjustments as needed. The processor can adjust the rate of the solids discharge to ensure the separator does not back up with solids, the volume of water being added to ensure oil clarity, the heavy- and light-phase back pressures to fine-tune the separation zone on the discs, and the rate

Figure 16.7

Clarified olive oil emerges from the vertical separator. *Photo:* Hector Amezcua.

of feed to ensure proper separation. The processor also can indirectly evaluate the cleanliness of the internal parts of the separator when assessing the clarity of the oil. If the internal parts must be cleaned, the processor can run an automated clean-in-place cycle if available or stop the separator and take it apart for cleaning.

Settling

The oil emerging from the vertical separator can be further clarified by 24 to 48 hours of storage, ideally in a stainless steel settling tank that allows for drainage from a steep conical bottom. Do not have the tank open to air and light, and maintain the temperature above 65°F (18°C) to aid settling. Drain settlings every 2 hours and remove any foam floating on the surface every 6 hours or as needed.

Filtration

Filtration removes suspended solids and moisture in the olive oil before storage, eliminating any possibility of further fermentation. There are many filtering methods, such as pumping oil through paper, cotton, cellulose, ceramic, polypropylene, or diatomaceous earth; using inert gas to separate the oil from the solids; or moving the oil over a membrane. A literature review (Ngai and Wang 2015) showed that the type of filtration method can have positive, negative, or neutral impacts on the oil's stability, phenolics, volatiles, appearance, pigments, sensory profile, and shelf life, with these impacts further influenced by the oil's initial chemistry, sensory profile, variety, and storage conditions (fig. 16.8).

Racking

Racking is an alternative to filtration. During racking, fine particles in suspension in the fresh oil settle over time at the bottom of the storage tank as mud (sediment and water), and the oil is pumped away from it to a clean tank. Proper racking typically takes 30 to 45 days and the oil is moved several times. Extra care must be taken to move the oil often enough to avoid fermentations but not so often that excessive aeration occurs, which hastens oxidation.

Storage

Ideal storage conditions slow down oxidation and prevent fermentative defects produced by any sediment. Olive oil is best stored in a stainless steel tank with a conical bottom to allow for drainage of settlings. The tank should be filled with oil with minimal headspace, and the oil should be held at 59° to 64°F (15° to 18°C) and regularly injected with nitrogen or argon to displace oxygen (fig. 16.9).

Compromising on these best practices reduces the shelf life of the oil. A large olive oil processor that evaluated the published literature concluded that plastic storage containers can reduce shelf life 10 to 70 percent compared to stainless steel tanks, temperatures above 64°F (18°C) reduce shelf life by 7 percent for each increment of 1.8°F (1°C), and shelf life drops by 25 percent in tanks without regular nitrogen flushes.

Packaging

Keeping the oil in proper bulk storage conditions and packaging it on demand extends the shelf life of the oil compared to packaging it all at once. A

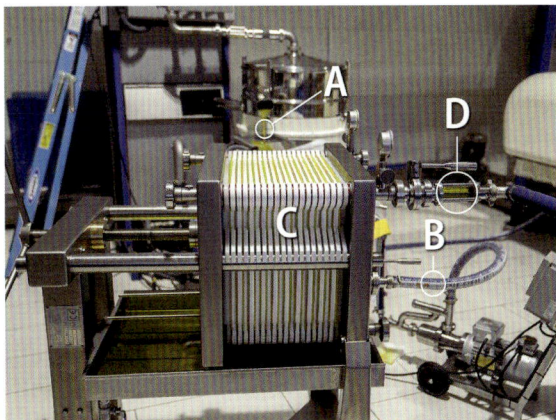

Figure 16.8

Cloudy oil from a vertical separator (A) is pumped (B) through cellulose plates (C) to produce clarified oil (D). *Photo:* Hector Amezcua.

Figure 16.9

Stainless steel tanks in temperature-controlled conditions preserve olive oil shelf life. *Photo:* Corto Olive Co.

review of the scientific literature revealed that the type of packaging influences the oxidation rate of the oil (Wang et al. 2014). Dark or opaque packaging minimizes the light exposure that hastens oxidation. To increase shelf life, oxygen in the container should also be limited, particularly in the headspace between the oil and the top of the package, although some headspace is necessary to allow for the natural expansion of the oil that occurs with increases in temperature. Some bottling equipment creates a vacuum in the package and/or injects inert gas into the package to eliminate oxygen in the headspace.

Bag-in-box packages, which include an inner bladder that collapses as the package is emptied, thereby keeping out air, have been shown to be superior for maintaining oil quality. Packages made of metal, such as tinplate, are desirable as they completely keep light out of the oil, as do coated-paperboard containers.

Dark glass has the desirable features of reducing light penetration and being nonpermeable, although there is some concern that oxygen enters the bottle through the cork or screw top. Clear glass exposes oil to light, but a label can be made to almost fully cover the bottle, or the glass can be given a protective coating to keep out ultraviolet light—there is evidence that consumers like to see the oil when deciding what to buy.

Polyethylene terephthalate (PET) is the primary type of plastic used in olive oil packaging. PET has the advantages of low cost and recyclability, and a dark-colored PET reduces light degradation of the oil. However, PET is porous, which allows humidity and gases to enter the package.

By-product management

The by-products of oil extraction are pomace and water. The three-phase system has separate outlets for pomace and water, while the two-phase system mixes the pomace and water. Another major by-product is the wastewater from washing equipment and general cleaning of the facility.

California regulators require processors to carefully manage raw pomace and wastewater because the high chemical oxygen demand of these by-products may pollute groundwater, soil, and waterways by depleting available oxygen. There is ongoing research to identify efficient ways to separate the water-soluble phenols from pomace and wastewater, as the natural antioxidant and antibacterial properties of the phenols have value as supplements in processed food and cosmetics.

Some processors use specialized equipment to remove pit fragments from the pomace, and either sell the pits or use them on-site as an energy feedstock. California research (Sedej et al. 2016) has found that filtered wastewater from the vertical separator produces a phenol-rich concentrate with potential uses in food processing and clean water for use in the processing facility or orchard.

A processor with high volume may consider investing in a separate processing line to provide a second oil extraction of the pomace. The oil will be of lower quality than the first extraction, but its bulk value may justify the processing expense. Residual oil in pomace also may be extracted with solvents and graded as olive-pomace oil, although California producers have not pursued this option.

Pomace and wastewater may be applied in the orchard within limitations approved by state regulators. Preliminary California research (Hodson et al. 2020) found that composted olive pomace likely has no phytotoxic effects on olive trees and may improve tree nutrition. The research found that raw pomace (not composted) does have phytotoxic effects but did not reduce growth of young olives in a greenhouse, raising the possibility that raw pomace could be used in the orchard to suppress weeds. The research also found that raw pomace immobilized nitrogen if applied without added fertilizer, but it also stimulated beneficial bacterial and fungal-feeding nematodes involved in nutrient cycling.

After oil extraction, large California processors store pomace in silos or allow it to partially evaporate in ponds or dry out on nonpermeable surfaces such as asphalt or tarps. As the pomace oxidizes and ferments, it produces an odor like the "fusty" defect in olive oil. After evaporation or drying, the pomace may be hauled from the facility for use as an animal feedstock. This option typically does not produce income for the processor, and it may have a cost, depending on the market price of other feedstocks.

California research (Cecchi et al. 2019) has shown that the pâté-type paste produced by a multiphase decanter can, when dried, fortify the beneficial phenols of processed food products such as pasta, bread, and granola bars. Consumer acceptance is high, according to the research, but food processors have yet to adopt this practice.

References

Cecchi, L., N. Schuster, D. Flynn, R. Bechtel, M. Bellumori, M. Innocenti, N. Mulinacci, and J. X. Guinard. 2019. Sensory profiling and consumer acceptance of pasta, bread, and granola bar fortified with dried olive pomace (pâté): A byproduct from virgin olive oil production. Journal of Food Science 84(10): 2995–3008. https://doi.org/10.1111/1750-3841.14800

Clodoveo, M. L. 2012. Malaxation: Influence on virgin olive oil quality. Past, present and future—An overview. Trends in Food Science & Technology 25(1):13-23. https://doi.org/10.1016/j.tifs.2011.11.004

Hodson, A., J. Milkereit, and L. Archer. 2020. Effects of raw and composted olive pomace on productivity and soil health in California orchards. Poster presented at 2020 California Plant and Soil Conference, American Society of Agronomy.

Ngai, C., and S. Wang. 2015. Filter or not? A review of the influence of filtration on extra virgin olive oil. UC Davis Olive Center.

Polari, J. J., and S. Wang. 2019. Hammer mill sieve design impacts olive oil minor component composition. European Journal of Lipid Science and Technology 121(10): 1900168. https://doi.org/10.1002/ejlt.201900168

Polari, J. J., D. Garcí-Aguirre, L. Olmo-García, A. Carrasco-Pancorbo, and S. Wang. 2018a. Interactions between hammer mill crushing variables and malaxation time during continuous olive oil extraction. European Journal of Lipid Science and Technology 120(8):1800097. https://doi.org/10.1002/ejlt.201800097

———. 2018b. Impact of industrial hammer mill rotor speed on extraction efficiency and quality of extra virgin olive oil. Food Chemistry 242:362–368. https://doi.org/10.1016/j.foodchem.2017.09.003

Sedej, I., R. Milczarek, S. C. Wang, R. Sheng, R. d. J. Avena-Bustillos, L. Dao, and G. Takeoka. 2016. Membrane-filtered olive mill wastewater: Quality assessment of the dried phenolic-rich fraction. Journal of Food Science 81(4):E889–896. https://doi.org/10.1111/1750-3841.13267

Wang S. C., X. Li, R. Rodrigues, J. D. Flynn. 2014. Packaging influences on olive oil quality. UC Davis Olive Center.

Suggested reading

Barranco, D., R. Fernandez-Escobar, and L. Rallo, eds. 2010. Olive growing. Translated by Susan E. Hovell, William A. Hovell. Pendle Hill, NSW, Australia: RIRDC. Originally published as El cultivo del olivo. Madrid: Junta de Andalucia, Consejeria de Agricultura y Pesca, and Ediciones Mundi-Prensa, 2004.

Index

Page numbers with an italic f, such as 17*f*, refer to figures, and those with an italic t refer to tables.